Rome, Pollution and Propriety

Rome, Pollution and Propriety brings together scholars from a range of disciplines in order to examine the historical continuity of dirt, disease and hygiene in one environment, and to explore the development and transformation of these ideas alongside major chapters in the city's history, such as early Roman urban development, Roman pagan religion, the medieval Church, the Renaissance, the Unification of Italy, and the advent of Fascism. This volume sets out to identify the defining characteristics, functions and discourses of pollution in Rome in such realms as disease and medicine, death and burial, sexuality and virginity, prostitution, purity and absolution, personal hygiene and morality, criminality, bodies and cleansing, waste disposal, decay, ruins and urban renovation, as well as studying the means by which that pollution was policed and controlled.

MARK BRADLEY is Associate Professor of Ancient History at the University of Nottingham. His main research and teaching interests are in the visual and intellectual culture of ancient Rome, and his recent work has been particularly concerned with exploring cultural differences in perception, aesthetics and sensibilities. His first book, *Colour and Meaning in Ancient Rome* (Cambridge, 2009), was longlisted for the 2011 Warwick Prize for Writing, and he has published widely in the field of Roman visual culture and the modern reception of antiquity. He is Editor of *Papers of the British School at Rome*, and is currently working on a book on *Foul Bodies in Ancient Rome*.

BRITISH SCHOOL AT ROME STUDIES

Series Editors

Christopher Smith
Director of the British School at Rome

Bryan Ward-Perkins
Chair of Publications and member of the Council of the British School at Rome

Gill Clark
Registrar and Publications Manager of the British School at Rome

British School at Rome Studies builds on the prestigious and longstanding *Monographs* series of the British School at Rome. It publishes both definitive reports on the School's own fieldwork in Rome, Italy and the Mediterranean, and volumes (usually originating in conferences held at the School) on topics that cover the full range of the history, archaeology and art history of the western Mediterranean.

Rome, Pollution and Propriety

Dirt, Disease and Hygiene in the Eternal City from Antiquity to Modernity

Edited by MARK BRADLEY

with Kenneth Stow

CAMBRIDGE UNIVERSITY PRESS
Cambridge, New York, Melbourne, Madrid, Cape Town,
Singapore, São Paulo, Delhi, Mexico City

Cambridge University Press
The Edinburgh Building, Cambridge CB2 8RU, UK

Published in the United States of America by Cambridge University Press, New York

www.cambridge.org
Information on this title: www.cambridge.org/9781107014435

First published 2012

Printed and Bound in the United Kingdom by the MPG Books Group

A catalogue record for this publication is available from the British Library

Library of Congress Cataloguing in Publication data
Rome, pollution, and propriety : dirt, disease, and hygiene in the eternal city from antiquity to
modernity / edited by Mark Bradley with Kenneth Stow.
    pages   cm. – (British School at Rome studies)
ISBN 978-1-107-01443-5 (hardback)
1. Urban sanitation – Italy – Rome – History.   2. Public health – Italy – Rome – History.
3. Pollution – Italy – Rome – History.   4. Quality of life – Rome – History.   I. Bradley,
Mark, 1977–   II. Stow, Kenneth R.
TD80.R65R66   2012
363.7309456′32 – dc23   2012010804

ISBN 978-1-107-01443-5 Hardback

For Mary Douglas
(1921–2007)

Mary Douglas. Photo reproduced by kind permission of the
British Academy.

Contents

Illustrations

Note on contributors

ALESSIO ASSONITIS received his doctorate in Renaissance art history from Columbia University. He is Director of the Medici Archive Project at the Archivio di Stato in Florence, a position endowed by the Florence J. Gould Foundation. He has published articles on Renaissance art, Medici history, Mendicant aesthetics, and the history of antiquarianism. His monograph on Domenico Ghirlandaio's follower, Bastiano Mainardi, came out in 2011. He is currently working on a book on the painter Fra Bartolomeo della Porta.

MARK BRADLEY is Associate Professor of Ancient History at the University of Nottingham. His main research and teaching interests lie in the visual and intellectual culture of ancient Rome, and he is the author of *Colour and Meaning in Ancient Rome* (Cambridge University Press, 2009) and editor of *Classics and Imperialism in the British Empire* (2010). He is also Editor of *Papers of the British School at Rome*, and general editor of a series of volumes on 'The Senses in Antiquity'. His recent work explores approaches to obesity in the ancient world, and he is currently working on a book on *Foul Bodies in Ancient Rome*.

PENELOPE J. E. DAVIES is Associate Professor in Art History at the University of Texas at Austin. Her work focuses primarily on public monuments of Rome and their propagandistic functions. She is the author of *Death and the Emperor: Roman Imperial Funerary Monuments from Augustus to Marcus Aurelius* (Cambridge University Press, 2000) and co-author of *Janson's History of Art*, seventh and eighth editions (2006, 2010). She is currently working on a book on the art and architecture of the Roman Republic, to be published by Cambridge University Press.

ELAINE FANTHAM studied Greats at Oxford and focused her research interest on literature and society from the middle Republic to the Flavian Principate. She taught at the University of Toronto until she moved to Princeton in 1986, retiring in 2000. Work on Ovid's calendar poem of religion and ritual has directed her recent interests to exploring traditional Roman religion, one of many rules and taboos, and as many escape hatches. Her most recent book (2006) is a biography of Augustus' scandalous and maligned daughter Julia.

DAVID GENTILCORE is Professor of Early Modern History at the University of Leicester. His principal interests lie in the social and cultural history of Italy, focusing on the relationships between different levels of society with regard to beliefs and practices, especially in the areas of religion, medicine and diet. Previous research has looked at attempts to control and regulate all aspects of the practice of medicine and healing, published as *Medical Charlatanism in Early Modern Italy* (2006), which was awarded the Royal Society of Canada's Jason A. Hannah medal. His current project, funded by a Leverhulme Trust major research award, explores the reception and assimilation of New World plants into Italy. *Pomodoro! A History of the Tomato in Italy* was published in 2010.

JUDITH L. GOLDSTEIN is Professor of Anthropology at Vassar College. She completed her PhD in Anthropology and Near East Studies (Princeton University, 1978), and has done field and archival research in Iran, Israel, France and Italy. She has published widely on such topics as aesthetics and modernity, cultural identity, social classification, and gender and consumer culture. Her publications include 'The Things They Left Behind', in Sarshar (ed.) *The Jews of Iran* (forthcoming); 'The Origin of the Specious', *Differences* 15.1 (Spring 2004); 'Realism Without a Human Face', in Cohen and Prendergast (eds) *Spectacles of Realism* (1995); and 'The Female Aesthetic Community', *Poetics Today* 14.1 (Spring 1993).

JOHN HOPKINS is an ACLS New Faculty Fellow at Rice University, and has been a postdoctoral fellow at the Getty Research Institute. His work deals primarily with architecture and viewer perception in the Roman world, especially as regards the early urbanization of the city. He is currently finishing a book on the architecture of Rome *c.* 700–450 BC and its impact on modern understandings of culture, society and politics in Rome through the early Republic.

DOMINIC JANES is Senior Lecturer in History of Art and Religion at Birkbeck College, University of London. He has research interests in the interactions of British textual, visual and material culture from the eighteenth century, as well the history of sexuality and the modern reception of classical antiquity and the Middle Ages. His most recent book is *Victorian Reformation: The Fight over Idolatry in the Church of England, 1840–1860* (2009).

JACK LENNON completed his doctorate in Classics at the University of Nottingham, and is currently Teaching Fellow at University College London. His PhD thesis examined aspects of religious pollution in ancient Rome, and he has wide-ranging interests in Roman religion and magic. He has published articles on the rhetoric of pollution in Ciceronian invective, pagan impurity in Christian polemic, and other aspects of pollution in Roman society, religion and culture.

KATHERINE RINNE is an urban designer and historian whose research is focused on water infrastructure in Rome from its foundation to today. She is author of *The Waters of Rome: Aqueducts, Fountains, and the Birth of the Baroque City* (2010). She teaches architectural design and urban history and theory at California College of the Arts in San Francisco, and is founder and editor of the cross-disciplinary online resource *Aquae Urbis Romae* (www3.iath.virginia.edu/waters).

MARTINA SALVANTE is an IRCHSS Postdoctoral Research Fellow at the Centre for War Studies at Trinity College Dublin. She works principally on Italian and European history, Fascism, gender and sexuality and disability, and her current research project focuses on the disabled Italian veterans of the First World War.

CELIA E. SCHULTZ is Associate Professor of Classical Studies at the University of Michigan. She has also taught at the Pennsylvania State University, Bryn Mawr College (from which she received her PhD), Johns Hopkins University, and Yale. She is the author of *Women's Religious Activity in the Roman Republic* (2006), and co-editor of *Religion in Republican Italy* (*Yale Classical Studies* 33, 2006). She is currently working on a study of Cicero's *De divinatione*.

KENNETH STOW is Professor Emeritus of Jewish History at the University of Haifa. He has twice been a visitor at the Institute for Advanced Studies, Hebrew University, Jerusalem, and most recently has been Bodini Research Fellow at the Italian Academy at Columbia University. He is founder and until 2011 editor of the periodical *Jewish History* and author of *Alienated Minority: The Jews of Medieval Latin Europe* (1992); *Theater of Acculturation: the Roman Ghetto in the Sixteenth Century* (2001); *Jewish Dogs: An Image and Its Interpreters – Continuity in the Catholic Jewish Encounter* (2006); and *The Jews in Rome* (1995, 1997).

TAINA SYRJÄMAA is Professor of General History at the University of Turku. Her main fields of study are nineteenth- and twentieth-century urban history, Italian history, and the history of tourism and consumer culture. Most recently she has studied the manifestation and production of 'progress' in world exhibitions. Her main publications include *Constructing Unity, Living in Diversity: A Roman Decade* (2006).

Preface

This volume examines the significance of pollution and cleanliness in the art, literature, philosophy and material culture of the city of Rome during antiquity and from the Renaissance through to the twentieth century. Dirt, disease and pollution and the ways they are represented and policed have long been recognized by historians and anthropologists as occupying a central position in the formulation of cultural identity, and Rome holds a special status in the West as a city intimately associated with issues of purity, decay, ruin and renewal. In recent years, scholarship in a variety of disciplines has begun to scrutinize the less palatable features of the archaeology, history and society of Rome. This research has drawn attention to the city's distinctive historical interest in the recognition, isolation and treatment of pollution, and the ways in which politicians, architects, writers and artists have exploited this as a vehicle for devising visions of purity and propriety. And yet, in spite of the volume of research into isolated instances of filth and cleanliness at Rome, there has been no comprehensive study of the history of pollution within the city. The challenge that remains, then, is to develop a more sophisticated analysis of developments over time in one geographical location, and to situate approaches to pollution in the city of Rome more broadly within cultural anthropology and the history of ideas.

This volume focuses on the theme of 'Pollution and Propriety' and the discourses by which these two antagonistic concepts are related. How has pollution in Rome been defined, and by what means is it controlled? To what extent is dirt culturally constructed (a position championed by Mary Douglas, but currently under challenge)? If dirt is dis-order/'matter out of place', how useful is it as an index of order or social and cultural system? How does Rome's own social and cultural history affect the way states of dirt and cleanliness are formulated? Does purity always accompany political, physical or social change? How different are pagan and Christian approaches to pollution and propriety at Rome, and do these approaches change over time from ancient to modern? Does Rome's reputation as a 'city of ruins' determine how it is represented? What makes images of decay in Rome so picturesque? And what do approaches to dirt at Rome tell us about contemporary value systems?

Rome, Pollution and Propriety brings together scholars from a range of disciplines in order to examine historical continuities in dirt, disease and hygiene within one environment, and to explore the development and transformation of these ideas alongside major chapters in the city's history, such as early Roman urban development, the Roman Empire, the early Church and the Renaissance, the Unification of Italy, and the advent of Fascism. This volume sets out to identify the defining characteristics, functions and discourses of pollution in Rome in such realms as disease and medicine, death and burial, sexuality and virginity, prostitution, purity and absolution, personal hygiene and morality, criminality, bodies and cleansing, waste disposal, decay, ruins and urban renovation. It also studies the means by which that pollution was policed and controlled. By combining literary and visual material on pollution, this volume integrates areas of academic inquiry that are normally separated in scholarly research in order to identify underlying cultural patterns. Its common theme is the enduring concern for purity in its various forms, as well as the concern for the environment that is evident throughout the history of the city.

This volume will be of interest to students and scholars working in archaeology, anthropology, art history, classics, cultural history and the history of medicine, as well as anyone interested in the history, society and culture of Rome. It provides a compelling context for examining general theoretical approaches to pollution and purity, which have experienced a resurgence of interest in academic and popular circles in recent years in the form of projects, publications and exhibitions. In doing so, the volume evaluates the applicability of these models to Rome, as well as using Rome as a test study for evaluating the models themselves.

Acknowledgements

It was a conversation with the late Keith Hopkins early in 1999, while I was struggling with my MPhil dissertation at Cambridge on ancient Roman laundries, that led to the conception of this project: the Romans had some filthy habits, Keith exclaimed in his inimitable manner, and somebody needs to do a proper study of what they thought about dirt and cleanliness. After the completion of my PhD (on an altogether cleaner Roman topic), I was determined to follow Keith's advice. Following my appointment at the University of Nottingham in 2004, a grant from the British Academy allowed me to spend a summer at the British School at Rome, plumbing the depths of the city's sewers, latrines, prisons, fulleries, tombs and all the less palatable features of Roman civilization – and it became clear that Roman dirt was a subject of interest to a wide range of scholars, and not just those working on antiquity. In 2005, I met Richard Wrigley, a regular at the British School at Rome and Professor of Art History at Nottingham, who had particular interests in health, disease and hygiene and their impact on the art and architecture of early modern Rome, and we could not pass up the opportunity to team together and get our hands dirty. Together we set out to organize a conference on continuities and differences in approaches to pollution and purity across the history of the city of Rome. A two-day conference ('Pollution and Propriety: Dirt, Disease and Hygiene in Rome from Antiquity to Modernity') was held at the British School at Rome on 21–2 June 2007, attended by nearly a hundred people and with speakers from all over Europe, America and Australia, from such diverse disciplines as archaeology, classics, history, literary studies, the history of art, the history of medicine, sociology and anthropology. In spite of its disciplinary and chronological range, the meeting quickly established a coherent and effective interdisciplinary dialogue around the central themes of the conference, and there was an impressive level of continuity in the arguments, ideas and material presented. The conference was only possible due to generous financial support from the University of Nottingham and the indispensable resources and facilities of the British School at Rome: in particular, I would like to thank Andrew Wallace-Hadrill, Elly Murkett, Geraldine Wellington and Peppe Pellegrino for making the conference run so seamlessly. Many of the conference's participants have contributed to this present volume, but those who have not are nonetheless owed a debt of gratitude for helping us to develop and enrich its approaches and ideas: Bob Arnott, Carlin Barton, John Bodel, Meredith Carew, Katy Cubitt, Val Curtis, Caroline Goodson, Adam Gutteridge, Gemma Jansen, Conrad Leyser, Pamela Long, Ann Koloski-Ostrow, François Quiviger, Renato Sansa and (most of all) Richard Wrigley. The volume itself has developed slowly, and I am grateful to all the contributors for their patience and encouragement across the last four years, and to Jack Lennon for his assistance with the copy-editing. Thanks are also owed to audiences at Austin, Lampeter, Liverpool, the London School of Hygiene and Tropical Medicine, Nottingham and Rome for their feedback on the project and its underlying methodology, themes and arguments. I also owe a debt of thanks to Michael Sharp and the staff of Cambridge University Press for their support of this project, and to two anonymous readers for their valuable and encouraging feedback. Finally, I must also reserve a special mention for Mary Douglas, who had agreed to be keynote speaker, but who passed away just a month before the conference was held. Her voice, approaches, ideas and arguments were nevertheless resonant throughout the proceedings, and offer intellectual coherence and unity to the volume that has emerged out of the conference.

Mark Bradley
University of Nottingham
March 2012

Introduction

MARK BRADLEY AND KENNETH STOW

In March 2007, Rome's Piazza del Popolo was desecrated with a provocative and topical type of dirt. For nine days, the Piazza was populated by approximately a thousand life-sized figurines, formed out of compacted, crystallized domestic and commercial waste (see figure 0.1). H. A. Schult's 'Trash People', a German exhibition dedicated to making a loud and highly visible statement about consumerism and environmental waste, had in fact already defiled Paris, Moscow, the Great Wall of China, the pyramids of Egypt and several other key landmarks around the world, and would go on to disrupt the skyline at Barcelona, New York and Antarctica.[1] Of course, this anthropomorphosed trash did not really constitute 'pollution': Schult's virtuous and politically correct mission to convince the world that (as he put it) 'we live in a time of garbage' and his argument that this 'social sculpture' operated as a 'mirror of ourselves' meant that his dirty exhibits – however 'out of place' they might seem – were thoroughly sanitized and legitimated, a powerful warning about what that world *might* become if it did not moderate its relationship with rubbish. 'Trash People', then, was a safe and institutionalized representation of dirt at its most global. Rome's 2007 encounter with universal trash, however, was by no means typical of the historical relationship between the city and its filth. Whereas 'Trash People' was a product of a modern global awareness imposed on the city from outside, Rome throughout its 2,500-year history has done a thorough job of formulating, evaluating and policing its own internal, culturally specific forms of pollution. This volume, then, is dedicated to exploring the history of pollution not within human thought (as many previous studies of the subject have done), but as part of the history, society, religion and politics of a single city. By doing so, it is hoped that the present study will both demonstrate the importance of pollution as a concept within the history of Rome, and provide a sensitive and nuanced case study for the construction and negotiation of pollution by culture.

This volume is a collection of fourteen essays and an Envoi addressing the development and transformation of ideas about dirt, disease and hygiene in

---

[1] See Schult (2002).

0.1. H. A. Schult, 'Trash People', Piazza del Popolo (March 2007). Photo: H. A. Schult.

several major chapters of the history of the city of Rome, both in antiquity and in the modern period. It also argues that pollution and purity were guiding factors in the organization of the city's religion, politics, literature, art and architecture. Its contributions, representing the collaborative efforts of classical, Renaissance and modern scholars, span approximately two thousand years of Roman history from the Republican period through to the early decades of the twentieth century. They are structured broadly in chronological order, although each chapter addresses particular themes in the history of the city: pagan ritual; urban development; early Church doctrine; plague; sanitation; immorality. Most of the contributions concentrate on discrete historical or discursive moments, but collectively they span some of the most significant chapters in the city's history and provide a wide-ranging analysis of the synchronous development of Rome's society, religion and culture and ideas about purity and pollution, order and disorder. They represent a range of approaches to the subject, but set out to integrate detailed studies of events, individuals, literature and visual culture within the broader theoretical framework of pollution and propriety. Some chapters examine, from a synchronic perspective, patterns in how the city's inhabitants integrated these concepts into their lives (Lennon,

Fantham); others examine specific events and developments (Gentilcore, Salvante), the characterization of particular groups (Stow on Jews, Janes on Victorian Protestants) or periods (Rinne, Syrjämaa), or shifts in practice over time (Davies, Hopkins); others adopt a primarily prosopographical approach (Assonitis) or explore the history of an idea (Bradley chapter 6 and Schultz on ancient punishment). Many of the chapters concentrate on literary approaches to pollution, others on archaeological, legal or visual material. In spite of the complex character and diverse definitions of 'pollution', this volume has mobilized Mary Douglas' core formulation of dirt as 'matter out of place' as a common and interactive theme that unites all the chapters. Above all, the volume as a whole and its constituent parts are dedicated to creating a constructive dialogue between disciplines by exploring a common and pervasive theme of interest that cuts across several areas of scholarly research and that appears as a recurring feature in the history of the city from antiquity to modernity.

One key theme that ties together many of the chapters in this volume is the idea that the discourse of purity and pollution has been used as a political weapon across the city's history, whether in urban zoning, forensic rhetoric, early Church doctrine, treatments of plague, sanitary reforms or the marginalization of minority groups; indeed this idea that purity systems have functioned as forms of social control across all human cultures is an essential component of theories proposed by Mary Douglas and other anthropologists about the role of cleanliness, dirt and danger within communities (see chapter 1, pp. 11–18). This is particularly evident in the sphere of religion, identified by many theorists as a central mechanism in the negotiation and policing of social boundaries and values: from pagan ritual to early Church law to papal sanitation measures, this volume demonstrates that throughout the city's history the religious authorities have appropriated ideas about cleanliness, pollution and purity to maintain and reassert social control. This is one important aspect of historical continuity: at every stage in its history, Rome has mobilized what Davies in this volume describes as a 'pollution-fighting infrastructure' (p. 74). Furthermore, a diachronic study of these themes reveals that a regime's claims to a state of purity or sanitation often involve developing discourses about the physical, moral or political decay and uncleanness of what went before. The restored and immaculate Rome of Augustus, for example, frequently imagined a late Republican city characterized by crumbling temples, polluted streets and political decay; early Christian writers bemoaned the impiety and ritual pollution of pagan religion; and the literature and imagery of Fascist Rome went to great lengths to compare and contrast the clean and renovated contemporary cityscape with the ruins and squalor of its recent past. The type of 'dirt', then, with which this volume is concerned is the stuff of language, discourse and representation: this volume does not explore how filthy the Roman sewers *actually* were, or how many people were actually carried off by the 1656 plague, but instead engages with the political and religious discourses that mobilized these phenomena as essential components of the city's value system – the expulsion of sewage as a powerful metaphor for purging Rome's criminals or the Church's moral quest to identify and eliminate the causes of disease.

Another theme that is central to thinking about pollution and purification is the organization of *space*: carrying weapons on a battlefield is normal, but carrying them inside the city can be highly inappropriate; sex in a marital bed is 'in place', but sex in a temple or church is polluting; corpses in a necropolis are where they should be, corpses on the street are not. The formula 'matter out of place', then, has necessarily put space and location high on this volume's agenda, and it is no accident that many of its chapters are concerned with the appropriate and inappropriate organization of urban space: sites of burial, sewage networks, disposal of criminals, the quarantining of plague victims, the zoning of minority groups or the renovation of key urban areas. Furthermore, throughout Rome's history we find an enduring concern with the negotiation, establishment and maintenance of boundaries: the sacred *pomerium* of the ancient city that distinguishes the internal and the external, the civic and the military, for example; or boundary walls dividing wealthy districts from poor districts; or spaces marked out for ritual activity, economic activities or particular social groups. Connected to this is the theme of 'racialization' that characterizes several episodes in the city's history, where groups perceived as marginal or threatening are compartmentalized socially or physically: the Jewish ghetto, for example (see Stow), or the political discourse that identified the city's male prostitutes as foreign rather than indigenous (see Salvante).

This volume, however, with its diachronic emphasis, demonstrates that these boundaries are negotiable: matter can be put in place not only by physically moving it, but also by reconfiguring or even renaming it. As Hopkins demonstrates (this volume), Rome's sewers could be both cleansing and polluting: they were constructed to drain Rome of its dangerous overflows, but (even as they were being built *c.* 600 BC), we are told, they were polluted by the crucified bodies of unwilling sewer-workers; by the early Empire, they were held up as one of the wonders of ancient Rome alongside roads and aqueducts, carrying an estimated 100,000 lb of human waste every day out into the Tiber and a focal point for religious activity to mark their purificatory role, but they were also where Nero is imagined washing (and so further contaminating) his bloodied hands after his nocturnal brawls; they were precisely the right place for depositing the body of the depraved emperor Elagabalus, but not that of St Sebastian, a striking example of 'matter out of place' dramatically represented in a painting by Lodovico Carracci (1612) commissioned for the Church of Sant'Andrea della Valle, where the saint's body had reportedly been recovered from the sewer (see cover image).[2] Whether feats of hygienic engineering or sites of contamination, the sewers were potent and versatile carriers of meaning across the city's history. And there are many other compelling examples of Roman filth reconfigured: in the hands of fullers or tanners urine could stop being human waste and become a detergent; bodies that had been in the ground long enough could lose their stigma as carriers of pollution (there is evidence for multiple burial grounds within the sacred boundary of the city); Christian corpses buried *ad sanctos* (next to the saints) were exactly where they should be; and executed criminals could be exposed in the Forum for righteous mutilation (see Bradley chapter 6). Furthermore, it is important to note that corpses on the street, dungheaps or fish-markets are not *per se* evidence of urban pollution: such sights and smells, particularly if they are part of an everyday experience, can be normalized and recalibrated. And alongside internal developments, a welter of external influences has shaped how dirt and cleanliness have been perceived within the city: the water technologies of the ancient Greek East, for example, ideas about purity circulated by resident Jews (Stow), or the religious and moral values of Victorian England (Janes). Pollution, at least in the terms this volume envisages it, is a shifting and organic component of political or cultural discourse rather than a static, objective phenomenon.

This volume is set out in two distinct halves: the first half concentrating on Rome in antiquity, and the second half on the city from the early Renaissance through to the twentieth century. We beg the reader's indulgence in omitting a long and important stretch of the city's history (late antiquity and the origins of Christianity, as well as the development of the medieval Church) for which the themes of pollution and propriety are pivotal: these periods have been the subject of some fascinating recent work (see pp. 28–33), and the volume's approaches and contributions have benefited greatly from these perspectives. However, the modest aim of the present work is to persuade the reader that the city's ancient and

---

[2] In fact, the painting's title is misleading if this was indeed the findspot of the body: Sant'Andrea della Valle lies close to the 'Giuditta' sewer, not the Cloaca Maxima (see figure 5.1).

modern history alike were informed and influenced by a common set of discourses about pollution and propriety, and that the shift from antiquity to modernity and from paganism to Christianity preserved some striking similarities in Rome's approach to dirt and the strategies it developed to fight it.

The volume begins by embedding the study of pollution and propriety in Rome within the theoretical and scholarly research context in which concepts of dirt, cleanliness and purity have been most important. Bradley's opening chapter considers the impact of Mary Douglas on approaches to pollution and the significance of her seminal work *Purity and danger* (1966) on scholarship in a wide range of disciplines, as well as addressing alternative theoretical approaches to the subject and the debates that have emerged out of them. It then discusses the importance of these themes for research on the society, religion and culture of antiquity – Greece and Rome, as well as ancient Judaism and early Christianity. The chapter then considers how some of these approaches have been developed by studies of medieval Europe and the modern West, as well as exploring some of the new directions adopted by this scholarship. It finishes by discussing various considerations for approaching dirt, cleanliness and pollution in the twenty-first century, and by doing so highlights the flexibility and malleability of these concepts, as well as their intellectual potential as indices of culture.

The first half of the volume then discusses aspects of the city of Rome in antiquity, concentrating on the late Republic and early Empire. Chapters 2 and 3 explore the various manifestations of pollution and purity in the early stages of Roman society, an exercise that is familiar within scholarship on ancient Greece, but which is under-represented in work on Rome (see chapter 1, pp. 19–22). The chapter by Lennon examines several basic aspects of ancient Roman life (birth, sex, blood, death) in order to demonstrate the pervasive significance of pollution and purity in pagan society, as well as the centrality of religion and ritual in creating patterns of belief and imposing system and order on the city's inhabitants. Following on from this, Fantham discusses the formulation of pollution and purification in one area of Roman religious ritual – purification and the avoidance of pollution on public holidays in the Roman calendar year – and the way this filtered across into the community's secular life; in particular, this chapter makes an important point about the intimate relationship between washing and ritual in ancient Roman thought, and the importance of water (and the Tiber) in achieving states of purification. These two chapters, then, critically assess the terms in which pollution and propriety could be formulated and described in the initial phases of the city's history, and set out some preliminary parameters for understanding and approaching these concepts in the rest of the volume.

The three chapters that follow examine the ways in which notions of pollution and purity helped give the ancient city physical shape, by exploring the establishment and negotiation of boundaries at critical stages in Roman historical development and considering how the very process of urban cleansing could be incorporated into the city's religious and symbolic system. These chapters draw together important recent work on Roman urban sanitation, its significance for the city's early political and legal developments, and its thematic and metaphorical currency within contemporary literature and rhetoric (see pp. 22–5). Davies' chapter discusses urban planning during the Republic, when the city was expanding faster than at any other period, and considers Roman attitudes to the quintessential taboo, death and burial, and how these attitudes dictated the relationship between 'inside' and 'outside', between clean and unclean, as well as how Republican politicians directed their professional careers to establishing and reinforcing urban sanitation and cleanliness. Hopkins then focuses attention on a single urban monument by examining the ambivalent representation of the magnificent ancient sewer (the Cloaca Maxima) – simultaneously a miracle of engineering for purging the city and a receptacle and focal point for the city's dirt and impurity. Bradley maintains the emphasis of the previous two chapters on topography and dirt removal, but extends it into the sphere of criminal behaviour and punishment by examining a critical part of the urban landscape in which the capital punishment of Rome's criminals was carried out, the Capitoline Hill complex (with its execution chamber, sewer channels and Tarpeian rock). He argues that criminality in legal rhetoric and across the full range of Latin literature became analogous to the literal pollution that so much of the city's infrastructure was designed to remove, and explores the creative representation of crime and punishment within the city's landscape. Like chapters by Davies and Hopkins, this chapter draws attention to the distinctive symbolic and intellectual currency attached to the process of waste disposal in ancient Rome. Collectively, these three chapters demonstrate how a seemingly narrow aspect of ancient urban development could permeate social, religious and cultural life, and point forward to themes in the city's more recent politics, religion and law that are explored by the volume's later chapters.

The final chapter on the ancient city, by Schultz, pursues similar concerns about the organization of space by examining strategies in pagan Rome for disposing of individuals whose actions or behaviour had threatened the religious establishment and made their presence polluting to the city. This chapter corroborates anthropological arguments about the role of religion as a mechanism for social control, as well as the thesis championed by Mary Douglas that purity and pollution were central discourses in this process (see p. 12). Schultz revisits a classic problem of Roman religious ritual – the live interment of transgressive Vestal Virgins – and considers the traditions of this practice alongside other expiatory rituals in Roman religion that involved expulsion and elimination, patterns of Roman pagan behaviour that would shape and influence early Christian approaches to heresy and dissidence.

Part II of the volume then shifts the focus on to the city in modernity, beginning with plague and bodily pollution in fifteenth-century Rome and scrutinizing the themes of pollution and propriety across each major phase of Rome's modern history. Chapters 8 and 9 address a theme that was critical to both ancient and medieval urban culture and the history of medicine: the effects of disease on politics and urban management, as well as the establishment's exploitation of disease as a tool of social and political control (for a discussion of some of these ideas, see pp. 28–31). Assonitis examines the treatises and sermons of the fifteenth-century Dominican friar Fra Girolamo Savonarola, who, drawing upon the language of contemporary medical culture, presented the Roman Church as a diseased body, and the city of Rome as a plague-stricken graveyard for pagan morals and behaviour. Assonitis focuses on a familiar theme in recent scholarship on urban history and the study of pollution: the somatization of the city and the use of anatomical and medical language to negotiate and evaluate its values and morals. Gentilcore then discusses the Roman plague of 1656 and the attitudes adopted by the authorities in treating and controlling the outbreak; by examining their willingness to bargain over the health of Rome's inhabitants with one particular Neapolitan 'alchemist', Gentilcore discusses the character of this epidemic in Rome (and in other Italian cities), and the quest for new and effective remedies to fight it. These two chapters consider the extent to which the outbreak of disease, and its effects and treatment, were discussed and evaluated within the city in comparable ways at different stages in Rome's history.

The next three chapters, which address the theme of sanitation and renovation, return to some of the ideas explored in earlier chapters on ancient urban management, by examining various efforts in the history of Rome from the early modern period to the late nineteenth century to fix Rome's prevailing reputation as a city of dirt, disease and corruption through the zoning of particular regions, the public supply of water and the renovation of buildings and districts. Chapters 10, 11 and 12 explore spatial aspects of cleansing and sanitation, and the association of particular areas with perceived sources of pollution within the local population. Stow begins by discussing the marginalization of Rome's ghettoized Jews in the late sixteenth century and the means by which this section of the Roman community dealt with poor urban sanitation; by doing so, he addresses the issue of urban zoning in early modern Rome, the stigmatization of minority groups, and the discourses of purity and pollution with which contemporary social issues were described. Rinne then surveys papal efforts in this same period to improve urban sanitation by renovating the city's water supply and infrastructure and thereby symbolically cleansing both the city and the Church of vice and corruption; these renovations and urban cleansing programmes, she argues, were a key part of papal propaganda to reinstate Rome as the rightful centre of Christendom. Finally, Syrjämaa explores the tensions generated by differing approaches to urban space and Rome's identity, both those formulated by outsiders and those proposed by the internal authorities, once Rome had become the nation's capital in the late nineteenth century; by examining a range of visual and literary evidence from nineteenth-century Rome, this chapter explores the intellectual currency of dirtiness and cleanliness in a period of intense social and political change. These three chapters, then, highlight the continuing significance attributed to programmes of cleansing and purification in Rome's modern history and consider the relationship of these programmes to the city's longstanding associations with dirt and pollution.

The final two chapters in part II consider two modern contexts in which the city of Rome has been associated with physical and sexual immorality: first, by Victorian commentators in mid-nineteenth-century England; second, by the Fascist legal and political discourses of 1920s Rome. Deviant sexual behaviour, and its association with crime, disease and immorality, has been recognized as a critical component of Western urban society, particularly in the cities of modern Europe (see pp. 31–2), and these final chapters integrate several themes in the history of Rome explored in earlier chapters. Janes discusses the city's representation as a site of physical and moral danger by religious figures of Victorian England for whom Rome had become an evocative lesson about the dangers of decay, corruption and the seductions of the flesh. Second, Salvante explores a case study within Rome itself, concerning the regulation of 'deviant' juvenile sexuality – specifically, male prostitution – and its identification with the city's physical and moral margins. The final part of the volume, then, considers pollution discourses within the city's more recent history and their role in formulating and shaping aspects of Rome's current urban identity. The volume closes with a short Envoi, in which Goldstein considers the role of dirt and cleanliness as a system of communication in modern crime fiction and investigative writing set in Italy, and the enduring significance of Mary Douglas' approach to pollution for considering all aspects of this system.

1 | Approaches to pollution and propriety

MARK BRADLEY

The aim of this chapter is to set the study of pollution and propriety in the city of Rome within the larger theoretical and scholarly research context in which concepts of dirt, cleanliness and purity have been most prominent. It will first address various theoretical approaches to the subject in anthropology and sociology, with particular focus on the contribution of Mary Douglas to scholarly debate on pollution, purity and danger across a range of disciplines. The chapter will then discuss the importance of these themes for research on the society, religion and culture of ancient Greece and Rome, as well as ancient Judaism and early Christianity, before moving on to address aspects of dirt, disease and hygiene in the urban centres of medieval Europe and the modern West. Finally, by exploring various considerations for approaching these themes in the twenty-first century, this chapter will demonstrate that pollution and propriety are flexible, negotiable and organic concepts which are not only embedded in the society's memory, experiences and modes of communication, but sit at the very heart of the community's value system.

Pollution in anthropology and sociology

The methodology adopted by this volume is most indebted to the work of Mary Douglas (1921–2007), the British anthropologist who put the study of pollution and purity on the map and who passed away just a month before she was due to speak at the 'Pollution and Propriety' conference in Rome. Her most celebrated book, *Purity and danger: an analysis of concepts of pollution and taboo*, first published in 1966 and released in multiple editions, examined how pollution is formulated and policed in a range of world societies and religions. Its core definition of pollution – or dirt – as 'matter out of place' was critically important: dirt is a by-product of a systematic classification of matter, inasmuch as ordering involves rejecting inappropriate elements.[1] If dirt is dis-order, dirt tells us about *order*, about

---

[1] Douglas (1966) 36. For a clear and concise summary of Douglas' contribution, see Moore (1997); Fardon (1999), esp. ch. 4, '*Purity and danger* revisited' (pp. 75–101). See also Wuthnow

social and cultural system. It also tells us about such things as morality, sexuality, law and communication. It helps us to understand what is included in that system, and what is excluded. Purity systems, the book argued, are there to maintain category boundaries and can be seen as a mechanism for social control. *Purity and danger*, then, was committed to the idea that dirt – what is polluted or unclean – is formulated and constructed by culture, and can tell us a great deal about a culture's value systems: 'in chasing dirt, in papering, decorating, tidying, we are not governed by anxiety to escape disease, but are positively re-ordering our environment, making it conform to an idea'. This extraordinary study has now been translated into more than fifteen different languages, and is recognized as one of the most influential and successful works of twentieth-century anthropology, and its author's recent death makes this an opportune time to re-evaluate her contribution to the field.[2]

Douglas was building on a diverse range of studies in theoretical anthropology. Most notably, she was committed to the theoretical framework laid down by the French sociologist Emile Durkheim, who had argued that religion was a cohesive mechanism developed to protect a threatened social order: religion defined and policed both the pure and the impure, and purity systems in turn embraced the very categories that shaped and governed a society.[3] Durkheim's approach had also been developed by the work of the Romanian philosopher Mircea Eliade on the sacred and the profane, concepts which informed Douglas' approach to pollution.[4] Douglas was influenced too by the work of Arnold Van Gennep, who had argued that the rites of passage that characterized all human societies involved transitional phases where subjects were in a position of marginality or liminality and were considered outside society and thus dangerous or 'sacred': such individuals, Douglas contended, were unclassifiable anomalies and were therefore regarded as both polluted and polluting.[5] However,

---

et al. (1984), esp. chs 8–11. For an online video interview conducted with Douglas by A. Macfarlane in 2006, see www.alanmacfarlane.com/ancestors/douglas.htm (accessed July 2011).

[2] So for example Campkin and Cox (2007), published shortly after Douglas' death. Douglas' influence has been evident across several distinct disciplines: see for example Shonfield (2001). For Douglas' obituaries, see R. Fardon in the *Guardian*, 18 May 2007 (at www.guardian. co.uk/news/2007/may/18/guardianobituaries.obituaries) and in *The Times*, 18 May 2007 (at www.timesonline.co.uk/tol/comment/obituaries/article1805952.ece) (both accessed July 2011).

[3] Durkheim (1912). On this relationship, see Fardon (1987). Other anthropologists concerned with religion who influenced Douglas' work include H. B. Tylor, J. Frazer and W. Robertson Smith.

[4] Eliade (1957).

[5] Van Gennep (1960), first published in 1909 as *Les rites de passage*. Van Gennep's approach was extended, shortly after the publication of *Purity and danger*, by V. Turner in *The ritual process:*

Douglas also managed to integrate this theoretical methodology effectively with cross-cultural analysis, an approach developed through her ethnographic fieldwork among the Lele of Zaire and influenced by her mentor at Oxford, the comparative anthropologist Edward Evans-Pritchard. This approach to pollution allowed for much more flexibility and diversity between groups and communities than Durkheim's concept of a sweeping 'social consciousness'. *Purity and danger*, then, developed a methodology that was sensitive to local variations but also pointed to patterns and rules that recurred in all human societies. After *Purity and danger*, Douglas continued to pursue research in the field of religion, sociology (particularly concerned with the symbolic significance of foodstuffs) and – most recently – aspects of literary composition, but, apart from the occasional article, did little to develop further her ideas on pollution.[6] *Purity and danger* has nevertheless remained a cornerstone of Western thought about pollution, and it has influenced (and continues to inspire today) a wide range of research in sociology, anthropology, classics and various other disciplines.[7]

Douglas' ideas, for example, were critically examined by V. Valeri in his important discussion of taboos among the Huaulu forest hunters of Eastern Indonesia (*The forest of taboos*, 1999), which examines death, sex and feeding as sites for pollution, arguing ultimately (as Douglas herself put it in her review of the book) that 'taboos reinforce the oppositions that construct the universe'; Valeri, however, moved beyond Douglas' theoretical framework by directing attention to the significance of corporeality and the physical constitution of the subject for formulating ideas about pollution and propriety.[8] This was not the first critical reappraisal of Douglas' theories in cultural anthropology: A. Meigs had already attempted to develop a new theory of pollution in her study of communities in Papua New Guinea, focusing on rituals and rules of feeding and (like Valeri) arguing that the

---

   structure and anti-structure (1969), which examined rituals among the Ndembu, as well as ideas about classification, liminality, hierarchy and 'communitas' (an unstructured community in which all members are equal). See also Turner (1979).

6 See for example, Douglas (1991) on witchcraft and leprosy; (1998) on bodily purity. Douglas took some steps towards modifying her approach to Leviticus: see Douglas (1999), esp. introduction; cf. Bendlin (2007) 189.

7 In classics, the most important application of Douglas' theories was made by R. Parker's *Miasma* (1983) (see below pp. 19–21); see also Beard (1980), who applied Douglas' ideas about social classification in religion to understanding marginal identities and interstitiality in the cult of Vesta. In a later article (1995, cited much less often than Beard 1980), she revised her initial line on Vestal Virgins by rejecting structural anthropology as an effective approach to understanding religious meaning.

8 Valeri (1999), reviewed by Douglas (2000). For a good theoretical discussion of the state of thinking on taboo and its relationship to religious systems, see Zuesse (1979).

body can function as a map for formulating and expressing moral concerns.[9] Most anthropologists, however, have readily accepted and extended Douglas' approach to their individual studies, particularly those concerned with religious rites and death rituals.[10]

Recent years have also seen a growing interest in interdisciplinary research on dirt, cleanliness and pollution. Several general studies of the subject, many of them released straight into paperback, attest to modern preoccupations with the less palatable features of civilization. W. I. Miller's *The anatomy of disgust* (1998), for example, analyses the emotional function of disgust in life in the modern West and draws upon history, literature, psychology and moral philosophy to demonstrate that disgust and responses to dirt perform essential roles both in self-preservation and (following Mary Douglas' line) in the creation and maintenance of social hierarchies.[11] Miller, Professor of Law at Michigan and author of an earlier book on the theme of humiliation (1993), surveys a wide range of material and phenomena from ancient and medieval through to modern Europe as part of a compelling history of emotions in the West. Particularly successful is the book's treatment of the role of aesthetics and the senses in the formulation of disgust, as well as its integration of ideas and material that are normally kept distinct. Although the premise of this investigation is indebted to Douglasian theory, some of his conclusions are ambitious (e.g. that democracy is dependent on a shared sense of disgust) and unpersuasive (e.g. that misogyny stems from male disgust at semen), and his sweeping approach to the subject, cutting across chronological, geographical and generic boundaries – while thought-provoking – is less sensitive than it might be to cultural differences in the formulation and policing of pollution.[12]

Even more wide-ranging is V. Smith's 2007 monograph *Clean: a history of personal hygiene and purity*, which examines approaches to cleanliness from prehistoric times, through to ancient Greece and Rome, early Christianity, medieval and early modern Europe, and the twentieth and twenty-first

---

[9] Meigs (1978) and (1984). Cf. Miner (1956), which discusses the rituals and ceremonies developed by this North American community to police the physical and moral dangers posed by the various uses of their bodies.

[10] See for example Goldschmidt (1973); Watson (1982) and (1988); Woodburn (1982). More recently, see Barrett (2008) on Aghor pollution.

[11] Research on the social significance of disgust has also found a home in more popular domains. Channel 4, for example, ran a series in August 2000 called *Anatomy of Disgust* which examined the various functions of disgust in psychology, society politics, religion and art. See Leyerle (2009) esp. 352–5 for an application of Miller's ideas to an example of early Christian 'rhetoric of disgust'.

[12] Further on disgust, see Menninghaus (2003); Nussbaum (2004); Rindisbacher (2005). For earlier approaches, see Knapp (1967); Ignatieff (1995).

centuries. Smith, honorary fellow at the London School of Hygiene and Tropical Medicine, is committed to a highly interdisciplinary method that integrates a range of historical, archaeological, religious, sociological and biological material and argues that – in spite of all the diverse cultural manifestations of cleanliness – there are a number of basic continuities in approaches to cleaning bodies that have dictated the historical development of human hygiene. The resulting narrative is engaging and thought-provoking, although the book covers so much ground that it is difficult to reach a detailed understanding of any of the individual periods it surveys. A similar approach, discussing comparable material, is taken by K. Ashenburg's *Clean: an unsanitised history of washing* (2008), although Ashenburg adopts the much safer line that cleanliness is culturally formulated and packaged: 'the way a culture approaches and achieves cleanliness always says something interesting about that culture'.[13] At the other end of the spectrum, studies such as J. Scanlan's *On garbage* (2005) have argued that what counts as waste is both fluid and culturally negotiable.[14]

Many of the ideas explored by these studies have in recent years been communicated to the public domain in the form of radio and television programmes and exhibitions: most recently, the Wellcome Collection hosted an exhibition in London called 'Dirt: The Filthy Reality of Everyday Life' (24 March–31 August 2011) which collected over 200 artefacts spanning visual art, documentary photography, film and literature, scientific instruments and items of everyday life, ranging from Victorian London to Nazi Germany, and from seventeenth-century Holland to futuristic New York. It is telling that, still after nearly fifty years, Mary Douglas' words and ideas were pivotal to the themes and arguments of the exhibition and its accompanying literature and events.[15]

This question of whether dirt and cleanliness are culturally or biologically determined is perhaps the most significant – and interesting – subject of current debate surrounding the study of pollution. Douglas was firmly committed to the idea that dirt is an anomaly that is constructed by culture

---

[13] Ashenburg (2008) 295. See also Shove (2003), which adopted a cultural rather than a universalist approach to cleanliness.

[14] See also Thompson (1979), a smart and sensitive approach to the complex identities of waste. Cf. Logan (1995), which examines the history of dirt (in the sense of soil, compost and manure) from antiquity to the modern West and its relationship to culture. For a more sophisticated diachronic approach, concentrating on the history of 'public health' (including disease, public order and morality) in the West from antiquity to the modern world, see Porter (1998).

[15] For the exhibition, see www.wellcomecollection.org/whats-on/exhibitions/dirt.aspx (accessed July 2011). The splendidly illustrated exhibition catalogue (Cox *et al.* (2011)) collects six essays on a range of related themes such as personal hygiene, domestic cleaning and urban sanitation.

rather than determined by human nature, and this is the approach that has proven most attractive and persuasive to scholars working within the humanities, to whom pollution offers a compelling window onto the values, aesthetics and priorities of the specific communities, cultures and periods they are studying. Others, however, have argued that disgust, cleansing and other responses to dirt are biological problem-solving tools which have evolved over many centuries, instinctive protective mechanisms that guard the body from disease and harm. This is the premise underlying V. Smith's *Clean: a history of personal hygiene and purity* (see above), but it has been most forcefully championed in recent years by V. Curtis, Director of the Hygiene Centre at the London School of Hygiene and Tropical Medicine. Curtis, whose approach is informed by epidemiology and anthropology and is geared towards improving the understanding and practice of hygiene in underdeveloped countries, has argued that 'hygiene and disgust originated well before culture and history', *contra* Douglas *et al.* who claim that the local cosmology, or world order, comes first, with dirt as its by-product.[16] Recently, several studies in psychology have set out to demonstrate a basic correlation between cleanliness and moral judgement: most recently, S. Schnall, J. Benton and S. Harvey (2008) conducted experiments which they claimed provided evidence that those who had been physically cleansed or psychologically exposed to notions of cleanliness were intuitively less likely to make severe moral judgements, a study that has important implications for thinking about the role of purification in human society.[17] Because the focus of the current volume is the cultural history of a single city, however, this debate about the natural or cultural origins of cleanliness and dirt is not one that is actively engaged, although most of the volume's contributors are necessarily committed to the idea that approaches to dirt, disease and hygiene are indicative of specific values, trends and developments at Rome.

A concept that has grown out of Douglas' theories about 'matter out of place', and which has grown popular in recent decades, is that of the 'abject' – something that appears familiar (a human body, for instance) but which has been 'cast out' or marginalized due to some physical, social or political quality or process: for example, a corpse, a convict or a prostitute. The abject evokes strong reactions such as disgust, and is often resolved

---

[16] Curtis (2007), esp. 663. See also Curtis (1998); Curtis and Biran (2001). The idea that religious ritual practice could be determined by biological factors has also been championed by Burkert (1996).

[17] Schnall *et al.* (2008); cf. Wheatley and Haidt (2005); Zhong and Liljenquist (2006).

by the creation of 'spaces of abjection' in which it can be situated and normalized: a cemetery, a prison or a red light district. This approach to pollution, articulated most clearly in J. Kristeva's *Powers of horror: an essay on abjection* (1982), owes a great deal to Douglas' ideas about dirt, purity and spatial organization, but allows for a greater level of ambivalence in society's relationship to pollution and provides a more subtle model for exploring the relationship between space and politics. The abject has also become a familiar term in modern art and architectural theory, and has sometimes been connected with Freud's concept of the 'uncanny'.[18] Aspects of abjection and cognate themes are explored in the present volume particularly in the chapters by Stow (on the Jewish ghetto) and Salvante (on male prostitutes in 1920s Rome).

Although *Purity and danger* already allowed for a flexible and diverse approach to pollution that could apply to different societies and cultures (and this is one of the principal reasons for its success as a methodology), the recent scholarly trend has been to focus on the internal mechanisms of purity systems within specific communities, and the impact of those systems on a community's activities, beliefs and traditions. This is the methodology adopted by the present volume, although it hopes to develop its approach in two innovative ways: first, it sets out to demonstrate that the modes of communication developed by the community – the literature, art, architecture, laws and so on – shaped and constructed that community's approach to pollution and propriety as much as they reflected them (anthropological and sociological studies of the pollution have too often focused on the latter); second, the city of Rome offers a unique opportunity to study the development of purity systems over time. Ideas about pollution and purity are not static: this much is now taken as axiomatic (even if a great deal of twentieth-century scholarship has overlooked the flexible and organic character of these concepts). A rich and varied body of material from two and a half thousand years of Rome's history is available to help us to understand the complex diachronic processes at work in formulating, developing and negotiating ideas about dirt, cleanliness, purity and pollution and their significance in one community's society, religion and culture. This volume also aims to integrate areas of academic inquiry that are normally separated in scholarly research. Studies in the history of pollution, purity, sanitation and

---

[18] See Campkin and Cox (2007) 4–5; Robinson (2000); Sibley (1995). The abject has also become a familiar term in modern art and architectural theory: see for example Lahiji and Friedman (1997). On the 'uncanny', see Kelley (2004); and Vidler (1992) on the architectural uncanny.

propriety – particularly in connection to a single locality such as Rome – are usually confined to separate disciplines: classics/ancient history; medieval studies; or modern history, sociology and anthropology. The following section, therefore, will outline some of the research in these various areas that has informed and influenced the contents, themes and arguments of this volume.

Pollution in antiquity

Issues of pollution and purity in classical antiquity have received sporadic attention in the early years of the twenty-first century.[19] Besides the conference on which the present volume is based, M. Goodman and R. Parker co-ordinated a seminar series at New College Oxford in 2002 on the theme of 'Purity and Pollution in the Religions of the Mediterranean World in Antiquity', at which Mary Douglas gave the opening address and which examined Greek, Roman, Jewish and early Christian approaches to pollution. As the online report by J. Kirkpatrick demonstrates, this series of seminars exposed the topical and controversial character of pollution and purity in (and between) ancient religions, the difficulties of identifying common rules, and the fundamental complexities of definition.[20] These seminars took as their premise the contentious idea that purity systems are a defining feature of many *pre-scientific* religions, and that the study of pollution is therefore a peculiar concern of pre-modern historians. While it is certainly true that scientific knowledge has often been at odds with approaches to pollution that are rooted in ritual and tradition, the two are by no means mutually exclusive, and (as Douglas herself has demonstrated) groups and individuals in the modern West continue to subscribe to a basic set of rules about order, cleanliness and propriety which have little to do with scientific knowledge about pathogenicity and hygiene. Nonetheless, various ancient cultures – Egyptian, Jewish, Greek and Roman included – appear to have a deep-seated and pervasive concern with purity and pollution, and this concern has been recognized as being at the heart of their religious systems.[21]

---

[19] On pollution in non-classical antiquity, see for example Bleeker (1966) on guilt and purification in Egypt. More generally on scapegoating and violence in early religions, see Girard (1977).

[20] See www.classics.ox.ac.uk/faculty/oxprinceton.asp (accessed July 2011).

[21] For a wide-ranging philosophical study of the concept of 'purity', considering ancient Greek, Jewish and early Christian approaches (alongside those of the modern West, which, she argues, tend to misunderstand or ignore the concept), see Mullin (1996).

The role of purity and pollution as key concepts in the Jewish world-view has received some detailed attention, and indeed was an important focus for Douglas' *Purity and danger*. A distinction has sometimes been drawn in Jewish studies between 'ritual purity' (that is, the set of rules identified in Leviticus, Numbers and Deuteronomy regarding things like purification rituals and dietary prescriptions) and 'secondary' concepts of purity that permeated the moral domain, particularly concerning sexuality, idolatry and murder (where purity has sometimes been described as 'metaphorical') – although, as J. Klawans has persuasively argued, it is not clear that the two can necessarily be treated separately.[22] Focusing more on formal Jewish ritual pollution, some of the issues discussed by Douglas were explored in M. Poorthius and J. Schwartz's wide-ranging interdisciplinary volume *Purity and holiness: the heritage of Leviticus* (1999). More recently H. Harrington's *The purity texts* (2004) gathered the relevant Dead Sea Scrolls on the subject of ritual purity and analysed this data as part of a coherent Jewish ideology about the relationship between purity, pollution and social system: Harrington identified five key 'impurities' on which there is remarkable congruence among the Qumran texts as well as other ancient Jewish texts – corpses, leprosy, bodily discharges and outsiders. These themes chime well with those subjects identified as pollution in a wide range of world cultures and religions.[23]

As with Judaism, pollution has long been recognized as a central concern for Greek and Roman society, religion and culture. The most thorough and comprehensive study of pollution in the classical world remains R. Parker's *Miasma: pollution and purification in early Greek religion* (1983), which (in spite of its title) is a wide-ranging and important examination of the significance of pollution and purity across the full range of Greek public and private life, history, mythology, philosophy and law.[24] Parker considers an impressive quantity of literary and epigraphic material, and sets it against

---

[22] Klawans (2000), which argues against the idea that moral impurity is metaphorical, and therefore secondary to ritual purity. For an earlier study, see Neusner (1973). Arguing against the secondariness of metaphor as an index of value, see Lakoff and Turner (1980). Further, see below pp. 103–4.

[23] On ritual purity, see Milgrom (1976) and, on Leviticus, Milgrom (1992). For a more specific aspect of Jewish pollution, see Wasserfall (1999); cf. McCarthy (1969) and (1973); Meyer (2005). Some work has been done on the relatively sparse emphasis on purity and pollution in Christian doctrine: see for example Kazen (2002).

[24] Building on the ideas and arguments of E. Rohde's classic study *Psyche: Seelencult und Unsterblichkeitsglaube der Griechen* (1894) and L. Moulinier's *Le pur et l'impur dans la pensée des Grecs d'Homère à Aristote* (1952); cf. Fehrle (1910). More recently, see Chaniotis (1997); Lupu (2004). Cf. Dyer (1969) on Apolline purification rituals.

the broader background of anthropological and sociological studies of pollution. *Miasma* is particularly influenced by the work of Douglas, and Parker is committed to the principle that pollution – a 'kind of institution' (p. 120) – is the 'by-product' of culturally defined order: 'a culture's beliefs about pollution [do not] derive from anxiety or a sense of guilt. They are rather by-products of an ideal of order' (p. 326). Parker even manages to demonstrate that this definition was itself on the lips of various Greek thinkers such as Plato, for whom pollution was a 'science of division'. Pollution, Parker's argument goes, necessarily existed in domains of life which required system or order: law, morality, religion, nature. Furthermore, Parker utilizes the work of Van Gennep to approach the transitional and pollutive status of birth and death – alongside sexual intercourse, defecation, commerce, etc. – in Greek thought (and especially religious contexts), and *Miasma* also contains sections on sexual activity, bloodshed and sacrilege. Parker argues that pollution was an expression of disorder in a society without a fully developed legal system. However, as *Miasma* itself demonstrates, pollution and purification were not straightforwardly primitive strategies for dealing with anomaly: religious rituals and magic continued to be influential in antiquity alongside medicine and law as a means of dealing with crisis, misfortune or disease; for example, Parker examines the role of the scapegoat in Greek communities as a means of communal purification, as well as Eleusinian and Dionysiac rites of purification and their connection with Greek concepts of salvation and the afterlife. By examining a number of different cultural and literary registers in which pollution and purification are visible, and a diverse range of linguistic categories through which it was expressed, Parker demonstrates that ancient Greek 'pollution', far from being a single homogeneous concept, was varied and flexible – although he does identify a unifying formula (not dissimilar to that laid down by Douglas) by which pollution points to alarm at the breach of a natural order. By privileging this technical interpretation, then, Parker plays down the significance of emotions such as disgust, revulsion or anxiety in the formulation of pollution, and it is perhaps in the history of such emotions that the most important work remains to be done. Nevertheless, *Miasma* is an extraordinary empirical study of the history of a concept and its various manifestations in an assortment of Greek literary contexts; it has great interdisciplinary appeal and is an essential starting point for the study of pollution and purity in antiquity.

It is a testament to *Miasma*'s influence on the classics that pollution and purification are increasingly to be found as standard topics of discussion in general studies of the ancient world, particularly religion. Recently,

A. Bendlin contributed a chapter to *A companion to Greek religion* (2007) on 'Purity and pollution', which provides a clear and comprehensive discussion of approaches and ideas explored by Parker, with a discussion of linguistic categories, religious rituals, forensic law and tragedy.[25] Bendlin argues that 'purity' and 'pollution' – both of which he considers to be the opposite of normality rather than of each other – represent 'two powerful religious categories by means of which Greek religion enforces a religious world-view upon the daily lives of ordinary Greeks' (p. 178). Bendlin presents little evidence that ideas about Greek pollution have developed much beyond *Miasma*, although the chapter does show some critical engagement with Douglas' formulation of pollution as a by-product of order, which Bendlin argues offers a limited approach to the topic. Furthermore, Bendlin goes further than Parker in suggesting sets of underlying rules about pollution and purity in Greek society and religion rather than (as Parker) patterns formulated in extant literature, although Bendlin does lay considerable emphasis on the significance of Attic tragedy (and especially Sophocles' *Oedipus Tyrannos*) as evidence of Greek approaches to pollution. More generally on pollution and purity in the ancient world, but again with an emphasis on religious ritual, sections of S. Johnston's edited volume *Religions of the ancient world* (2004) address the definition, practice and theory of pollution and purification in Near Eastern, Greek, Roman, Jewish and early Christian religious systems.[26] In spite of the differences evident between these systems, it is not difficult to identify common features in their formulation of pollution and purity: water, fire, blood, semen and excrement, for example, all appear to be components of purity systems with similar sets of functions and associations (further on these substances, see Lennon, this volume).[27]

Pollution in ancient Rome has received rather less attention than that in Greece, not least because Latin lacks a single overarching term such as *miasma* with which to define it (see Lennon, this volume). It has been argued recently that Roman religious life was largely unconcerned with pollution and purification, and that such associations were principally Greek imports; however, while there are clearly some examples to corroborate this latter point, such a totemic argument cannot stand.[28] 'Ancient Rome' was, from a very early stage in its formation, the product of many different

---

[25] For a shorter summary, see Bradley (forthcoming a).   [26] Attridge (2004).
[27] See also Lennon (2010c) on menstrual blood in ancient Rome.
[28] See Kirkpatrick (2003), who focuses on the absence of purity content in lustration rituals (at least as the Romans themselves conceived them); it is not clear that a propitiatory explanation, plausible as that might be, is so distinguishable from a purificatory one.

cultural influences – Greek, Etruscan, Latin, Phoenician – and its religion, society, literature and art integrated a diverse range of rituals, traditions and associations. Throughout its history Rome was at a busy crossroads between communities, and it is evident that one of its virtues as a case study for a volume of this kind is that the city itself was a prominent meeting-point of cultures, both synchronically (as a cosmopolitan centre) and diachronically (as a focus of transition from pagan to Christian, for example). Furthermore, it is clear that aspects of dirt, purification, transgression and immorality have been prominent in several areas of research in recent studies of Roman society, culture and literature.

The early twentieth century saw a growing interest in taboo in Roman religion, particularly in the wake of J. Frazer's influential comparative study of folklore, magic and religion in ancient and modern cultures, and there have been some sporadic attempts in recent years to address specific aspects of religious pollution and purity at Rome, focusing particularly on rituals of 'lustration' performed to purify the city, the fields and the army, as well as various priestly figures, sacrificial prodigies and sacred water.[29] As Parker has done for Greek pollution, Lennon and Fantham (this volume) demonstrate the centrality of religion and ritual for the formulation of systems of purity and pollution in ancient Rome.[30] This said, it is important not to compartmentalize the religious and the secular in pagan life, and it is unsurprising to find that ideas about pollution pervaded many other aspects of Roman society and thought. The idea of moral and social transgression in the Roman world has received some detailed attention: C. Edwards' *The politics of immorality in ancient Rome* (1993), for example, examined contemporary preoccupations with decadence, sexuality and depravity in the literature of the early Empire and argued that these discourses were the product of social and political tensions and conflicts, an approach that chimed well with Douglasian definitions of pollution.[31] Ancient Roman dirt mattered.

Particularly prominent among studies of marginality in ancient Rome in recent years is research on waste disposal and its importance for the organization of cities. X. Dupré Raventós and J. Remolà's edited volume *Sordes*

---

[29] Frazer (1890). See for example Wülker (1903); Warde-Fowler (1911a) on Latin *sacer*; Burriss (1929) and (1931) on 'taboo'; MacBain (1982); Tatum (1993). More recently, see Ziolkowski (1998–9) on Roman urban cleansing; Beard (1980) on Vestal Virgins (and above n. 7); Edlund-Berry (2006) on sacred water. On 'lustration', see Bradley (forthcoming b).

[30] For a detailed study of the significance of ritual pollution and purity in Cicero's invective against Clodius in the mid 50s BC, see Lennon (2010a).

[31] See also Kaster (1997); cf. Richlin (1983). On linguistic aspects of Roman sexual behaviour (focusing on transgression), see Adams (1982).

urbis: la eliminación de residues en la ciudad romana (2000) discussed the disposal of various types of waste (excreta, food, bodies, artefacts, rubble, submerged wreckage and so on) in communities in Roman Italy and in the imperial peripheries, and makes some important observations about the relationship of waste disposal to myth and ritual, its representation in literature and art, strategies for recycling waste, and shifts in practice over time.[32] Studies such as this one draw attention both to the extraordinary efforts in place in Roman cities to eliminate waste (the scale and extent of the Great Sewer of Rome, the Cloaca Maxima, for example) and to the remarkable differences that existed between ancient and modern sensibilities about dirt (waste in the streets, or the use of human urine in laundries).[33] They also serve as an important corrective to the traditional line on Roman sanitation adopted by those such as A. Scobie, who argued that Roman cities – and especially the metropolis itself – were dangerous, unhealthy and unpleasant places to live, and that Romans were by and large insensitive to the benefits of systematic urban sanitation.[34] The sophistication of Roman approaches to waste control and its relationship to urban organization has also been explored by E. Gowers' seminal article on representations of the landscape of Rome – focusing on the Capitol and the Cloaca Maxima – as a metaphor for the body, with the former representing the head of the city, and the latter its bowels.[35] A number of these ideas are explored in this volume by Davies (on the political ramifications of urban cleansing in the Roman Republic) and by Hopkins (on the practical and symbolic complexities of the Cloaca Maxima).

In connection with this research, scholars in the last thirty years or so have examined the cultural importance of baths and bathing in the Roman world, and some of this research has explored the ideological role of cleansing in these establishments (alongside, paradoxically, their practical lack of hygiene and sanitation).[36] In literature, an urbane, elegant individual was sometimes described as *lautus* ('well-washed'); conversely, an unkempt appearance and filthy clothes signalled those on the margins of Roman society – something that became institutionalized as a political statement for Romans on trial, in mourning or having suffered some defeat or disaster.[37] In addition, the last twenty years have seen pioneering research by Roman archaeologists into water supply, sewers and latrines: the activities of Roma

---

[32] Cf. Magdelaine (2003).   [33] On the latter, see Bradley (2002).

[34] Scobie (1986).   [35] Gowers (1995).

[36] See for example Yegül (1992); Fagan (2002); Jackson (1990). On bathing and washing in the Greek world, see Ginouvès (1962).

[37] In particular, see Blonski (2008).

Sotterranea (an institution dedicated to the excavation of subterranean Rome) and K. Rinne's cross-disciplinary online resource *Aquae Urbis Romae* (examining the full range of hydraulic features across the history of the city) are of particular importance.[38] There has also been some archaeological and historical interest in the phenomenon of communal latrines, as well as possible evidence for shared sponge sticks in place of toilet roll: such phenomena, it has been argued, point to a very different approach to propriety and hygiene.[39] Much of this work, however, is relatively specialized and remains to be properly integrated into Roman cultural history. There have also been some significant efforts to address basic concepts of disease in the Roman world, in terms of both medical knowledge and religious doctrine.[40] A related area in which dirt, disease and purity have featured heavily is the ancient history of medicine, and recent research within this field is extensive: one study which is particularly relevant to the present volume is H. King's edited volume *Health in antiquity* (2005), which contains essays on disease (both actual and represented), responses to disability, urban 'salubriousness' and spiritual health.[41]

One category of waste disposal that has attracted a great deal of scholarly attention is the disposal of the dead in the city of Rome and other Roman communities (on some aspects of this, see Lennon, this volume). One important shot across this territory, published in the same year as *Sordes urbis*, is V. Hope and E. Marshall's *Death and disease in the ancient city* (2000), which collected essays on death and burial, disease, medicine and urban development in ancient Rome as well as a range of Greek city-states. Particularly important here is J. Patterson's topographical study of urban boundaries (physical, ritual, economic and legal) and the marginalization of 'dangerous' or unhygienic activities at the fringes of the city; Patterson makes a compelling case for the practice in ancient Rome of what might be described as 'urban zoning', a highly significant phenomenon of city planning in a range of modern contexts.[42] From rather different

---

[38] For the latter, see www3.iath.virginia.edu/waters/ (accessed January 2012). *Aquae Urbis Romae* has published some important online articles relevant to the present volume: see for example Hopkins (2007). Some of the research on drainage and sanitation has recently been synthesized in Koloski-Ostrow (forthcoming).

[39] On Roman toilets, see Neudecker (1994); more recently, Hobson (2009).

[40] See for example André (1980); Nutton (2004). On approaches to dangerous spaces, see Borca (1997). More generally, see Jackson (1988).

[41] This volume complemented a large number of seminal studies of Greek approaches to health and disease, such as Lloyd (2003), which examined Greek intellectual approaches to disease from Homer to Aristotle (and beyond), and raises important questions about ancient representations of disease.

[42] On one type of marginalized group in ancient Rome, see Whittaker (1993).

perspectives, V. Hope and J. Bodel discuss burial and the treatment of corpses in ancient Rome, the former using literary evidence to argue that the treatment of the corpse reflected elite concerns about public honour and dishonour, the latter developing a sophisticated argument – principally on the basis of archaeological and epigraphic evidence – that the disposal of the dead was a more ambivalent activity than those elite concerns might suggest.[43] Beyond this volume, Hope herself has published extensively on this topic: *Death in ancient Rome: a sourcebook* (2007), which demonstrates the intellectual currency of death, mortality, funerals, burial and the afterlife across an extensive range of literary, epigraphic and visual source material; and, most recently, *Roman death: the dying and the dead in ancient Rome* (2009), a wide-ranging study of the ways in which ideas about death permeated Roman life.[44] P. Davies, in *Death and the emperor: Roman imperial funerary monuments from Augustus to Marcus Aurelius* (2000), examined the architecture and topography of emperors' funerary monuments in the early Principate and their connection with imperial propaganda and cosmic rites of passage, and made an important statement about the relationship of the physical architecture of death to Roman concepts of social life and politics.[45]

Complementing these historical, archaeological and visual approaches to death in ancient Rome, C. Edwards' *Death in ancient Rome* (2007) explores representations of death in literature from the late Republic to early Christendom, arguing that (high-profile, 'unnatural') death in the Roman world operated as a compelling spectacle that formulated and communicated information and ideas about the character of individuals, political regimes and morality.[46] Public spectacles of death in Rome, in the form of public executions and gladiatorial combat, have also received some detailed attention: D. Kyle's *Spectacles of death in ancient Rome* (1998), for example, addresses – with an interesting comparative perspective – topics such as violence, capital punishment, gladiatorial spectacles, Christian martyrdom and disposal of bodies, as well as more theoretical aspects of the subject

---

[43] These chapters are followed by a rather generalized essay by H. Lindsay (2000) on the concept of 'death-pollution', burial and the purification of the family in pre-Christian Rome. See also Bodel (1994).

[44] See also Hinard (1987) and (1995); Davies (1999).   [45] Cf. Bodel (1999).

[46] For a similar approach, with a useful comparison between pre-Christian and Christian approaches to death and the afterlife, see Bernstein (1993). Cf. also Clark (1998) on late antique martyrdom, virginity and resurrection. Most recently on the ritualization of death in ancient Rome and its literary and visual representations, see Erasmo (2008). The classic study of Roman death is Toynbee (1971); cf. Shaw (1996). On Greek death, see Sourvinou Inwood (1995).

such as the ritualization of death and social purgation (further on this, see Bradley, this volume, chapter 6).[47] The disposal of marginalized outcasts – criminals, gladiators, captured enemies and so on – was in the ancient city the subject of spectacle, ritual and literary ekphrasis. The connection of these processes with human sacrifices and scapegoating has received some detailed attention.[48]

A related area of study in ancient Rome which finds expression within this volume is crime and punishment, on which there have been several recent studies principally concerned with Roman law. The development of a rigorous legal system to police transgression, it has been argued, is evidence of a deep-seated cultural concern to preserve the integrity of the community and remove any anomalies that were perceived to threaten it.[49] O. F. Robinson's *Penal practice and penal policy in ancient Rome* (2007) is a wide-ranging study of six examples of criminal behaviour and legal penalties from the mid-Republic through to the late Empire, discussing Roman juristic, moral and philosophical approaches to punishment, deterrence and retribution and their development over time.[50] In the same year, J. Harries published *Law and crime in the Roman world* (2007), which explores the moral and social aspects of crime, as well as its legal ones, and argues that legal practice, in spite of its conservative traditions, was sensitive and responsive to social changes. Roman criminal law, therefore, can be seen as a concerted, systemic effort to police internal pollution in the form of anomalous behaviour, where anomaly represented an organic, developing phenomenon which was in tune with current concepts of propriety and order.[51] The more conceptual aspects of crime and punishment in ancient

---

[47] See also Mustakallio (1994) on capital penalties in early Roman historiography and Cloud (1971) on laws governing *parricidium*. Most recently, see Charlier (2009) on violent death and Zimmermann (2009) on extreme forms of violence in antiquity (the chapters by Zimmermann, Rohmann, Huttner and Krause address Roman imperial themes).

[48] See esp. Bremmer (2007) on human sacrifice; Bremmer (1983) on Greek scapegoats. Cf. also Burkert (1983) for an anthropological approach to Greek sacrifice; Eckstein (1982) on human sacrifice in Republican Rome; Rives (1995) on human sacrifice from paganism to Christianity; Green (2001), Maccoby (1982) and Parry (1982) on human sacrifice more generally. On sacrifice in antiquity, see most recently Mehl and Brulé (2008); more generally, see Hubert and Mauss (1964).

[49] Cf. Jońca (2004), who argues that homicide was punished privately by the victim's relatives in archaic Roman society and that 'blood revenge led to chaos, disorder and seriously undermined the state authority' (p. 51). On the role of pollution in stigmatizing murder in contemporary Athenian law, see Arnaoutoglou (1993).

[50] Building upon a number of ideas already developed by Bauman (1996).

[51] Connected with this, violence in ancient Rome is a phenomenon that has attracted significant scholarly attention: see Lintott (1999b) on Republican Rome; Drake (2006) on late antiquity.

Rome have also been brilliantly explored by G. Thome's 'Crime and punishment, guilt and expiation: Roman thought and vocabulary' (1992), and an important collection of essays by J. L. Voisin (1984) has addressed the symbolic significance of capital punishment in ancient cities.[52] One example of public punishment specific to Rome, the phenomenon of *damnatio memoriae* (a loosely defined process in which a condemned emperor's memory was desecrated through such public gestures as the desecration of his images and the annulment of his laws), has also been the subject of recent scholarship.[53] All these studies can be linked to general studies of *transgression*, which has been a popular focus for research in recent years; while transgression is most clearly represented by criminal behaviour and criminal law, these studies have demonstrated that transgression in all its manifestations is a prominent feature of literature, art and culture.[54]

Another area of transgression in Roman life which has attracted a great deal of scholarly attention is that of female sexual behaviour. Sexual impropriety, particularly among aristocratic women and – most conspicuously – among Vestal Virgins was a popular topic for invective and moral debate among the educated elite of ancient Rome (further, see Schultz, this volume). The rules regarding Vestal chastity, the penalties for losing that chastity, and the connection between the priestesses' personal sexual behaviour and the purity and integrity of the state have generated a large number of classic discussions about the symbolic relationship between gender, sexuality and religion in ancient Rome.[55] More broadly, work on female prostitution in Rome has examined some of the taboos attached to the profession, while also drawing attention to its pervasiveness in Roman life and an extraordinary disregard for the zoning and policing of the activity in Roman cities.[56] R. Flemming (2000) has also explored the relationship between women's bodies, sexual behaviour and the development of Roman medicine. Other sexual offences, such as incest and *stuprum* (debauchery), have also received some attention.[57]

---

[52] Cf. Wallace-Hadrill (1982) on propriety and sin in Augustan Rome.

[53] See esp. Varner (2004).

[54] C. Jenks' *Transgression* (2003) has offered a wide-ranging interdisciplinary study of transgression in the modern West, building on the ideas of Durkheim, Douglas and Freud. A classic study of transgression in all its manifestations is P. Stallybrass and A. White's *The politics and poetics of transgression* (1986).

[55] Starting with Beard (1980); revised in Beard (1995) (see above n. 7); Mustakallio (1992) on Vestal cruelty; Staples (1998) on gender classification in Roman religion; Parker (2004) on Vestal chastity and state security.

[56] See for example Flemming (1999); cf. Edwards (1997).

[57] Cornell (1981) on the *crimen incesti*; Fantham (1991) on *stuprum*.

Pollution and purity in ancient societies, then, have in fact been extensively studied from a number of different angles, and a substantial amount of material is available in diverse areas of ancient life: religion, morality, law, literature, art and architecture. Furthermore, it has long been recognized that multiple aspects of ancient Roman society were permeated by concerns with the maintenance of proper boundaries and the avoidance of pollution: this volume will address several of these aspects, particularly in the context of religious ritual, urban development and criminal behaviour. In all of these aspects, it is clear that ancient approaches and responses to purity and pollution are intimately connected with the values and priorities formulated and exercised within the community. Some of the most characteristic features of ancient Roman society – rigorous rituals, baths, sewers, necropoleis, the arena, female behaviour, criminal legislation and so on – were governed by concerns with pollution and propriety, and it is through an integrated study of these concepts, this volume contests, that a valuable contribution can be made to the history of ideas within the city of Rome.

Pollution beyond antiquity

It is perhaps unsurprising, then, that studies of pollution and purity in the urban centres of medieval Europe and the modern West have continued to explore similar themes and issues to the scholarship engaging with classical antiquity. In part, this is the inevitable consequence of universal human patterns: human waste and dead bodies cause disease; hygienic measures promote sexual reproduction; and so on. But it is also important to note that the patterns, practices and ideologies of Greece and Rome continued to be influential across the Western world, and what counted as 'clean' (and therefore what constituted 'dirt') for the ancients had great currency for later societies.

One prevailing concern of medieval communities, exploited at great length by the political and religious authorities to bestow legitimacy on their regimes, was the control of disease and the exercise of 'sanitary' measures. A series of devastating plagues, most notably those of the 1340s and 1350s, has been identified as one of the defining features of medieval Europe, and perhaps one of the triggers behind the reconstruction of society that led to the Renaissance. Some studies of disease, therefore, have examined medieval society *en bloc*. A. Foa (1990), for example, has examined European responses to the outbreak of syphilis at the end of the fifteenth century and argued that traditional associations of disease with sexual impurity governed political and literary responses to this outbreak; she has also demonstrated that the scapegoating of marginal groups – both Jews and new immigrant Amerindians – at this time revealed a pattern of response to 'disease' that stretched back to antiquity. Others have focused on the impact of plague in specific parts of Europe.[58] Out of this approach to disease and healing, it has been argued, medieval Europe saw the emergence of popular heresy as an institutionalized response to deviancy, and forms of persecution characteristic of the period which targeted Jews, lepers and other minority groups.[59]

Such approaches to disease and its perceived carriers, and attempts to define, police and prevent that disease, continued to be central concerns of Renaissance society and the politics, law and culture of early modern Europe: Stow (this volume), for example, examines strategies for marginalizing Rome's ghettoized Jews in early modern Rome.[60] There has been a particular interest in the scatological in the Middle Ages: most recently, S. Morrison's *Excrement in the late Middle Ages: sacred filth and Chaucer's fecopoetics* (2008) explores the material and symbolic history of fecal matter in late medieval England with particular emphasis on Chaucer's *Canterbury Tales*, the symbolism of the grotesque body and developing urban policies of waste disposal, and closely follows the theoretical models laid out by Douglas (on 'matter out of place') and Kristeva (on abjection). Other aspects of the urban environment which were perceived as a source of pollution, illness and disease have also attracted attention: M. Jenner ((1995) 551), for example, has examined interpretations of smoke pollution and noxious airs in seventeenth-century London and their effects on health, reproduction and social behaviour, as well as the use of these phenomena in literature as a metaphor for contemporary politics.[61] Jenner ends his article by referring to a claim in M. Douglas and A. Wildavsky's *Risk and culture* (1982) that 'Generally, pollution ideas are the product of an ongoing political debate about the ideal society' (p. 36); pollution, the argument goes, is intricately connected to broader cultural issues and values. The chapters in this volume by Assonitis and Gentilcore demonstrate that the representation of, and

---

[58] See for example Slack (1985) on Tudor and Stuart England; Carmichael (1986) on Renaissance Florence; Sonnino and Traina (1982) on the 1656–7 plague at Rome.

[59] See in particular Moore (1987). See also Barber (1981) on minority groups opposed to Christendom in the early fourteenth century.

[60] See for example Cipolla (1981) on seventeenth-century Italy; Slack (1988) and Carmichael (1993) on early modern Europe. More generally on the flexible and negotiable character of disease in Western history and culture, see Crawford (1914); Sontag (1979) on the metaphors of sickness.

[61] Further on concepts of polluted airs in the early modern period, see Cipolla (1992).

response to, plague at various stages in Rome's history attest precisely to such developments in the city's ideals and politics.

Particularly from the Renaissance onwards, European cities began to emulate ancient Rome in the public supply of water.[62] Hand in hand with this came the provision of baths and the culture of bathing. Bath-houses and spas sprang up across Europe, and were used by the powers that be, alongside street-cleaning, urban renovation, the provision of laundries and various sanitary measures, as evidence of a clean and healthy nation (on this theme in sixteenth-century Rome, see Rinne, this volume); this continued into the twentieth century, and the provision of public bathing facilities and the ideology of bathing continue to function in Western cities as prominent factors in urban management and political manifestoes.[63] One important and sophisticated study of bathing and washing, and shifts in approaches to cleanliness between the medieval and modern period, is G. Vigarello's *Concepts of cleanliness: changing attitudes in France since the Middle Ages* (1988). Vigarello elegantly refutes the traditional argument that saw the history of cleanliness as a straightforward progression from the filth and dirt of the medieval period to the hygienic and sanitary standards of the modern West. Water, he argued, was perceived as a dangerous substance in sixteenth- and seventeenth-century France, and a medium for the transmission of disease; other methods of cleansing were sought such as the frequent changing of linen or the vigorous scrubbing of the body with a towel. By contrast, medieval France saw a proliferation of bath-houses and regular bathing, connected with preparations for festivals and rituals. Post-plague, eighteenth-century France 'rediscovered' bathing as part of a growing taste for amenities and luxuries, although this was checked by waves of ascetic preferences for corporeal discipline. Vigarello neatly demonstrates that the history of cleansing (at least in France) was part of a complex succession of moral and scientific discourses, and this approach offers a useful point of entry into thinking more broadly about the development of ideas about pollution and purity in Western history.[64]

---

[62] The relationship between water and health in modern Europe has been explored by Goubert (1989). See also Jenner (2000) on water in early modern London.

[63] In general, see Porter (1990) and Wilkie (1986). On baths in Renaissance Italy, see Palmer (1990); cf. Thorndike (1928) on medieval and Renaissance baths and street-cleaning. On their importance in the development of American urban culture, see Williams (1991). On the social and political ramifications of laundry provision in sixteenth-century Rome, see Rinne (2001–2). On Rome, see San Juan (2001), esp. ch. 4 'Water's overflow'.

[64] Cf. Louz and Richard (1978), who examined French proverbs to demonstrate a prevailing peasant lore that the retention of a layer of grease on the skin protected individuals from cold. For some of these issues in eighteenth-century England, see Jenner (1998).

As one of the by-products of disease and plague, death and funerary practice from the Middle Ages onwards have been important topics for historical study. In 1974, the French medievalist P. Ariès published an influential book (*Western attitudes toward death: from the Middle Ages to the present*) which examined changing approaches to death in the West, focusing particularly on the differences between attitudes in the Middle Ages and those of modern Europe.[65] His principal argument was that death was normalized in earlier ('primitive') periods, when it was a regular and visible part of life and communities were more in tune with religious doctrines about the afterlife, whereas in later periods, death – and all its accompanying rituals, monuments, spaces and responses – started to become the subject of horror, indignity and revulsion, with hospitals, graveyards, bodies, funerals, coffins and so on treated as abject phenomena to be shielded from public view and contemplation.[66] This was a big claim and, in spite of the evidence Ariès presents, it is evident that such sweeping generalizations cannot apply straightforwardly to the diachronic study of individual locations such as Rome; nevertheless, such an approach to the subject can be illuminating in helping us to understand the relationship between death and pollution, as well as shifts in sensitivities and sensibilities over time. Since Ariès, there has been a proliferation of studies of death from anthropological, sociological and theoretical perspectives, and it is clear that the way societies responded to death, organized the rites of passage, and integrated the memories, spaces and bodies of the dead into their world was frequently permeated with concerns about (or concepts of) pollution and purity; in this volume, for example, Janes explores the evaluation of Roman catacombs by visitors from Victorian England as a focus for the city's dead and decayed moral values.[67]

Another widely examined aspect of the society of the Middle Ages which was governed by concerns about purity and the avoidance of pollution was marriage and sexuality. J. Brundage's influential book *Law, sex and Christian society in medieval Europe* (1987), for example, examined the

---

[65] This study was followed by Ariès (1981).

[66] Some of these arguments are pursued, albeit in a rather more focused and sophisticated form, by McManners (1981), which examines the growing ambiguity, melancholy and fear associated with death in early modern France. On mortuary practices in late medieval and early modern Italy, see Strocchia (1992). See also Laqueur (1983).

[67] See Younger *et al.* (1999), who examine the complexities and ramifications (from various disciplinary perspectives) behind definitions and evaluations of death. For a comparative anthropological approach to death (emphasizing diversity of practice), see Metcalf and Huntington (1991); see also Hockey *et al.* (2001). On cultural responses to death in Britain, see Gorer (1965); more generally, see Hallam *et al.* (1999); Grainger (1998); Richardson (1988).

creation of legislation about sexual conduct in medieval Church law which institutionalized prevalent beliefs about and responses to a range of sexual behaviour (marital sex and adultery, prostitution, homosexuality, incest, masturbation, bestiality and so on), much of which itself drew on the sexual morals prominent in classical antiquity. This book, examining a thousand years of Church history (AD 500–1500), makes an important case for the continuity of classical ideas about sexual purity both from antiquity to the Middle Ages and from Church law to the secular ideologies of the modern West, and demonstrates that the control of sexuality is a complex and evolving historical phenomenon which has been an essential component of the exercise of authority and power in the Western world. This theme has been explored in a wide range of contexts: K. Stow, for example, has examined the relationship between sexuality and holiness in the Jewish family in the Middle Ages, and the significance of Jewish laws about menstrual purity.[68] On the theme of masculinity, C. Leyser (1998) has discussed the threat posed by masturbation and wet dreams to notions of sexual purity in Europe in this period. Homosexuality in the Middle Ages has also received its fair share of attention: as well as its widespread condemnation across medieval Europe, K. Gade has shown that homosexual behaviour was the subject of discrimination and marginalization in Old Norse society and literature.[69] Like disease, sexuality and perceptions of deviant sexual behaviour were prominent factors in the persecution of minority groups across Europe.[70]

In spite of the relative lack of interest in pollution and purification across Christian doctrine compared to pagan and Jewish systems (see above, p. 19), it is clear both that existing concepts about purity contributed to various aspects of the formation of Christian thinking, and that the development of various Christian schools of thought has influenced approaches to purity in European communities.[71] One such school, which has its origins in late antiquity, is asceticism; comparable to Christian concepts of chastity, this religious lifestyle advocated abstinence from various types of worldly pleasures with the aim of achieving physical and spiritual purity, and rejected elements which were perceived as polluting or sinful in various ways, such as sex, alcohol, various foodstuffs and material luxuries. The Roman Catholic Church embraced asceticism most conspicuously in the form of monasticism, and purification, atonement and the avoidance of

---

[68] Stow (1987) and (1992).   [69] See Gade (1986); cf. Gade (1989).

[70] See for example Payer (1993) and Richards (1991).

[71] See for example Leyerle (2009) 338 on the overlap between Jewish and Christian approaches to purity; cf. 344 on contemporary 'scriptural caricature' of Jewish obsessions with purification.

sin were often professed as reasons for pursuing an ascetic lifestyle. From the Christian authors of late antiquity to Francis of Assisi, and through to such revolutionary figures as Stalin, Hitler and Mussolini, asceticism has often been evoked as a rationale for creating a system of values in which the pure and the impure could be effectively formulated and policed.[72] And from the earliest surviving records of Christian thinking, it is evident that the discourse of pollution was employed as a compelling means of alienating undesirable figures, philosophies and behaviours: recent studies, for example, have examined attacks by early Christian writers on traditional Graeco-Roman practices through allegations of blood pollution in pagan ritual, or through associations of waste and excrement to mobilize disgust among Christian congregations towards extravagant and non-Christian behaviours.[73]

The engagement of other types of religious literature and thought with aspects of pollution has received a great deal of attention in recent decades. S. Greenblatt, for example, has explored – among other aspects of the abhorrent and the filthy – scatological imagery in principally religious invective on carnivalesque practices in seventeenth-century Western culture.[74] Other studies of the relationship between cleansing and Christian notions of purity include K. Thomas' important article 'Cleanliness and godliness in early modern England' (1994), which examines the coterminous development of the two concepts from Tudor and Stuart England through to the nineteenth century, and addresses the significance of various religious doctrines (including asceticism) for shifts in practice.

While histories of dirt and cleanliness since the Middle Ages have focused principally on diachronic shifts and developments, several important studies focusing on the modern world have emphasized geographical variations, both within and between communities. B. Campkin and R. Cox's edited volume *Dirt: new geographies of cleanliness and contamination* (2007), for example, is a collection of essays on spatial and geographical aspects of pollution in a range of communities in the modern world (Europe, Asia and America), focusing mainly on twentieth-century material but with two

---

[72] See Finn (2009) on Graeco-Roman asceticism; Brakke (1995) on Athanasius and asceticism; Rousseau (2010) on asceticism in the late antique Church; cf. Brown (1988) on early Christian sexual renunciation. Specifically addressing the theme of purity, see Luckman and Kulzer (1999). On the creation of sainthood in the medieval period, see Stouck (1999).

[73] See Lennon (2010b) on early Christian accusations of blood pollution in the rituals of Jupiter Latiaris; and Leyerle (2009) on the use of the imagery of refuse, filth and excrement in the *Homilies* of John Chrysostom to attack decadent and consumptive lifestyles and 'false teachers'.

[74] Greenblatt (1990). The excrementality of Carnival was a theme originally explored by Bakhtin: see Lachmann (1988–9). See also Jenner (2002) on scatology in Restoration England.

discussions of pollution in Victorian London.[75] Its contents are structured under three spatial headings: 'Home' (discussing boundaries and matter out of place in the household), 'City and Suburb' (discussing urban zoning, sexuality and sewers, with an emphasis on the city of London) and 'Country' (on associations of pollution with the rural workforce and contaminated food). This volume's geographical focus makes it a welcome addition to existing studies of pollution in modern culture, not least because Douglas' core definition of dirt as 'matter out of place' necessarily privileges the study of space and spatial boundaries in approaching and formulating pollution.[76] Other research has focused on the more technical aspects of spatial organization in cities and its relationship to sanitation: one example is M. Melosi's *The sanitary city: urban infrastructure in America from colonial times to the present* (2000), which examines urban sanitary systems across the United States from colonial times until the end of the twentieth century.[77] Melosi's book complemented existing studies of the importance of cleanliness in American culture, such as S. Hoy's *Chasing dirt: the American pursuit of cleanliness* (1996) and J. Sivulka's *Stronger than dirt: a cultural history of advertising personal hygiene in America, 1875 to 1940* (2001).[78] There have also been a number of important studies of urban pollution in underdeveloped countries.[79]

More recently W. Cohen and R. Johnson's edited volume *Filth: dirt, disgust and modern life* (2004) has surveyed – against the background of sociological and anthropological theory – a much broader range of materials dealing with literary and cultural constructions of dirt and cleansing (sewers, the bourgeois *toilette*, labourers and foreigners, disease, sexuality) in modern contexts, focusing on London, Paris and their colonies in the nineteenth and early twentieth centuries.[80] This volume, then, opens a number of diverse windows onto the role of filth in all its manifest forms in social

[75] Further on Victorian London, see Allen (2007); cf. Anderson (1993) on Victorian approaches to 'fallenness'; Maynard (1993) on Victorian sexuality and religion; Bashford (1998) on purity and pollution in Victorian medicine.
[76] This volume builds upon ideas explored in earlier studies of space and marginality in human geography, such as Shields (1991) and Sibley (1995).
[77] Cf. Melosi (1981); Otter (2004) on nineteenth-century London.
[78] Cf. Bushman and Bushman (1988) on the early history; Vinikas (1992) on hygiene in American advertising; most recently, see Brown (2008). More broadly on marginal identities, class and city life in urban America, see Sennett (1996).
[79] See for example Neves (2004) on Brazil.
[80] The sewers of Paris and London have received a great deal of scholarly attention: see for example Reid (1991), Radford (1994) and Gandy (1999) on Paris; Halliday (1999) on Victorian London. Cf. Pike (2005a) on subterranean Paris and London. For an older study of sanitation, see Reynolds (1943); cf. Wright (1960).

management and the formation of identities in colonial and postcolonial Europe.[81] Indeed, the suggestion that pollution 'could provide a powerful integrative theme' within modern urban history has been made several times, and M. Jenner (1997) has argued that disease, slums, waste disposal and other less palatable features of urban life are, following the principal tenets of Douglas' *Purity and danger*, critical for understanding the organization of space and the formation of values in urban communities.

A snapshot of how some of these issues played out in the urban landscape of seventeenth-century Rome is provided by R.-M. San Juan's *Rome, a city out of print* (2001), which examines the visual imagery (maps, posters, guidebooks, etc.) of the time and the representation and urban organization of street activities, the Jewish ghetto, hospitals, public punishments and other marginalized activities. This imagery could be highly ambivalent: as Syrjämaa shows (this volume), the grubby ruins and backstreets of 'Roma capitale' in the late nineteenth century could be subject to a highly picturesque re-imagination, while Janes (this volume) demonstrates that this same landscape for Victorian contemporaries evoked the lessons of decay and corruption. Some of these themes were also creatively explored in G. Orwell's first full-length work, *Down and out in London and Paris* (1933), which, written from a playfully autobiographical stance, explored the experience of poverty in early twentieth-century Paris and London. Others have focused on the ideological role of public health, hygiene and cleanliness in the colonial peripheries: A. Bashford's *Imperial hygiene: a critical history of colonialism, nationalism and public health* (1994), for example, examines hygiene, borders, ethnicity and foreign bodies within British colonialism and White Australia between the mid-nineteenth and mid-twentieth century, and studies the role of segregation as a tool of imperial control.[82]

The marginalization, humiliation and elimination of various minority groups, and the use of systems of purity and shame with which to characterize these groups, has continued to be a major focus for sociological research. The stigmatization of Jews in the nineteenth and twentieth centuries has been the subject of extensive work, and (particularly from the end of the nineteenth century) the identification and segregation of homosexuals as a site for deviancy, disease and immorality has been a topic for debate and controversy.[83] Studies of society's responses to sexuality have

---

[81] Public health and urban management have been prominent areas of research: see for example Anderson (1995).

[82] Cf. Bashford and Hooker (2001) on contagion; Strange and Bashford (2003) on isolation.

[83] The scholarship is extensive. See for example Boswell (1989), with an emphasis on the formulation of law out of social prejudice.

been at the forefront of research on the relationship between pollution, purity and the body. Recently, M. Nussbaum (*Hiding from humanity: disgust, shame and the law*, 2004) has argued that legislation about sexual behaviour and pornography has conventionally been dictated by disgust, a social convention (the author argues) that is a primitive and inappropriate foundation for law.[84] The historical extent of this has been most evocatively explored through studies of sexuality (and especially prostitution) in Victorian London.[85] Much emphasis has been placed in recent research on the relationship between sexual behaviour and urban 'zoning', with prostitution and gay life relegated to particular parts of the city in order to preserve the order and perceived integrity of the community at large.[86] Accordingly, the final part of *Rome, Pollution and Propriety* ends with a chapter (Salvante) examining the condemnation and segregation of male prostitution in Fascist Rome, and the role of this activity as a foil to moral and civilized Italian life.

Along with such ethnic and sexual marginalization, the formulation of social outcasts based on profession has received some detailed attention: K. Stuart, for example, has examined dishonourable professions in early modern Germany (focusing on executioners and skinners, as well as bailiffs, gravediggers, bathmasters and so on), arguing that their stigmatization – in terms of legal rights, religious associations and literary representation – was a form of ritual pollution rooted not in universal anthropological rules but in the specific social context of city guildsmen defending their own unique claims to honourable trades; this practice, she argues, was gradually phased out in the eighteenth century with various social and economic developments in Germany.[87] Indeed, this emphasis on the *localization* of purity systems, and their negotiation within particular social and historical contexts, represents one important development in recent scholarship on pollution. As with studies of earlier European communities, the elimination (and sometimes the reintegration) of marginalized groups in the modern West has been an important subject for research in sociology and criminology: the identification of 'crime' and the construction of a culture of 'shame', and the negotiation and organization of these concepts in law, language and

---

[84] Cf. Olyan and Nussbaum (1998) on sexuality and human rights in American religious discourse. On homosexuality and sexual distaste, see Barsani (1987).

[85] For example: Walkowitz (1980); Koven (2004). Cf. Houlbrouk (2005) on queer London in the early twentieth century. Other studies, such as S. Jeffrey's *The idea of prostitution* (1997), have attempted to identify universal patterns of prostitution and social responses to the activity. Cf. Henriques (1966); O'Connell Davidson (1998).

[86] See for example Ingram *et al.* (1997); Hubbard (1999); Koven (2001).

[87] Stuart (1999). Cf. Danckert (1963).

literature, have continued to be central factors in modern approaches to pollution and purity.[88]

Other modern studies have tended to focus on the body as a pivotal feature in the formulation of dirt. R. Longhurst's *Bodies: exploring fluid boundaries* (2001) examines – from the perspective of human geography – the concept of the 'body', its material and metaphorical associations, the fluidity of somatic boundaries and the significance of these approaches for politics, gender and sexuality.[89] Other studies have focused on anomalous body types, such as deformity and obesity, or the branding of bodies with stigmata or tattoos.[90] Furthermore, many studies of cities from the perspective of pollution and cleanliness have focused on the analogy of the city to the human body.[91] The centrality of the body and bodily features and products in understanding wider social, political, linguistic and intellectual aspects of human life has been provocatively discussed by the French psychoanalyst D. Laporte's *Histoire de la merde: prologue* (1978), which employs theory from Nietzsche, Freud, Marx, Foucault and others to connect the development of waste disposal in Western Europe to the formulation of social, political and sexual identities and the dialogue between the individual and the state.[92]

Addressing similar themes, and perhaps the closest precedent to *Rome, pollution and propriety* in terms of its approach, methodology and scope, is A. Hopkins and M. Wyke's *Roman bodies* (2005), a collection of seventeen interdisciplinary essays on various aspects of the body in the culture, literature and art of Rome from the early Empire through to the early modern period.[93] This remarkable and diverse collection examines, as the authors

---

[88] See for example Braithwaite (1989); cf. Gatrell (1994) on execution in early modern England. Cf. Morrison (1986) on shame; Jenkins (1998) on child molesters in America.

[89] See also Shildrick (1997) on the idea of 'leaky bodies'; cf. Shweder (1991) on menstrual pollution. On the dirtiness of masculinity, see Widding Isaken (2002); cf. Hershman (1974) on hair, sex and dirt. For a collection of essays on 'bodies and boundaries' in antiquity, see Fögen and Lee (2009). For a more general approach to the role of the body in thought and culture, see Shilling (1993). For a legal perspective on the use of bodies, see Nussbaum (1999).

[90] See Haslam and Haslam (2009) on obesity; cf. Bradley (2011) on obesity in Roman art and culture. Cf. Robertson (1996) on bodily deformity and disgust. Specifically on ancient approaches, see Garland (1995); on Roman satirical representations of distorted bodies, see Miller (1999); more generally, see Winkler (1991). On the significance of tattooing across Western history as a sign of marginality and deviant identities, see Jones (2000); cf. Jones (1987) on Graeco-Roman tattooing.

[91] See for example Grosz (1992) and Pile (1996). More recently, see Gilbert (2004), which examines the use of maps in the Victorian period to identify urban spaces affected by cholera outbreaks and elevates the metaphor of the city as body. For a comparable study of body–city imagery in ancient Rome, see Gowers (1995), and above p. 23.

[92] Cf. McLaughlin (1971).

[93] Cf. Montserrat (1998), focusing generally on the body in Graeco-Roman thought.

put it (p. 8), 'the multiple and sometimes contradictory ways in which bodies have shaped the city and the city has shaped the bodies of its inhabitants'. By emphasizing the pivotal, organic role of the body in shaping the history of the city (under the headings 'Empire', 'Church' and 'Religion and science'), it makes claims that are similar to this volume's argument that pollution and purity were integral to the development and representation of the city throughout its history. Indeed, many of its essays concentrate specifically on bodily pollution or purification or on somatic defilement: three on circumcision, two on decapitation, as well as chapters on lameness, burial, dissection and saintly bodies. Like the present volume, Hopkins and Wyke demonstrate the importance of localization for the history of ideas: bodies, like concepts of dirt and cleanliness, may be human universals, but they are shaped, packaged and negotiated by and within the communities in which they function.

Dirt, particularly in modern contexts, has often been approached from the angle of the senses: as well as matter that *looks* out of place, bodily or environmental smells, disturbing or distressing noises, and food or drink that offends the tastebuds have all been the subject of various studies concerned with pollution. The social and cultural importance of smell from antiquity to the present, for example, has been examined in C. Classen *et al.*'s *Aroma: the cultural history of smell* (1994); this book explores the significance of odours both in the Western world and in a range of non-Western societies.[94] More specifically, A. Barbara and A. Perliss' *Invisible architecture: experiencing places through the sense of smell* (2006) examines – from the perspective of theoretical architecture – the development of the olfactory sense across the centuries as a tool for recognizing, evaluating and responding to the environment.[95] Several studies have focused on the particular role of smell in specific cultures: for example, A. Corbin's *Le miasme et la jonquille: l'odorat et l'imaginaire social XVIIIe–XIXe siècles* (1982) discusses the

---

[94] For a more scientific approach to odours, see Engen (1982); cf. Engen (1991). On approaching smells from an archaeological perspective across a range of historical periods and locations, concentrating on live animal trade, tanning, sea-purple dyeing industries and meat processing, and making some important points about the cultural relativity of smell, see Bartosiewicz (2003); cf. Lilja (1972) on smells in Greek and Roman verse. See also Leyerle (2009) esp. 353 on the use of early Christian rhetoric about bad smells to 'sharpen the senses' towards inappropriate behaviours; cf. Potter (1999) on connections between scents, social status and economic practices in Rome from the early Empire to late antiquity. I am currently editing a volume on *Smell* for Acumen's forthcoming 'The Senses in Antiquity' series (see Bradley (forthcoming c)); cf. also Drobnick (2006) for a wide range of cultural perspectives on smell within human societies. More generally on senses, see Classen (1993).

[95] On the olfactory dimensions of eighteenth- and nineteenth-century Rome, from an art-historical perspective, see Wrigley (forthcoming).

history of French attitudes towards smell from the eighteenth century to the present and their relationship to various social and political developments. Like smell, taste – particularly in the form of food and consumption – has represented another important area in which the body's contact with the external world can be understood in terms of pollution and propriety, and in which these terms can be appropriated and negotiated according to culture.[96] Modern epidemic outbreaks connected with the consumption of particular contaminated foods has arguably exacerbated this sensitivity to appropriate and inappropriate foodstuffs.[97]

It is often argued (and occasionally overstated) that the discovery of pathogens has transformed our notion of the dirty, as well as what it means to cleanse or purify. There is little doubt that the microscopic and the invisible have featured more prominently in concepts of pollution formulated across the past hundred years than they did previously.[98] J. Amato's *Dust: a history of the small and the invisible* (2000) is a smart demonstration of the valency of tiny foreign bodies in modern thought, and the development of this phenomenon from earlier understandings of cosmic constituents. The effect of this scientific knowledge on approaches to domestic cleaning has also been widely studied.[99] It has also been argued that over-cleaning and under-exposure to pathogens can actually reduce the body's resistance and make us more susceptible to illness and disease, a counter-hygienic line of thought that has contributed to such contentious recent phenomena as 'swine-flu parties' and probiotic drinks.[100] One thing is clear: dirt, cleanliness, pollution and purity have been flexible, negotiable and hotly contested topics for well over two thousand years of Western history, and that is still the case today.

This cursory outline of the wide-ranging research contexts out of which this volume has been produced, then, is evidence of the highly interdisciplinary character of the subject. Pollution and purity have been objects of interest to classicists, archaeologists, medieval and modern historians, anthropologists, sociologists, criminologists and scholars working in a number of other disciplines. This does not mean that the subject is ill-defined and incoherent: purity and pollution represent states of cultural order and

---

[96] See for example Warde (1997). For approaches to food and edibility in antiquity, see Durand (1989).

[97] See for example Waddington (2006).

[98] Though see Conrad and Wujastyk (1999) for approaches to contagion and infection in pre-modern societies; cf. Nutton (1983).

[99] See for example McClary (1980) and Berner (1998). On germs and domestic cleaning, see Thomas (2001); in America, see Palmer (1989) and Tomes (1998).

[100] See for example Schaub *et al.* (2006).

disorder, elicit strong gut responses from participants in this culture, and are frequently bound up with the organization of space and the maintenance of system, and at least since Mary Douglas, they are concepts shared and understood across disciplines and in multiple contexts of human society. Particularly in ancient, medieval and early modern contexts, these concepts are often inextricably connected to religious world-views and form a critical part of moral discourse, and the role of political authorities in defining and policing standards of cleanliness continues to be a central tenet of modern urban management in the twenty-first century. The challenge that remains is to explore the history of these themes in a single geographical location, in order to address their development over time and the continuities and differences between approaches to pollution in various contexts from antiquity to modernity.

PART I

Antiquity

2 | Pollution, religion and society in the Roman world

JACK LENNON

It has now been over twenty-five years since Robert Parker's seminal work, *Miasma* (1983), laid down solid foundations which should have allowed the study of pollution across the ancient world to flourish. Proceeding systematically through various human actions and events in Greek society, including various rites of passage, as well as crimes including murder and sacrilege, Parker meticulously examined the nature, causes and effects of pollution on both gods and men, assessing the threats and dangers perceived to society throughout. Sadly, however, the subject has failed to expand significantly beyond the boundaries laid down in Parker's investigation. Nowhere is this stagnation more prominent than in the study of pollution in the Roman world, which still lacks a single, coherent work from which others may develop. The result is a scattered series of ideas, typically viewed in isolation regarding a single specific occurrence of pollution. A significant factor in this situation is likely to have been the lack of an obvious point in Rome's religious vocabulary from which to start. Rome has no '*miasma*', no single term to classify pollution, although as Gabriele Thome has noted, 'it is not lacking in conceptual equivalents'.[1]

Despite lacking any single term corresponding with *miasma*, Latin offers a number of verbs which denote the acts of staining, fouling or otherwise 'polluting' (e.g. *polluere, inquinare, foedare, funestare, scelerare, maculare, contaminare*). Yet these terms cannot stand alone as indicators of specific forms of pollution, and so, most crucially, we must rely on the context of their individual application, as well as reactions to them, when drawing conclusions about what was considered 'impure' or dangerous in Roman ritual and society. These actions posed the greatest threat when directed against a place, object or person connected with the 'sacred', which by its very designation was made vulnerable to damage from the profane world

---

[1] Thome (1992) 73–98 at 77. For some discussions of the nature of individual forms of pollution, or their removal, within Roman society, see Burriss (1929) 142–63; Latte (1960) 47–50; Ogilvie (1961) 31–9; Bodel (1994); Ziolkowski (1998–9) 191–218; Patterson (2000) 85–104; Lindsay (2000) 152–73; Nasta (2001) 67–82; Attridge (2004) 71–83; Beck (2004) 509–11; Lennon (2010a) 427–45.

of everyday existence.[2] Nero damaged the sanctity of the Aqua Marcia's shrine as well as polluting (*polluere*) its healing waters by swimming in it.[3] Such damage could cause imbalance in the gods' goodwill (*pax deorum*), resulting in 'offence' (*scelus*), which has been identified as a vital factor, since it may be both the result of pollution and a form of danger attached to a wrong-doer, which might infect those around him or her.[4] This infection would spread and eventually come into harmful contact with the divine.[5] It was no coincidence that, when setting out his laws governing religion, Cicero's first instruction was that citizens should 'approach the gods with purity, bringing piety, and leaving gifts' (*ad divos adeunto caste, pietatem adhibento, opes amovento*).[6]

Douglas' central tenet was that dirt and pollution were 'matter out of place', requiring both 'a set of ordered relations and a contravention of that order' (further, see chapter 1 pp. 11–13).[7] In this picture, the body is considered a symbolic microcosm of society, and in both cases that which exists on the edge or boundary is considered dangerous.[8] Thus bodily emissions stemming from liminal orifices are imbued with power. Others take this a step further, arguing that such emissions (blood, semen, sweat, etc.) are considered to be 'dying' once separated from the living body, or when the body loses control over their flow.[9] Equally dangerous were those beings considered to be liminal, or out of place, within the established sphere of nature. This included physical beings, such as deformed births, as well as portentous events, such as rains of blood. While not instantly

---

[2] Douglas (1966) 8–10; Parker (1983) 18–31; Thome (1992) 77–8; Bendlin (1998) and (2007) 184–9; cf. Attridge (2004) 72–3, who specifically views pollution as 'that form of "dirt" that prevented participation in the realm of the sacred'.

[3] Tacitus, *Annales* 14.22. On the traditional purity of the Aqua Marcia see also Statius, *Silvae* 1.5.24; Strabo, *Geographia* 5.3; Pliny the Elder, *Naturalis historia* 31.24.

[4] This led Latte to conclude that *scelerare* and *polluere* had, in essence, the same contaminating result; Latte (1960) 48 n. 1; Wallace-Hadrill (1982) 24–9; Thome (1992) 77. The links drawn by Wallace-Hadrill between *scelus* and 'sin' require careful navigation, due to the overtly Christian connotations of 'sin', which does not entirely correspond in terms of its infectious danger of pollution; see Davies (1999) 182; Attridge (2004) 71–8.

[5] Thus, Horace's famous ode commenting on the civil wars, linking them to neglect of the gods' temples: Horace, *Carmina* 3.6.

[6] Cicero, *De legibus* 2.19; *Tusculanae disputationes* 72; Festus s.v. 'pura vestimenta'; Liebeschuetz (1979) 48–50; Dyck (2004) 290–2.

[7] Douglas (1966) 8–50 at 44.

[8] For example Gowers (1995) 23–32. This idea has been used particularly in studies of the roles in Roman religion of Vestal Virgins, in whose bodies scholars have read various degrees of significance: Beard (1980) 12–27; Mustakallio (1992) 56–63; Beard *et al.* (1998) I.51–4; Parker (2004) 563–601; Wildfang (2006) 51–64. Further, see Schultz, this volume.

[9] Meigs (1978) 304–18; Carson (1990) 135; Lugones (1994) 459; Mullin (1996) 514–5; Hallam *et al.* (1999) 26–34. See Cicero, *De divinatione* 2.27 on the proper placement of bodily fluids.

recognizable as pollutions themselves, such events were symptomatic of divine anger in Roman religion. Alternatively, we see in Cicero's speech *On the responses of the haruspices* that such portents could also be thought to have resulted from the pollution of religious rites (*ludos minus diligenter factos pollutosque . . . quando sunt ludi minus diligenter facti? quando aut quo scelere polluti?*).[10] Again the issue of *scelus* is present as a pivotal factor in pollution, this time as an offence to the gods.[11] Most importantly, the typical response was to nullify the danger through acts of ritual purification (frequently expressed by the terms *lustrare* and *expiare*) which, Thome argues, naturally 'presupposes the concept of pollution'. Purification thus restored balance in the *pax deorum*, re-establishing order.[12] In Roman religion, purification was frequently transportive, sending the offending person or object 'away' beyond the boundaries of Roman, or even human, society. This usually involved the use of the Tiber to remove beings that went against nature (hermaphrodites, parricides, etc.), or forms of dirt which Rome could not afford should come back to endanger either society or the gods.[13] By far the most important way to restore balance was sacrifice, and all such rituals were subject to equally stringent rules regarding purity and behaviour which controlled every aspect of the sacrificial process.

Birth and death

As with every culture, the processes of birth and death represented two of the commonest sites for pollution in Roman society. Just as in ancient Greece, both events threw the household into confusion, forcing family members to adjust social boundaries.[14] Both events also resulted in a member of the family losing control of his or her body, with the resulting emissions being viewed as contagious and threatening. The physical dangers to which pregnant women were exposed are well known, and even after birth, both mother and child were considered vulnerable. This danger received considerable attention in the anthropological theories of A. Van Gennep, who

---

[10] Cicero, *De haruspicum responso* 21.
[11] Festus s.v. 'impietus' succinctly reinforces the connection: '*impietus, sceleratus*'.
[12] Thome (1992) 78.
[13] On the use of rivers and the sea in transportive purification, see Le Gall (1953) 67–110; Kyle (1998) 213–241; Lindenlauf (2004) 416–33; Edlund-Berry (2006) 162–80; Barrett (2008) 42–5; see also Hopkins, this volume.
[14] Parker (1983) 52; Hertz (1960) 37–53; Shaw (1996) 100–38; Mullin (1996) 511–12, 520; Bendlin (2007) 182–3.

interpreted the ritual and physical separation of mother and child from ordinary society as one of the pivotal 'rites of passage'.[15] Such a ritual separation protected the mother and child during their liminal phase of reintegration, as well as protecting the wider society from the pollution which clung to them (a pollution which Parker believed helped Greeks 'define, and so limit, the period of danger and anxiety'). As is so often the case, the regulations regarding the Greek rites of separation are better attested in the ancient evidence.[16]

Evidence for Roman beliefs and practices is sketchy at best, but one important note has survived in Festus: *Lustrici: dies infantium appellantur, puellarum octavus, puerorum nonus, quia his lustrantur atque eis nomina inponuntur* ('*Lustrici* refer to those days, the eighth for girls, the ninth for boys, on which the children are purified, and assigned their names').[17] The need for ritual lustration coinciding with official acceptance into the family marks a logical rite of transition for both mother and child, and implied that a degree of pollution surrounded them up until their full reintegration into the family. St Augustine also records that both were protected immediately after the birth by a form of ritual sweeping on the threshold, ostensibly to protect them from the ancient Italian god Silvanus.[18] The act of purification warded off those very real physical dangers which threatened immediately after birth, whilst also ritually combating the equally real pollution caused by emissions accompanying labour. Beyond those dangers faced by the mother and child, the pollution incurred by birth to a household appears negligible. The father of the future emperor Augustus was able to attend the Senate's meeting during the revolt of Catiline on the same day as his son's birth, albeit having been detained by the birth itself. Similarly, Gellius reveals that it was acceptable to attend a home where a birth had taken place immediately, without fear or precaution.[19]

---

[15] Van Gennep (1960) 41–64.

[16] Pregnant women had to avoid temples for forty days after confirming their condition, but afterwards might attend them freely. Following the birth the contagious nature of the pollution lasted a matter of days, and pollution was probably limited to those family members who had been under the same roof; Parker (1983) 48–65; Censorinus, *De die natali* 11.7; Pliny the Elder, *Naturalis historia* 7.41, who suggests the embryo was thought to move first on the fortieth day.

[17] Festus s.v. 'lustrici'. The practice of naming girls and boys on the eighth and ninth days is also mentioned by Plutarch in his *Quaestiones Romanae*, and Macrobius, with a variety of potential solutions: Plutarch, *Quaestiones Romanae* 102; Macrobius, *Saturnalia* 1.16.36; Rose (1924) 210–1; Rüpke (2007) 231.

[18] St Augustine, *De civitate Dei* 6.9; Burriss (1929) 152–3; Bailey (1932) 38. Cato, *De agricultura* 83 notes that women may not attend the country rites of Mars Silvanus. Warde-Fowler viewed both deities as an offshoot of the same primal deity; Warde-Fowler (1911b) 131–4; Burriss (1925) 221; Dumézil (1966) 34–5, 616–17.

[19] Suetonius, *Divus Augustus* 94; Aulus Gellius, *Noctes Atticae* 12.1; Dixon (1988) 238–40.

Those purificatory actions which warded off the threat of death during this period were also visible in those rituals immediately following bereavement. As Bodel has noted, there was a marked importance in religious rituals to combine with genuine, practical issues of hygiene in the disposal of corpses, particularly in heavily populated urban settings.[20]

Death instantly made family members polluted, regardless of physical proximity to the corpse, the closest relations typically incurring the greatest degree of impurity.[21] At the moment of death the nearest family member came into direct contact with the body, by collecting the last breath with a kiss and subsequently closing the deceased's eyes, for Pliny reveals that it was considered *nefas* to look upon the eyes at the point of death.[22] The deceased was called by name three times (*conclamare*), then the body was washed, rubbed with perfumes, dressed in white robes and left to lie in state with the feet towards the door. Each of these actions contributed towards easing the transition of the spirit from the house and apparently prevented its return.[23] Conversely, surviving family members made their pollution apparent through public demonstrations. The house itself displayed branches of cypress to warn outsiders of the taint of death.[24] H. Lindsay suggests that this measure was designed to warn those members of society particularly at risk from death-pollution, most notably priests, although pregnant women were probably also advised to avoid the house during this period.[25] The family also marked themselves out by dressing in mourning garments, typically black, or otherwise dark, although J. Heskel also suggests the possibility of white garments which had been deliberately sullied with dust or dirt. Deliberate physical pollution, including refraining from washing, therefore demonstrated the family's metaphysical pollution

[20] Bodel (2000) 128–51.

[21] Hertz (1960) 27–34; de Visscher (1963) esp. 32–9; Toynbee (1971) 43–61; Hopkins (1983) 217–34; Prieur (1986) 18–45; Lindsay (1998) 69–74; Corbeill (2004) 67–106; Hope (2007) 85–6 and (2009) 71–4.

[22] Pliny the Elder, *Naturalis historia* 11.150; Virgil, *Aeneid* 4.684; Toynbee (1971) 43–4. The closing of the eyes may have protected the family from the departing spirit – '*in oculis animus habitat*': Pliny the Elder, *Naturalis historia* 11.145–6; Plautus, *Miles gloriosus* 1260–1; Cicero, *De legibus* 1.9.27; Evans (1935) 59–60; McCartney (1952) 187–8.

[23] Servius, *Ad Aeneidem* 6.218; Persius 3.98–106; Pliny the Younger, *Epistulae* 5.16; Lucian, *Dialogi mortuorum* 11–15; Burriss (1931) 72; Toynbee (1971) 44, 228 (fn. 119); Hope (2007) 97–9.

[24] Horace, *Carmina* 2.14.23; *Epodi* 5.18; Pliny the Elder, *Naturalis historia* 16.40, 16.139; Festus s.v. 'Cupressi'; Servius, *Ad Aeneidem* 3.64.

[25] Lindsay (2000) 155; Tacitus, *Annales* 1.62. The *flamen Dialis* was also forbidden to touch a corpse or enter a graveyard, although he could attend a funeral: Aulus Gellius, *Noctes Atticae* 10.15.23–5. On the susceptibility of pregnant women to pollution, see for example Seneca the Elder, *Declamationes* 4.1; Pliny the Elder, *Naturalis historia* 28.80.

openly.[26] Once the corpse had been carried out to the cemetery, which by law had to be situated beyond the sacred boundary of the *pomerium*, the deceased's heir was tasked with ritually sweeping the threshold, just as was done following a birth. Following deaths, however, this is more likely to have represented a purification of the household, which coincided with driving away the spirit of the departed.[27] On the family's final return from the grave they were purified with fire and water in the ritual of *suffitio*, before their reintegration was complete.[28]

To leave a body without proper burial could pose a serious threat to the community. Horace envisioned the words of an unburied corpse, addressing a nameless passer-by, imploring him to perform his burial rites. The spirit warns the stranger not to ignore his request and thus endanger the stranger's children by an inexpiable offence.[29] Three handfuls of earth are specified as the minimum requirement to put a spirit at ease, but amongst the poor and destitute, even this could not always be guaranteed. In such cases the task fell to specialists, undertakers and morticians. Such men were based at the sanctuary of Libitina, where all funerary equipment was made available,[30] and Bodel has argued that they lived perpetually under the shadow of pollution.[31] The surviving law code from the Italian town of Puteoli states that corpse bearers and executioners must enter the town only on official business, and during their journey were required to wear brightly coloured clothing, and to ring a bell as they proceeded (conditions that

---

[26] Heskel (1994) 141; Cicero, *In Vatinium* 30–2; Cicero, *Pro Sestio* 144–5; Lucan, *De bello civili* 2.333–7; Petronius, *Satyrica* 42.1; Diodorus Siculus 1.91; Parker (1983) 41; Bodel (2000) 142.

[27] Festus s.v. 'Everriator'; Ovid, *Fasti* 2.23–6; Frazer (1929) II.279–83. Frazer notes the possibility that the sweeping was performed by a religious officer, since Ovid mentions the presence of a *lictor*.

[28] Festus s.v. 'Aqua et Igni'. The location of the *suffitio* ritual is uncertain, but the marked similarity with the use of fire and water in Roman marriage ceremonies suggests the threshold of the household is most likely.

[29] Horace, *Carmina* 1.28.2.10–14 (*neglegis immeritis nocituram postmodo te natis fraudem committere?*).

[30] Varro, *De lingua Latina* 6.47; Dionysius of Halicarnassus, *Antiquitates Romanae* 4.15.5. Julius Obsequens 6 and 10 record instances of plague so severe that Libitina was overwhelmed (*pestilentiae Libitina non suffecit*). The likely location of the temple is thought to be beyond the Esquiline gate adjacent to the Roman necropolis: Bodel (1994). Similarly, the gate through which the dead were removed from the arena was the Porta Libitinaria; Cassius Dio 73.21.3; Ville (1981) 376–7; Kyle (1998) 156, 164–71.

[31] Bodel (2000) 135–44 examines the various tasks associated with disposal, noting that while some cultures made strict distinctions between the various duties regarding the dead, less distinction was made in Rome by the late Republic. He stresses that physically polluted should not automatically mean religiously impure, and no degree of uniformity can be assumed. In particular, profit from misfortune also played a major role in the social segregation that accompanied the profession.

Rüpke assumes represent a simplified form of Rome's own laws).[32] This was stressed particularly when dragging corpses away with a hook, a practice retained also in the arena, and suggestive of a degree of deliberate avoidance of the most impure corpses, the *noxii*, or otherwise publicly detested figures (further on this, see Bradley, this volume, pp. 107–18).[33]

Across rituals surrounding birth and death we see the apparent need for isolation for the protection of one group or another during periods of 'disorder'. Purification corrected the imbalance caused by these natural pollutions. In other areas, however, the extra dimension of human action, even deliberate intent, could cause greater fear of pollutions incurred within proximity of the gods.

Sex

In examining Greek attitudes towards sexual pollution, Parker stated that 'profane life is, necessarily, sexual'.[34] For Mullin and Carson, this pollution stemmed from the idea that 'impurity is mixture and sex is seen as mingling', while Meigs interpreted bodily emissions as 'dying' once separated from the body.[35] Varro saw the term *pueri*, referring to boys under fifteen years old, as stemming from the word *purus*, a form of purity based on sexually inactivity.[36] The rejection of sex in religion therefore served to separate it from the everyday activities of mortal men, and Virgil and Ovid reveal that even animals selected for sacrifices were marked out from birth and kept away from profane usage, specifically manual labour and breeding, both of which 'spoiled' the otherwise perfect offerings.[37]

In the majority of cases, however, sexual abstinence appears to have been temporary, and centred on specific religious events or rituals. Ovid imagined the pious king Numa receiving prophecies from Pan within a sacred grove, after dutiful sacrifices and a period of dietary restrictions and

---

[32] Rüpke (2007) 232–3; Bodel (1994) 72–81; Hinard and Dumont (2003) esp. 57–68; Bodel (2000) 142–3 is more sceptical, warning against assuming uniformity between towns.

[33] The emperor Vitellius was one such victim. The hook was a means of both avoiding direct contact and also showing disrespect towards socially prominent victims. The treatment is identified by Kyle as being employed particularly against those considered *hostes*, and Kyle adds that 'when hooks . . . are mentioned, dumping in the Tiber can be assumed', a further means of symbolic purification, transporting the body beyond the boundaries of the city; Suetonius, *Vitellius* 17; Cassius Dio 61.35.4; Kyle (1998) 218–20. Further on the death of Vitellius, see Bradley, this volume, p. 117.

[34] Parker (1983) 91.　　[35] Mullin (1996) 513; Carson (1990) 158–9; Meigs (1978) esp. 312–13.

[36] Censorinus, *De die natali* 14.2.

[37] Virgil, *Georgics* 3.157–61, 4.538–43; Ovid, *Fasti* 4.335.

sexual abstinence,[38] while, much later, the emperor Severus Alexander was thought to worship his family *Lares* each morning, unless he had been with a woman the night before.[39] Ritual abstinence became a motif for Republican poets, complaining of spurned advances because of the demands of Oriental mysteries, particularly those of Isis; Tibullus' Delia insisted she must sleep in a pure bed (*purus torus*) before ceremonies.[40]

Examples of the pollutive and damaging nature of sex within religion occur predictably in rites concerned with agricultural fertility, or gods connected with crops. Ovid reveals that during rites to Ceres Roman women were required to abstain from sex for nine nights (again apparently the standard period of isolation for religious purposes),[41] while Tibullus gives a famous description of the agricultural festival, the Ambarvalia, in which the farm's lands, animals and workers were purified, and the names of Bacchus and Ceres were invoked. Only those polluted by sexual contact were forbidden to approach the altars.[42] Bee-keepers were specifically ordered by Columella to be chaste, as were those who oversaw the household stores and cupboards, or who were involved in baking, cooking or other tasks involving the preparation of food.[43] The exact reasons behind these taboos are uncertain, although with agricultural festivals a sympathetic connection may exist – sexual activity amongst worshippers might draw away fertility from the fields.[44] With regard to those handling food, Meigs' theory that undesirable emissions might gain access to the body seems more probable.[45] Beneath each of these ideas, however, will have been the constant desire to keep the sacred and profane worlds firmly separated.

This was most certainly the pretext for the senate's brutal suppression in 186 BC of the cult of Bacchus, which was thought to have involved nightly orgies in which young citizens were corrupted by the sexual depravity (*stuprum*) of a foreign religion (although Livy curiously suggests ten days of sexual abstinence were required before initiation, possibly extending the existing poetic motif of abstinence surrounding Eastern mysteries).[46]

---

[38] Ovid, *Fasti* 4.651–8 (*unus abest Veneris, nec fas animalia mensis ponere, nec digitis anulus ullus inest*).

[39] Scriptores Historiae Augustae, *Alexander Severus* 29.2.

[40] Tibullus 1.3.23–6; Propertius 2.33.1–6. See also Propertius 4.8.81–8.

[41] Ovid, *Metamorphoses* 10.431–5. Similarly for the wife of the priest of Jupiter, cf. Ovid, *Fasti* 6.227–33.

[42] Tibullus 2.1.11–12 (*vos quoque abesse procul iubeo, discedat ab aris, cui tulit hesterna gaudia nocte Venus*); Pascal (1988) 523–36.

[43] Columella, *De re rustica* 9.14.3, 12.4.2.

[44] Columella, *De re rustica* 2.22.5 shows awareness that not everyone followed these restrictions.

[45] Meigs (1978) 311–12.

[46] Livy 39.8–18; Pailler (1988) 195–245; Turcan (1996) 300–7; Briscoe (2008) 230–91. On *stuprum* and sexual misconduct, see Fantham (1991) 267–91.

Similarly when a Roman matron was debauched in the temple of Isis, the emperor Tiberius acted swiftly and decisively against the corruptive influence of the cult.[47] Most famous, perhaps, was the scandal caused by Publius Clodius' intrusion on the women-only rites of Bona Dea, allegedly to seduce Julius Caesar's wife.[48] The rites were performed by the Vestals, whose sanctity (closely tied to their chastity) may have been compromised as much as that of Bona Dea. This certainly represented a serious violation of religion, and Clodius was tried *de incesto* by the senate after the Vestals and pontiffs judged the offence to be *nefas* ('unspeakable').[49] Throughout the Republican and imperial periods, sex was supposed to be kept firmly away from sacred space and ritual. The harsh reactions that occurred in those instances where these rules were broken indicate a firm resolve to maintain the show of dignity and respect to the gods.

This scene also illustrates that *incestum* did not necessarily signal incest, but could equate to various impurities thought to be worthy of divine condemnation.[50] Livy used the term to signify general uncleanness as a priest of Diana asks a Sabine preparing to sacrifice a calf if he will risk doing so while still impure (*inceste sacrificium Dianae facere?*).[51] Cicero, too, implies a wider religious dimension when he stipulates that the pontiffs should execute those guilty of *incestum* – the issue of whether this referred to incest or less specific sexual crimes is a subject of some debate.[52] Nevertheless incest was definitely also viewed in terms of impurity, which could be dangerous. Tacitus suggests that at first Claudius and Agrippina hesitated to marry, since marriage to a niece was incestuous, and might cause 'public misfortune' (*malum publicum*).[53] Juvenal also mocked the marriage laws of Domitian, who was polluted (*polluere*) by incest with his niece.[54] Incest was a capital crime; those found guilty (aside from emperors), faced execution by being hurled from the Tarpeian rock, a fate reserved for those guilty of

[47] Josephus, *Antiquitates Judaicae* 18.65–80; Rogers (1932) 253; Moehring (1959) 298–300.

[48] Cicero, *Epistulae ad Atticum* 1.12.3; *De haruspicum responso* 37–8, 44; Velleius Paterculus 2.45.1; Plutarch, *Cicero* 28; Cassius Dio 37.45.1–2; Bettini (1995) 224–35, esp. 228.

[49] The rites were re-performed and the danger nullified. Cicero would attack Clodius for this wilful act of pollution for the rest of his life; Cicero, *De domo sua* 105; *De haruspicum responso* 4–5, 8–9; *Pro Milone* 72–3, 85–6.

[50] For example Horace, *Carmina* 3.3.19; Livy 45.6; Shaw (1992) 269–70 explores the issue of *incestus* as the antithesis of *castus*, with wider reference to crimes prohibited by state law.

[51] Livy 1.45.6.   [52] Cicero, *De legibus* 2.22; see Dyck (2004) 317–18 with bibliography.

[53] Tacitus, *Annales* 12.5.

[54] Juvenal 2.29–33. The emperors Gaius, Claudius and Nero are all reported to have had similar incestuous relationships: Suetonius, *Gaius Caligula* 24, 36; *Nero* 28; Tacitus, *Annales* 12.8; cf. Liebeschuetz (1979) 41–3; Moreau (2002), esp. 29–36, 43–52. On reports of cultural incest in Roman Egypt, where the practice appears to have been most widespread, see Hopkins (1980) 303–54; Parker (1996) 362–76; Ager (2005) 1–34; Heubner (2007) 21–49; Remijsen and Clarysse (2008) 53–61.

particularly heinous offences.[55] Thus religion reinforced those conventions set in place to protect the Roman society from the corrupting influences they deemed to have overrun various barbarian nations.

Blood

In terms of pollution, blood is perhaps the most varied and powerful of all natural substances, in many ways matching the threat death was felt to pose to society. Whereas the pollution caused by death is invisible, however, blood is a highly visceral substance, pressing itself upon each of the senses. Its vivid colour is shocking to the eye; its smell is pungent and distinctive, affecting humans and animals alike; its taste is metallic and bitter; its touch stains many substances and fabrics irrevocably, and whilst it is warm when fresh from the wound, it cools and congeals quickly, a process which leads to changes in all of the above factors as the blood appears to die visibly once separated from the living whole.[56] In this respect, the pollutions of blood and death are, to a degree, inseparable, and in cases of murder contain an extra dimension of disorder, since the shedding of blood here represents an assault on the stability of the social order.[57]

The idea of blood taboo in Rome was originally dismissed by Fowler, who believed that continuous exposure to animal sacrifice, warfare and blood sports in the arena had 'eliminated the various chances that might arouse it'.[58] However, these areas should be recognized as some of the most rigidly controlled events in Roman society, carefully separated both physically and ritually from everyday life. When these parameters were breeched, these areas typically signified disorder, at which point the blood could be dangerous. The obligatory blood sacrifices were carefully controlled, and the blood was intended for the altar only. Those occasions on which animals broke free, spattering onlookers with blood, were viewed as terrible omens,[59] while corrupted blood from sacrificial wounds revealed the anger of the gods.[60] Similarly, numerous portents occurred throughout Roman literature in which blood appeared out of place and even nature, such as

---

[55] Tacitus, *Annales* 6.19.1; Robinson (1992) 194–5; Moreau (2002) 341–3. For further discussion of the Tarpeian rock, see Bradley, this volume, chapter 6.

[56] Robertson Smith (1927) 40, 606; Burriss (1929) 144–50; James (1962) 27, 60–1; McCarthy (1969) 166 and (1973) 205–10; Douglas (1975) 108–15, 248; Parker (1983) 104–44; Roux (1998); Mullin (1996) 512–15; Attridge (2004) 71–9; Meyer (2005) esp. 1–16.

[57] Mullin (1996) 512.   [58] Warde-Fowler (1911b) 33, 180–1.

[59] Livy 21.63.13–4; Lucan, *De bello civili* 7.170; Suetonius, *Divus Iulius* 59; *Gaius Caligula* 57.

[60] Lucan, *De bello civili* 1.609–37.

in rain or streams.[61] In such cases the disorder of *pax deorum* was made known through disorder in the natural world.[62]

Beyond sacrificial parameters, blood appears to have been damaging to the sanctity of both the gods and their precincts (although the true threat posed was to Roman society, upon which the punishment rebounded).[63] Aeneas illustrates his piety by refusing to touch the images of his household gods while stained with the blood of battle,[64] and Seneca's Amphitryon begs Hercules not to offer sacrifice until he has cleansed himself of the blood of his enemy (*nate, manantes prius manus cruenta caede et hostili expia*).[65] The layers of pollution caused by bloodshed are perhaps best illustrated by the story surrounding the murder of King Servius Tullius by his daughter, Tullia. The king having been attacked on Tullia's orders, Livy describes how she proceeded to drive her carriage over the body of her father, spattering her carriage with blood which she carried back to her home, thus polluting her household gods. Her husband Tarquinius Superbus was described as equally stained (*maculare*) by his part in the deed.[66] An alternative version offered by Ovid states that, thus polluted, she dared to enter her father's temple and was shunned by his image.[67] In this case murder is augmented by being both parricide and regicide combined. In defending Sextus Roscius from charges of parricide, Cicero stated that the blood of a parent stained so deeply that it penetrated the soul and could never be removed (*magnam possidet religionem paternus maternusque sanguis; ex quo si qua macula concepta est, non modo elui non potest, verum usque eo permanat ad animum*).[68] Tullia's crime stains not only her, but also her husband. Contact with the sacred causes her pollution to rebound upon herself, as she is shunned by her household gods and her father's temple statue. Even profane space is irrevocably tainted by the deed, as this act

---

[61] For example Livy 24.10. For full lists, commentary and bibliography on prodigies in the Republican period, see MacBain (1982); Engels (2007) 283–714.

[62] This idea is interestingly refuted in Cicero, *De divinatione* 2.60, which argues for rational explanations for any such inexplicable event.

[63] Parker (1983) 145–6.

[64] Virgil, *Aeneid* 2.717–20. See also Servius, *Ad Aeneidena* 6.8. Aeneas states that until he has washed in a running stream (*flumen vivum*), to do so would be *nefas*. Compare with Hector's similar actions in Homer, *Iliad* 6.300–18.

[65] Seneca, *Hercules Furens* 919.

[66] Varro, *De lingua Latina* 5.159; Livy 1.48.5–7; Dionysius of Halicarnassus, *Antiquitates Romanae* 4.39; Ovid, *Ibis* 363; Valerius Maximus, *Factorum ac dictorum memorabiliuem libri* ix 9.11.1; Cicero, *De republica* 2.46; Festus s.v. 'vicus Sceleratus'. For a similar argument see Cicero, *Pro rege Deiotaro* 15.

[67] Ovid, *Fasti* 6.609–24. On the Porta Scelerata see Coarelli (1988) 409–14.

[68] Cicero, *Pro Sexto Roscio Amerino* 66–8.

was accepted as the aetiological explanation behind the street named Vicus Sceleratus, which was well known even in Ovid's day. Yet it is not simply abstract pollution caused by the crime which plays a central role in the reports of the writers, but blood – the physical essence of the crime, which spreads the misfortune like a disease.

The role of blood in murder-pollution played a pivotal role in Parker's study of the Greek world, and the issue is equally complex in Rome.[69] The crucial issue is the extent to which pollution was a key consideration in cases of murder, and in both cultures it appears more prominent in myth (although it certainly appears elsewhere). In discussing the roles of social violence, Girard suggested that the original purposes of murder-pollution, and the emphasized expectance of family-administered 'vengeance', were to offer direction in societies without established judicial process and authorities.[70] The Roman Republic, however, was definitely not without established legal codes and practices, and acts of murder in self-defence or of those engaged in robbery are deemed in the xII Tabulae, not to have been committed 'without pollution', as was the judgement in Plato's Laws, but rather to have been done 'justly' (si nox furtum factum sit, si im occisit, iure caesus esto).[71]

From a strictly legal perspective, therefore, pollution may be said to have been superseded by ius. Parker believed a similar situation emerged in Greece, where pollution acted as a 'shadowy spiritual Doppelgänger of the law', reinforcing legal decrees and expressing the feelings of social disruption brought about by murder without the need for family vendettas.[72] The idea of the polluted killer continued to exist into the late Republic, and though it does not appear in legal statutes it was used repeatedly as a tool of skilled orators, who used it to sway the opinions of their fellow citizens and, specifically, the jury, the 'surrogate avengers' tasked with protecting the city.[73] It was not coincidence that Cicero reminded his audience of Clodius' blood-stained (cruentus) hands when defending his murderer, T. Annius Milo, or when refuting Clodius' ability to offer a shrine on the site of his home.[74]

---

[69]   Parker (1983) 104–44; Carawan (1993) 235–70; Sealey (2006) 479–81.
[70]   Girard (1977) 1–38. Girard used this idea expressly to support his theory of inherent violence, while Parker suggests a more balanced combination of the two factors of pollution and retribution; Parker (1983) 116–18; Arnaoutoglou (1993) 109–37; Jońca (2004) 44–51.
[71]   xII Tabulae 8.12; Plato, Leges 874a–e; Cicero, Pro Milone 8–9; Aulus Gellius, Noctes Atticae 20.1.7; Macrobius, Saturnalia 1.4.19.
[72]   Parker (1983) 116, 121; Jońca (2004) 46.   [73]   Parker (1983) 124–6; Girard (1977) 15–17.
[74]   Cicero, Pro Milone 20; De domo sua 108. For the opposite view of Milo as polluted by Clodius' murder, and the image of the bloodstained hand in epic, see Lucan, De bello civili 2.480, 3.135–6.

Cleansing the city

Forms of natural pollution or otherwise unavoidable forms of contamination springing from daily life in Rome could not be avoided, and so the city itself underwent ritual purifications at various interludes and in varying locations. These could be specific or wide-ranging, affecting both the populace and the fabric of the city itself. The simplicity of such pollution is encapsulated in Horace's rebuke concerning the city's temples and statues, soiled (*foedare*) by smoke and grime – a physical symbol of a people's religious and moral decay.[75]

Periodic rituals of purification are commonplace in Roman religion, the most notable being the *lustrum*, which occurred every five years, and the Secular Games, which were supposed to be performed once in a century. The *lustrum* was conducted by the censors, themselves paradigms of moral uprightness and purity.[76] Originally a lustration of the armies, the ritual was based on the Campus Martius, and involved the sacrifice of a bull, a sheep and a pig (*suovetaurilia*), which were ceremonially led around the site before sacrifice.[77] The same practice was used in the agricultural lustrations of the Ambarvalia, which also required leading the victims around a specified area intended for purification, although Ogilvie asserts that the use of fire as a further purificatory agent was limited to the *lustrum* alone.[78] These devices were taken even further in the Secular Games, as Zosimus records the directive to distribute torches along with sulphur, both symbols of ritual purification.[79] These were notably distributed by the emperor Augustus in his revival of the games in 17 BC.[80] Ritual prayers were also offered by two groups of twenty-seven men and women, all of whom still had both parents living.[81] Living parents were a requirement for prospective Vestals too, and groups of twenty-seven appear throughout Republican records in rituals of purification and expiation, typically in the wake of disturbing prodigies.[82]

Numerous other forms of purification in the city, whether naturally occurring or otherwise, made special use of the Tiber as a means of removing religiously dangerous or offensive materials. Marriage in the first part of June

---

[75] Horace, *Carmina* 3.6.1–4.
[76] Varro, *De lingua Latina* 6.86–7; Festus s.v. 'Minuebatur populo luctus'.
[77] Ogilvie (1961) 31–9 and (1969) 89; Liebeschuetz (1979) 96–8; Rüpke (1990) 145–7.
[78] Strabo, *Geographia* 5.2; Tibullus 2.1.1–24; Ogilvie (1961) 38–9.
[79] Zosimus 2.5.1–5; Cooley (2003) 266–79.
[80] Cooley (2003) 272; *Corpus Inscriptionum Latinarum* 3.2323.29–36.
[81] Zosimus 2.5.5. Cf. Aulus Gellius, *Noctes Atticae* 4.9, 5.17; Festus s.v. 'Patrimi et Matrimi'.
[82] Livy 38.36.4; Julius Obsequens 1.27a. 34.

was deemed inauspicious until the dirt swept and cleaned from the temple of Vesta (*purgamina*) was ritually removed in the Tiber, during which period the *flamen Dialis*' wife was forbidden to comb her hair or have sexual intercourse with her husband.[83] The need to protect all things connected with Vesta from the profane world appears extremely pronounced as a result – even the water used for the cleansing of the sanctuary was drawn from a sacred spring and when carried back could not be placed upon the ground at any point.[84] Running water therefore secured both purity and purification for the cult.

The Vestals also oversaw a more widespread purification of the city through the ritual of the Argei, which has remained a point of considerable contention for over a century.[85] While the purifications in June removed physical dirt, the Argei ritual, performed on 14 May, had more widespread aims. In it, the Vestals carried twenty seven rush-puppets, which had been placed in the individual districts of the city, to an ancient bridge (Pons Sublicius), and ceremoniously threw them into the Tiber. The rite was attended by the college of pontiffs as well as the wife of the *flamen Dialis*, who wore mourning dress with her hair unkempt.[86] The positioning of the festival on an even-numbered day is curious, and must be viewed in the context of the preceding festival days of the Lemuria (9, 11 and 13 May), in honour of the spirits of the dead.[87] Thus the propitiation of the dead was immediately followed by purification of the city, at which point the doors of the temples, which had been kept closed during these periods, could be opened again without fear of contaminating the gods' space with the pollution of death.[88] The positioning of the Argei puppets around pivotal areas of the city suggests they served as a form of scapegoat, absorbing the danger which would otherwise harm the populace. In order for life to return to normal, these items needed careful disposal. The Tiber presented the ideal method of removal to the sea, which could itself absorb such pollution without contamination.

---

[83] Ovid, *Fasti* 6.225–9; Frazer (1929) 4.166. This purification of Vesta's sanctuary may have deliberately preceded the Vestalia, which took place on 9 June.

[84] Servius, *Ad Aeneidem* 9.339.

[85] Warde-Fowler (1911b) 54–6, 105, 321–2; Wissowa (1912) 60, 283–4, 420–1; Rose (1924) 98–101; Latte (1960) 412–4; Maddoli (1971) 153–66; Nagy (1985) 1–27; Pötscher (1998) 225–34; Ziolkowski (1998–9) 191–218; Graf (2000) 94–103; Rüpke (2007) 178–9.

[86] Varro, *De lingua Latina* 5.45; Ovid, *Fasti* 5.621–62; Plutarch, *Quaestiones Romanae* 32, 86.

[87] Bayet (1973) 98–9; Nagy (1985) 1–27 places particular emphasis on the placement of the ceremony adjacent to the Lemuria.

[88] Ovid, *Fasti* 5.485–6.

Conclusion

Pollution within Roman society and religion must inevitably be viewed as a wide-ranging discourse, and not a singular phenomenon. It could appear in any context and was expressed in numerous ways across ritual, language and literature, occurring within epic and poetic verse as well as political rhetoric and history. Its dominant place within Roman religion is demonstrated by the considerable unease seen on those occasions when rites were deemed to have been improperly performed, or deliberately violated. Such fears could be played upon for literary effect, particularly in the field of rhetoric. Cicero appears a master in this, marking out enemies such as Catiline as out of place in decent Roman society, branding him a plague (*pestis*), which had spread, unchecked, like a disease through the Roman state, and begging him to 'cleanse the city' (*purga urbem*) by departing.[89] Cicero would continue to use accusations of physical and ritual impurity against his enemies throughout his career, to significant effect.[90] After Cicero the writers of the new Augustan age drew upon the horrors of the recent civil wars to present the late Republic as a period of deep social and religious corruption. Beck notes that the possibility that this was primarily the product of the Augustan administration is irrelevant, the crucial issue being that 'a narrative of collective sin, divine punishment, and expiation entered Roman discourse, however engineered'.[91] This is accurate to a degree, particularly regarding the literary fields in which pollution emerged in the late Republic. It must be noted, however, that the nature of the regulations governing rituals from the city's farthest antiquity suggests that pollution never simply 'emerged' into Roman society. Though the means of expression and interpretation changed over time, there may never have been a period in which pollution was not to exert influence over the Roman psyche. In many of the most important areas of daily life there appeared a concerted effort to separate the human world from those spaces and occasions reserved for the gods. Therefore, the Romans' relationship with physical and religious purity should not be viewed as static. The sources of the late Republic do more than merely reveal existing feelings regarding pollution and purification. Rather, they shaped

---

[89] Cicero, *In Catilinam* 1.10, 1.33, 2.1; Dyck (2008) 122. Further, see Bradley, this volume, pp. 103–4.

[90] Achard (1981) 110–40, 519–20. For similar examples see Cicero, *Post reditum in senatu* 14; *De domo sua* 23, 26, 35, 112, 139; *De haruspicum responso* 24, 42; *Pro Sestio* 20; *Pro Milone* 18, 20, 85; *In Pisonem* 95; *Pro Fonteio* 31; *Orationes Philippicae* 1.17, 2.6, 2.86, 11.29, 12.15.

[91] Beck (2004) 510.

and adapted them to serve new purposes and express new ideas. Yet at all points the sentiment remained that purity was intrinsically good, and to be associated with piety and 'correctness', while pollution could be used to mark out anything fundamentally wrong, which by necessity made it incompatible with the divine.

3 | Purification in ancient Rome

ELAINE FANTHAM

It can be instructive to chase pollution and purification in the subject indices of a university library, but it is not useful for our field. The bare terms call up major ecological studies of the pollution (and purification) of air and water: add the qualifier 'religious' or 'ritual' and the screen swims with texts in Arabic, or studies of Mosaic law. Anthropologists can take their starting point from Mary Douglas' 1966 study of what endangers purity: students of the Greek world have the marvellous guide by Robert Parker (1983), but despite the many excellent new books on Roman religion,[1] students of Rome can only extrapolate from those well-lit fields to an obscurer scene. One reason for this obscurity may be the relative secularization of Roman thinking during the period from approximately 60 BC to AD 100 when the most detailed Roman discussions originated. Educated Romans of the elite were intensely interested in state religion, largely because Rome's state priests were not theologians or holy men but active politicians. Another related factor is the absence of (for example) dietary taboos, so familiar to us from the Pentateuch and Islam.

The main Roman sources must be read sceptically – as sceptically as Horace's politically inspired wishful thinking in the Secular Hymn of 17 BC commissioned by Augustus: 'No chaste family is polluted by fornication', *nullis polluitur casta domus stupris*. Sexual pollution, defined as violation of married or marriageable women, was only one aspect of pollution at Rome, not usually treated as of significance but brought into prominence by Augustus with his new morality laws and ideology.

Horace's official hymn of praise had to affirm the successful impact of Augustus' moral legislation, but it strikes a note utterly different from other contemporary poetry, love poetry indifferent to issues of marriage and marriageability. And surely this offers an extraordinary contrast with the honest Nuer whose attitude to pollution Mary Douglas reports in *Purity*

---

[1] To mention only a few, Beard *et al.* (1998); Scheid (2003); Rives (2007): none of these index pollution or purification. Older books, from G. Wissowa's *Religion und Kultus der Römer* (1912) to K. Latte's *Römische Religionsgeschichte* (1960), and G. Dumézil's *Archaic Roman religion* (1970) (= *La religion romaine archaique*, Paris, 1966), took more interest and are still excellent sources.

and danger. The Nuer believed that adultery pollutes and physically withers the victimized husband. Adultery had become a matter of indifference to the elite of Augustan Rome and, to judge from Catullus, even in modest Italian towns. Of the two obvious taboos of pollution by murder of a fellow citizen or the violation of his marriage (wife, son or daughter), the first had long been transferred from the moral and religious sphere of pollution to the standing courts *de vi* and *de sicariis et veneficiis*, while the latter only troubled simple folk, and Augustan husbands had to be prodded into prosecuting their errant wives by the threat of being prosecuted themselves for pandering. In this huge and sophisticated city there was no real equivalent of the belief in *miasma*, which gave rise to Robert Parker's splendid study of pollution in Greek thought.

This is probably the best moment to review the actual verb *polluere*, and its surprisingly restricted use in Latin. It seems to be derived from **por +luere*, and connected with the same root as *lutum*, mud, and various compounds of *luere* denoting unclean water, flood waters, swill and sewage: here perhaps we come closest to the aspect of pollution which is employed elsewhere in this volume. The *Oxford Latin dictionary* gives the verb's primary meaning as 'to make foul or dirty, soil or stain', but its examples all come from the imperial period. The reference of most interest to my argument, (2) 'to make ceremonially impure, pollute', is found as early as Cicero's first speech, as are the third and fourth senses, (3) 'to violate, degrade by immoral action' and (4) 'to defile with illicit sexual intercourse'. Romans were in fact very selective in their use of the word. Three enormously prolific authors, the orator Cicero, the historian Livy and the encyclopedist Pliny the Elder, use *polluere* only around eight to twelve times each in their many volumes, usually with overt religious reference. Love elegists transfer the application of the verb to the violation of sexual fidelity; only the virgin's dreaded loss of chastity before marriage (*cum castam amisit polluto corpore florem*) in Catullus' wedding hymn 62 reflects the strict morality of the Nuer: whereas Propertius' three examples extend condemnation to the violation of illicit unions as well as family.[2] The political historian Sallust first applies *polluere* to sacrilege 'stealing from temples and polluting everything sacred or profane' (*delubra spoliare, sacra profanaque omnia polluere*, at *Bellum Catilinae* 11.6–7) but then uses the word to reflect the extreme snobbishness of the elite condemning the accession of a new man, Cicero, to the consulship:

---

2 Propertius 2.34.5: 'that god (Cupid) pollutes both kinsmen and friends'; 3.20.26: 'the man who has violated our pledged altars and polluted our marital rites with a new couch'; more ambiguous is 4.9.8: 'Cacus polluted Jupiter by his theft (of cattle)'.

'they believed the consulship was as it were polluted', *quasi pollui consulatum credebant* (23.6).[3] Epic poets too hold back; the word occurs only four times in the *Aeneid*, for the unwitting tomb-violation of a murdered prince denied proper burial (*Aeneid* 3.61), for the fouling of the Trojans' food by evil Harpies (3.234) and for the violation of a peace treaty (7.467).[4] For Lucan, the poet of civil war, almost every one of his eighteen examples reflects the shedding of citizen blood.

My interest in pollution and taboos was provoked by working on Ovid's versified calendar, the *Fasti*, which includes both general comments on means of purification and individual records of protective instruments and rituals. The main ritual act of purification – the *lustratio* – had many forms at Rome, but was mostly unrelated to the calendar; hence we discover what little we know of its forms from other sources. As the verb suggests, *lustratio* was thought of spatially; typically you would process and make offerings while walking around the place or body of men to be protected, and do it pre-emptively: you purified the army before campaigning, but not on its return. Certain elements – notably fire and water (cf. *Fasti* 4.785–90 and 6.291f.) – and natural growths – wool, laurel, myrtle, *verbenae*, even sacrificial ash – were used in this and related rituals to cleanse men and beasts from the pollution of past offences, and immunize them against future faults. Not all of these *februa* concerned the ordinary Roman: Ovid's examples[5] come mostly from priestly practice, and we can set aside the many priestly taboos of *flamen* and *flaminica Dialis* as extreme. I will also pass over here the criminal pollution of kin murder, mentioning only Pliny's report at *Naturalis historia* 15.119 that Roman and Sabine soldiers supposedly used myrtle to cleanse themselves of bloodshed after their conflict under Romulus (this was not a civil war between kin but war between *adfines* or

---

[3] Compare Livy's report (10.23.10, purporting to be from the late fourth century BC) of a conservative speaker declaring that patrician cults would be polluted by the admission of plebeians, as would the rites observed by augurs and pontiffs once plebeians could hold these offices.

[4] The fourth instance, of the harsh agony of Dido's great love subjected to pollution (*duri, magno sed amore dolores / polluto*, 5.6) raises too many issues to offer an undisputed interpretation.

[5] I quote Frazer's (1929) 1.53–5 translation of *Fasti* 2.21–30: 'Our Roman fathers gave the name of *februa* to instruments of purification; even to this day there are many proofs that such was the meaning of the word. The pontiffs ask the King (*rex sacrorum*) and the Flamen for woollen cloths, which in the tongue of the ancients had the name of *februa*. When houses are swept out the toasted spelt and salt which the officer gets as means of cleansing are called by the same name. The same name is given to the bough which, cut from a pure tree, wreaths with its leaves the holy brow of priests. I myself have seen the flamen's wife begging for *februa*: at her request a twig of pine was given to her. In short anything used to cleanse our bodies went by that name in the time of our bearded forefathers.'

cognati – men related by marriage) at the shrine of Venus Cloacina.[6] In general I have treated the requirement of washing or sprinkling the body with water as implying prior pollution, though obviously there is a strong component of hygiene involved. We have a very useful survey of purification by water by Ingrid Edlund-Berry (2006), though I shall disagree with some of her inferences. She is interested in sacred springs, also pools and rivers, and the evidence for such springs, both epigraphic and archaeological, is extensive. But I shall only deal with the various practices of using natural water sources as evidence that Romans recognized the need for purification after the shedding of bodily fluids, whether by water or – and this is a major alternative – by fire.

Even Romans living blameless lives were still considered as ritually polluted by the crucial life events of birth and death. According to Cicero's *De legibus* (2.55), when a Roman died his family and members of the household were considered *funesti*, polluted by death, for a nine-day period and could not leave the house.[7] This at least was observed by correct Romans as late as Pliny the Younger, who asks a family friend (a woman at that) to visit him at his house because he is in the mourning period and cannot leave home. This and other death-related practices have recently been well discussed by several scholars in the volume *Death and disease in the ancient city*.[8] J. Bodel writes as an expert on disposal of the dead, including the destitute; H. Lindsay with an eye to the more formal practices in good families. Corresponding to this period is the nine days for which a newly delivered mother must wait and remain confined to the home after childbirth, the same period necessary before the child is formally recognized by its father. Babies were known to be terribly vulnerable in those days, and there are folktales of vampire witches destroying them, but was the mother's state one of pollution? I am going to treat it is as such because she too was required to wash at the shrine of Cloacina, which was positioned alongside the Cloaca Maxima in the Forum.

Apart from the impact of birth and death let me introduce a pattern of ordinary behaviour which incurred pollution that could not be washed away.

---

[6] On *verbenae* and other sacred plants in Pliny, see *Naturalis historia* 22.5, 25.105, and the discussion of sacred plants by Koves-Zulauf (1972). Pliny himself cites both myrtle (cf. Cato, *De agricultura* 82 and 133.2) and laurel for use in cleansing, and reports (*Naturalis historia* 15.138) that laurel was used as a *suffimentum* (fumigant) to purge the slaughter of the enemy by soldiers returning from a Roman triumph.

[7] With Cicero's discussion of funeral rites in *De legibus* 2.55 cf. Varro, *De lingua Latina* 5.23, Livy 2.8 and 7.47.10; also Festus 18L and Aulus Gellius 4.6.8.

[8] Hope and Marshall (2000). See esp. chapters by Bodel ('Dealing with the dead in ancient Rome') and Lindsay ('Death, pollution and funerals').

Virgil and other agricultural writers report a Sabbath-like proscription of work on holy days: on *Feriae*, or *dies festi*, no new work could be undertaken, only emergency repairs to (for example) the fencing around grazing animals, or the clearing of blocked irrigation ditches.[9] Otherwise the peasant or landowner who had worked unaware of the holy day (this assumes that calendars were not generally available) could expiate his offence by sacrifice, but if he had intentionally violated the holy day nothing could atone. Did he pollute the *feriae*, as some sources report,[10] or himself? A different type of potential pollution on the home farm is mentioned by Columella; he declares (12.4.2–3) that the preparation and preservation of produce should not be handled by anyone sexually active; but how did this work? Columella required drinking vessels and food to be handled by boys below puberty or virgins – or at least someone sexually abstemious.[11] This is so unrealistic that we have to assume a much more common dilution of such taboos; that all sexual activity in some sense pollutes, but can be cancelled out either by washing, or by the lapse of time. And this restriction of sexual activity is so common in connection with sacred space or certain types of ritual – such as the nights of chastity required at times by Bacchus or Ceres[12] – as to need no further comment.

Let us go back to washing. Still water from a cistern was not sufficient; it had to be running water, and if this were carried from a distance the vessels used should not be rested on the ground. Hence the round-bottomed *futtiles* of the Vestals. In most cases the citizen is urged either to wash or to sprinkle himself, often by dipping tree-branches in holy water. I know of very few occasions on which expiation of an offence (or of a portent which was construed as marking divine anger in response to a past offence) actually requires sea-water, but this was spelled out after Nero's great fire of Rome, when the matrons were expected to go to Ostia and fetch water from the nearest seashore and sprinkle it on the image and temple of Juno.[13]

---

[9] See Virgil, *Georgics* 1.268–72, and Rome's earliest agricultural writer, Cato, *De agricultura* 2.4, setting aside those times for cleaning jobs.

[10] Macrobius, *Saturnalia* 1.16.7–10 reports that such action polluted both the *feriae* and the worker.

[11] Koves-Zulauf (1972) notes from Pliny, *Naturalis historia* 12.54 that incense growers had to avoid contact with women or corpses during harvest.

[12] Chastity and sleeping apart (*secubare*) for Bacchus; Ovid, *Fasti* 2.327–32; for Ceres, Ovid, *Amores* 3.10. 1–4; for Isis, Propertius 2.33.1–4.

[13] Tacitus, *Annales* 15.44: 'Juno was to be propitiated by the married women first on the Capitol, then by the nearest part of the sea, from which they drew water and sprinkled both the goddess's temple and image.'

I mentioned branches, and we must pay attention here to Ovid's account of these and other natural products that were seen as purifying agents, a kind of ritual first aid kit. When Ovid explains the name February as derived from ritual purifiers he instances hanks of wool (wool has a number of religious and protective uses, such as for the headbands binding the hair of a matron or priest). We need not read it as averting pollution, but it is notable that one custom made the new bride sit on a woollen fleece at the wedding ceremony, and we find priests and citizens performing the rite of incubation resting on fleeces.[14] Besides wool Ovid mentions the branch of an (unidentified) pure tree to wreath the priest's temples; certainly Roman priests used bunches of herbs, *sagmina* or *verbenae*, which were to come with the earth still on their roots, and it seems ordinary citizens gathered rosemary, myrtle and laurel branches to carry at festivals and triumphs. We think of the laurel as the wreath held over the head of the triumphant general and the honorific adornment of Augustus' palace doorway,[15] – as in Greece it was associated with Apollo and the athletes' prize in the Pythian games, but laurel was ubiquitous in Roman cult. The ladies of Augustus' family are shown carrying branches in the processional frieze of the great Altar of Peace; the Roman *curiae* (headquarters of religious districts) had bunches of laurel set before their doorways,[16] which were replaced at the beginning of each year, and the citizens used sprigs of laurel to sprinkle the soldiers returning in triumph to Rome. Although Masurius Sabinus' antiquarian work is lost we have it on Pliny's authority that Masurius explained the laurel as purifying the soldiers after the bloodshed of battle.

But Pliny recognizes that this is just one man's opinion, and given the eagerness of antiquarians to track down rituals and responses to pollution we must be careful not to put too much faith in, for example, Ovid's interpretation of ritual acts in his lively tales. When he describes the behaviour of the people at Rome's birthday, which was also the prehistoric shepherds' feast, the Parilia, Ovid is combining the old shepherds' cult of Pales, god or goddess of the flocks, with less ancient practices of the common city folk. Can we tell apart the common measures of precautionary hygiene and actual acts of propitiation and purification? When the shepherd first sprinkles his sheep pen and sweeps the ground with a twig-broom before fumigating with sulphur and fragrant herbs, this seems like an essential act

[14] Cf. Virgil, *Aeneid* 7.92–5 and Ovid, *Fasti*, 4.652–4.

[15] Cf. Ovid, *Fasti* 3.135–44, listing the *curiae* and temple of Vesta; 4.953–4 on the laurels of Augustus' house; see also *Tristia* 3.1.33–48, and Wiseman (1998a).

[16] Laurel: see Pliny, *Naturalis historia* 15.126, quoting Masurius Sabinus, and 127–38 for detailed comment.

of spring cleaning before the new mating season (it is 21 April); but he is also instructed to deck the pen first with garlands and boughs. And Ovid mixes in with the shepherds' practical cleansing the highly ritualized acts of the common people who obtain from the Vestal Virgins a personal ration of a mysterious compound: ashes from the tail of the October horse, sacrificed six months earlier, and from the foetus of a pregnant cow, sacrificed at the Fordicidia seven days before, bulked out with ash from beanstalks (*Fasti* 4.731–4). The people are supposed to set up rows of bonfires and leap across them while being sprinkled with laurel – experiencing water and fire together, as Ovid explains with his own series of interpretations of water and fire as two elemental forms of purification.[17]

Here surely the poet's first explanation represents Roman thinking: 'Devouring fire purges all things and cooks the dross out of metals, therefore it purges the shepherd and his sheep.' Again, after Ovid's shepherd has made a simple offering to Pales (without wine or blood), the poet has composed a delightful prayer for him which begins with a request for forgiveness for any trespass by himself or his flock (and here we can imagine mud and pollution) under a sacred tree or on graves, or for muddying pools sacred to the nymphs or taking refuge in a shrine: such intrusions of the human on the sacred were clearly seen as pollution. Even if this prayer is the poet's invention it seems a natural choice. I take a different line when in May Ovid invents a ritual for the merchant or huckster praying to Mercury as god of profit – and fraud. Ovid's merchant openly prays for the god's connivance not only in his past frauds but in future attempts, and dips laurel branches into his fumigated jug to sprinkle water over his goods (*Fasti* 5.675–90). Edlund-Berry treats this as a formal ritual; I see it as pure comic fiction.

Obviously rituals might as often, even more often, be taken up to avert future pollution as to atone for a past trespass, and were co-ordinated with a specific festival. One particularly complex and interesting sequence is the Vestals' cleansing of their temple for the feast called Vestalia in early June. What were they cleansing? We have usually assumed this was the accumulated dirt and dust of the year. A more subtle suggestion comes from L. A. Holland (1961): that the Vestals had to dispose of the straw and stalks gathered from a special plot of emmer wheat from which they made the sacred mixture of salted meal used in sacrifices. In either case this process imposed a taboo on marriage for the first half of June until the temple had been cleansed at the end of the festival week (*Fasti* 6.219–34 for

---

[17] On the Parilia and its rituals see the appendix volume of Frazer's major edition of *Fasti* (1929), or Fantham (1998) 721–806; cf. 785–805 for fire and water.

6 June, and 715–16 for the end at 15 June). And how was it cleansed? The debris was tipped into the Tiber to be carried out to sea. But the Tiber was sacred: was there an exemption for sacred waste? If the current was indeed strong enough to carry flotsam down to the sea many kilometres away, the river was still surely being violated? It is not as if the Romans did not know how dirty the river's waters were. Pliny, discussing edible fish, declares that the worst-tasting flesh comes from fish caught in the Tiber's waters as they passed through the city. The Romans had always disposed of monstrous births by casting them into the sea, and of offences such as Vestal incest by burying the offending girl underground. Here again we see an ambiguous act of averting pollution; protecting themselves from bloodshed by not actually killing the offender, and protecting the world above ground (but not the earth itself) from pollution by evil acts. It is perhaps in this double, even self-defeating action of trying to avert pollution by transferring the polluter or pollutant to the environment that we should see the most fertile opportunities for questioning.

It is generally assumed that river currents were strong enough to overcome polluting elements by (for example) sweeping bloodshed into the sea, but what if the river was too sluggish and the pollutant visibly remained? Even 1,300 years after Christianity set in, the women of Cologne bathed *en masse* in the Rhine on St John's Eve each year to wash themselves and their city clean, ensuring fertility for the coming year; thus the prevailing pattern simply associates pollution with physical contamination or dirt. Petrarch, our source, argues in *Familiares* 1.5 that you needed a physically strong current to carry away the pollution, and the current of the Tiber at Rome was too weak.

4 | Pollution, propriety and urbanism in Republican Rome

PENELOPE J. E. DAVIES

Early in his account of Vespasian's reign, Suetonius enumerates omens that presaged the future emperor's rule. Among them was the tale of how 'once when Vespasian was breakfasting a stray dog brought in a human hand from the crossroads, and threw it under the table' (Suetonius, *Vespasian* 5.4). The passage is one of many that Alex Scobie deploys in describing abject conditions in ancient Rome, where, so it might imply, a dog could readily find a human hand lying in the street (Scobie (1986) 419). In the ensuing debate on sanitary conditions in Rome, some scholars have seen a model city, far ahead of its early modern counterparts; others see an urban dystopia.[1] Like this anecdote from Suetonius, most pertinent evidence – literary and archaeological – dates to the imperial period. By contrast this essay focuses on the extent to which, if at all, concerns about pollution were at play in the Republican city's growth. It addresses Roman perception and management of pollution (street dirt and sewage, and air and water quality), rather than pollution as a cause of disease or morbidity, since Roman medicine did not develop a theory of bacterial or viral contagion (Morley 2005: 199). After presenting a brief picture of the city with regard to general cleanliness, I suggest some factors that might have affected pollution levels.

Despite a dearth of clear archaeological data, we might deduce the following about Republican Rome (figure 4.1). As the population increased – to an estimated 500,000 in the second century BC – housing must have grown densely packed and overcrowded.[2] Well drained and ventilated, hilltops were highly sought after, but elite and non-elite also developed less salubrious areas, such as valleys and places subject to flooding.[3] Streets were probably desperately crowded and noisy, and littered with the general

---

[1] For Rome as a model city, see Laurence (1997), who argues that the notion of Rome as an urban dystopia is chiefly an anglophone construction rooted in nineteenth- and twentieth-century interest in lower-density and suburban cities; cf. Morley (2005). Urban dystopia: Scheidel (2003); Mumford (1961).

[2] Boatwright (1998); Scobie (1986) 401–7.

[3] Cicero, *De republica* 2.11; Livy 5.53.4; Quilici Gigli (1995); Ammerman and Filippi (2004); Aldrete (2007) 46–50.

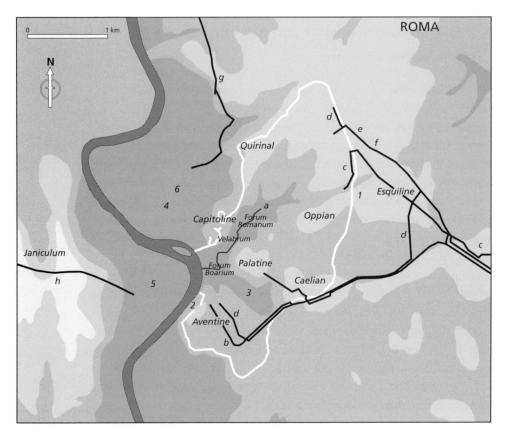

4.1 Map of Rome showing 1. Esquiline burials; 2. Emporium; 3. Circus Maximus;
4. Theatre–Portico of Pompey; 5. Gardens of Julius Caesar; 6. Baths of Agrippa, a.
Cloaca Maxima; b. Aqua Appia; c. Aqua Anio Vetus; d. Aqua Marcia; e. Aqua Tepula;
f. Aqua Julia; g. Aqua Virgo; h. Aqua Alsietina. Map: Penelope Davies.

detritus of everyday Roman living, such as animal and human faeces, dis-
carded food and even corpses of the destitute. Dungheaps were probably
commonplace.[4] By the early second century, modest bathing facilities provided a temporary refuge from crowded living quarters; though privately
owned, some baths, like those M. Junius Brutus inherited from his father
and held as an investment, were open to the public.[5] There is no evidence
at present for public latrines before the second half of the first century BC;[6]
on comparison with Herculaneum, Pompeii and Cosa, it is likely that any

[4] Cicero, *De lege agraria* 2.96; Scobie (1986); Bodel (2000).
[5] Cicero, *De oratore* 2.223–4; *Pro Cluentio* 141; Fagan (2002) 48.
[6] Neudecker (1994). A public latrine behind the Forum of Julius Caesar is most likely imperial;
see Morselli (1995) 305.

private latrines were in or close to kitchens. Liquid waste leeched away; a *stercorarius* or refuse collector removed solids.[7] Wood was the principal fuel for heating and Romans' minimal cooking needs, so the air must often have been heavy with smoke.[8] The gardens that had once been attached to dwellings for vegetable cultivation gradually dwindled out of existence, leaving the wealthy, with their urban and suburban gardens, as the sole possessors of open space, relative quiet and a modicum of fresh air. Most sacred groves had vanished by the end of the second century.[9]

The laws of the xii *Tabulae* dictated that burials should be outside the city, and there, beyond the fourth-century walls on the Esquiline, Rodolfo Lanciani found the apparent remains of Republican charnel pits, containing human and animal bones (figure 4.1: 1); elite tombs dating to the mid- to late Republic, some underground and some freestanding, lined major arteries out of the city.[10] Other than that, there is no evidence for zoning as we understand it in the twenty-first-century West, and residences stood cheek by jowl with industries that we might deem undesirable neighbours. Bone analyses suggest that though most animal slaughter occurred outside the city, at least some meat processing took place within (MacKinnon 2004); and on comparison with Pompeii it is likely that *fullonicae* (laundries) were located in residential neighbourhoods, and that fullers' pots in the streets collected urine for garment-cleaning from passers-by; unglazed, they must have leeched their contents into the packed earth.[11] Still, some areas, like the Forum Boarium or Suarium, the Velabrum and the Emporium district (figure 4.1: 2), were predominantly commercial; they clustered away from the civic centre in the Roman Forum and near the river, which was a useful depository for animal (and sometimes human) carcasses.[12] By imperial times the west bank was the place of craftsmen, but the potent stench of tanneries may already have been relegated there in the Republic.[13] This was

---

[7] Jansen (1991); Scobie (1986) 409.

[8] Horace, *Carmina* 3.29.11–2. Later Romans termed the air *gravioris caeli* (Frontinus, *De aquae ductu urbis Romae* 2.88) and *infamis aer* (Tacitus, *Historiae* 2.94). See Hughes (1994) 126–7, 163–4.

[9] Grimal (1943) 58. By Grimal's estimation the only remaining sacred groves in the early first century BC were the sanctuaries of Anna Perenna and Dea Dia, both on the outskirts of the city. See also Wallace-Hadrill (1998); Boatwright (1998) 72.

[10] Lanciani (1874) and (1897); Albertoni (1983); Wiseman (1998b); von Hesberg (1992); Barbera *et al.* (2005). For exceptions, see Frischer (1983).

[11] Bradley (2002); Scobie (1986) 414.

[12] Kyle (1995); Pisani Sartorio and Colini (1986); Aldrete (2007) 189–90. On human corpses, see Plutarch, *Gaius Gracchus* 17.5; Velleius Paterculus 2.6. On the separation of commercial from political zones, see now Hopkins (forthcoming).

[13] Leguilloux (2004); van Driel-Murray (2002); Hughes (1994) 126; Maischberger (1999).

true for at least some potteries, which consumed tremendous quantities of fuel; but others – manufacturers of terracotta tiles and architectural revetments – were located in the Velabrum.[14] Live animals to be slaughtered in sacrifice or displayed and fought in games in state festivals from 169 BC may have been kept in holding pens outside the city, as they were during the Empire.[15] This informal segregation does not suggest urban planning so much as a hierarchy of acceptable smells and wastes, calibrated according to the necessity of an industry or activity to everyday living.

There is no question that Republican Rome was more pleasant for the haves than the have-nots. Still, it is probably not high on the list of cities most contemporary Westerners would choose to relocate to, even with money. Why did Romans not do more to improve their standard of living? For one thing, some apparently unsanitary practices make sense in the context of ancient urban life. For instance, human and animal waste were a valuable commodity before the development of artificial fertilizers; similarly locating a latrine next to the kitchen allowed for easy disposal of food scraps.[16]

For another, as several essays in this volume stress, pollution is a relative concept (Douglas 1966). The very smells and sounds of industry that we protest today signalled progress in the nineteenth century (Thompson 2006). Putting aside the routine city-versus-countryside trope of Latin literature (on which, Morley (2005) 195–8) and comparing city to city instead, I suspect that Romans thought of their capital as clean; in fact, I suspect that was part of their civic identity. Writing under Augustus, the historian Livy reports that a group of Macedonians visiting Rome in 182 BC 'mocked . . . the appearance of the city, the public and private spaces of which were not yet embellished' (Livy 40.5.7). Rome's embrace of Hellenic forms and materials immediately thereafter confirms that Romans looked to cities of Greece and the East with admiration for their public buildings (on which, Winter (2006)); yet where those metropoleis could not compete with Rome was in water and waste management (as Strabo would later note: 5.3.8). As early as the sixth or fifth century BC, Roman engineers had developed a monumental drainage/sewer system centred on the Cloaca Maxima (figure 4.1: a), with later subsidiary branches in the Circus Maximus, the Campus Martius, and possibly the Aventine area.[17] Roman sewers were not fitted with traps, but

---

[14] Ammerman *et al.* (2008); Winter *et al.* (2009).

[15] Procopius, *De bello Gothico* 1.22.10; 1.23.13–23; Kyle (1995) 184; Jennison (1937).

[16] Varro, *De re rustica* 1.13.14; Columella, *De re rustica* 1.6.24; 10.84f.; 11.3.12; Scobie (1986) 408–9, 413–14.

[17] Scobie (1986) 416; Bauer (1993); Quilici Gigli (1995); Hopkins (2010).

4.2 Roman Forum and northern slope of the Palatine, showing remains of Republican houses. Photo: Penelope Davies.

the only known houses with waste systems linked to the public sewer (sixth-century houses on the lower north slope of the Palatine, used throughout the Republic[18]) stood just above the 20 m a.s.l. flood-plain (figure 4.2), and thus avoided the incursions of rats and surge of urban waste that minor rises of the Tiber might have caused at lower levels (Scobie (1986) 413; Aldrete (2007) 175; Hodge (1992) 334). Though the Cloaca Maxima would come to channel street waste into the Tiber, and though noxious vapours escaped from it into the air through large drainage holes in the streets, it was still unparalleled throughout the Mediterranean. As Pliny would later put it, 'If one carefully considers the abundance of water . . . the distances the water covers to reach the city, the arches constructed, the mountains tunnelled and the valleys levelled, one will admit that there has been nothing more marvelous in the whole world' (*Naturalis historia* 36.24.123).[19]

Unparalleled, too, were Rome's aqueducts. In the short space of 200 years, four major aqueducts were constructed to provide the city with a continuous

---

[18] Carandini and Papi (1999); Cifani (2008).
[19] For a late antique view, see Cassiodorus, *Variae* 3.30.1–2.

flow of water: the Aqua Appia of 312, the Anio Vetus of 272, the Aqua Marcia of 144 and the Aqua Tepula of 125 (figure 4.1: b–e).[20] Bath-owners and fullers could pay to pipe aqueduct water from overflow fountains into their establishments, and the state granted the same right to honoured citizens; but otherwise, scholars believe, these structures delivered water to public spaces.[21] For Frontinus, curator of aqueducts under Nerva and Trajan, vast quantities of water were key to salubriousness; he records a decree of the consuls of 11 BC 'that public fountains may flow as continuously as possible, day and night' (Frontinus, *De aquae ductu urbis Romae* 88, 103–4). These 'ruinously expensive' aqueducts cleansed the city by flushing streets with their non-stop run-off;[22] as Strabo put it in the Augustan period, 'Water is brought through the aqueducts in such quantities that rivers flow through the city and its sewers' (5.3.8). Their maintenance seems to have been principally the mandate of censors (and aediles or quaestors in years without censors), who oversaw water contractors, the *aquarii*, with a workforce of slave and free labour.[23] By Trevor Hodge's estimation, only Hellenistic Greek waterworks were in the Romans' league; but like their classical antecedents, those were built with terracotta pipes rather than masonry conduits and therefore delivered substantially smaller quantities of water (Hodge 1992: 19–47). Pergamon stood apart: probably built by Eumenes II (197–159 BC), its Madradag aqueduct ran for over three kilometres in a huge siphon, with a gradient four times the steepest Roman aqueduct's.[24] Contemporaries recognized it as a wonder, and it may not be coincidental that the Aqua Marcia of about two decades later was the first aqueduct of Rome designed with a deliberately monumental appearance, raised on arches for 7,463 paces, of which 6,472 were within the city's seventh milestone.[25] Whatever the case, by the time of Augustus, the Greek Dionysius of Halicarnassus could still state that in his opinion 'The three most magnificent works of Rome, in which the greatness of her empire is most apparent, are the aqueducts, the paved roads and the sewer works, on account of not only the utility of these constructions . . . but also their lavish cost: of which one instance serves as evidence, if one puts faith in Gaius Acilius, once when the sewers had been neglected and water could no longer flow through them,

[20]   See Evans (1994); Hodge (1992); de Kleijn (2001).
[21]   Frontinus, *De aquae ductu urbis Romae* 94.2–6; *Inscriptiones Latinae Selectae* 5779, for a
         first-century BC grant of water in a Campanian town; Hodge (1992) 5; Peachin (2004) 72–3.
[22]   Hodge (1992) 6; Eschebach (1983).
[23]   Plutarch, *Cato Maior* 19; Livy 39.44; Robinson (1992) 95–7.
[24]   Garbrecht (1979) and (1987); Fahlbusch (1977); Hodge (1992) 42–5.
[25]   Frontinus, *De aquae ductu urbis Romae* 1.7.

4.3 Paved section of the Sacra Via in the Roman Forum. Photo: Penelope Davies.

the censors let a contract for their cleaning and repair at a thousand talents' (3.67.5).

In fact, the full potential of these flushing waters must only have been realized after Romans began to pave their city streets with blocks of volcanic stone (figure 4.3). This enterprise was first tackled in 238, when the aediles L. and M. Publicius paved the Clivus Publicius near the Porta Trigemina, using fines levied on illegal lessors of public lands.[26] In 174, 'First of all the censors [Q. Fulvius Flaccus and A. Postumius Albinus] let contracts for paving the

[26] Varro, *De lingua Latina* 5.158; Ovid, *Fasti*, 5.275; Festus 276 L; Clivus Publicius (Coarelli (1993b)); also Staccioli (2003) 17–19. In 189, the censors also paved the Via Appia outside the city walls between the Porta Capena and the temple of Mars: see Livy 38.28.3.

streets in the city with flint and for laying substrata of gravel for roads outside the city and edging them with foot paths.... And they oversaw the pavement of the Clivus Capitolinus with flint... And outside the Porta Trigemina they paved the emporium with stone... And inside the same gate they paved the portico to the Aventine with flint.'[27] Although we are inclined to treat asphalt surfaces as something of an urban blight, descriptions of nineteenth-century London and Paris recall how much dirt unpaved city roads generate. 'Just a quarter of an hour of rain is enough to turn Paris back into... a city of mud', writes Giovanni Rajberti in 1857, 'a slush that threatens to send you tumbling at any moment, that ruins your clothes, that plasters you to the bone.'[28] In 1934, Samuel Cyril Blacktin complained that dirt 'dispersed constantly from clothing in walking; constantly and increasingly raised by wheels'; clothiers designed special garments such as veils, face-masks and shoe-guards to protect pedestrians and motorists from the inescapable dust.[29] The principal remedy was to spray streets with water – just as Roman fountains effectively did (Zardini (2006) 243). Emile Zola, Gustave Flaubert, Guy de Maupassant and Edmond and Jules de Goncourt all hailed the advent of asphalt with resounding praise, dubbing it the material *par excellence* for promenades (Schnapp (2003)). Even if the paving initiatives in Rome did not affect minor streets, the change to the physical environment as a whole must have been phenomenal.

It is likely that, regardless of our judgement of Rome, and regardless of how elite Romans compared their temples to those of Greek lands, in terms of pollution-fighting infrastructure they felt they were ahead of the game. Relatively speaking, improvement was not urgent.

Still, that is not to say that by their own estimation they could not have done better, especially in terms of air and water quality. In the earliest years of the Empire, Horace urged readers to ignore the smoke and din of Rome, if they could (Horace, *Carmina* 3.29.11–12); and Strabo recognized that Spanish metallurgists built tall chimneys for their furnaces so gas from ores would rise high into the air, since it was heavy and deadly.[30] Vitruvius, too, noted the effects of industrial air pollutants on those who worked in the vicinity: 'We can take an example from the lead workers, since their body colouring is overcome with pallor. For when lead is exuded as it is poured, its vapor settles in the joints of their bodies, and burning up from there it steals the strength of their blood from their limbs' (*De architectura*

[27] Livy 41.27.5–7; Wiseman (1993); Broughton (1951) I.404; Staccioli (2003) 19.
[28] Rajberti (1857) 68; Zardini (2006) 241–2.   [29] Blacktin (1934); Zardini (2006) 243.
[30] Strabo 3.2.8; Hughes (1994) 127.

8.6.11). When Vitruvius encouraged architects to consider the inhabitants' *salubritas* as an important factor in town planning, and advocated public parks to enhance the health of eyes and body through walks in the open air,[31] he was echoing the views of the fifth-century physician Hippocrates of Cos. Hippocrates' treatise *De aera, aquis, locis* (*On airs, waters and places*) was familiar to educated Romans, and in it he insisted that the natural environment had a formative influence on people and civilizations.[32] As for water, Vitruvius published methods for testing its purity, and expected results: 'If [the springs] are flowing and open, before drawing water is begun, one should observe and consider the constitution of the people who live around the springs, and if they have healthy bodies and glowing colour, and if their legs are unflawed and their eyes uninflamed, then the waters will have been proven outstanding' (*De architectura* 8.4.1).[33]

The last decades of the Republic saw dramatic shifts in Romans' standard of living, and this too suggests that they knew things could be better. As part of his Theatre–Portico complex on the Campus Martius, dedicated in 55 BC, Gnaeus Pompey opened up expansive formal gardens to the public (probably plots of land previously privatized to infuse money into an ailing treasury) (figures 4.1: 4 and 4.4), and Gaius Julius Caesar bequeathed his gardens on the Transtiberim to the people after his death.[34] Spacious public gardens had been known in South Italy and Sicily from the second century BC on; a famous example was at Tarentum, and the public garden of Croton was still famous in Petronius' time.[35] Pompey may have taken the idea from Hellenistic cities of the Greek east, such as Miletus, Priene or Antioch, where porticoes and promenades ornamented with greenery surrounded the theatres (Grimal (1943) 85) or from the *paradeisoi* of regal palaces in the East (Gleason (1994)). Whatever the source of the inspiration, the Campus Martius and Transtiberim gardens were probably controlled landscapes with shade trees and other formal plantings, which provided cool, fresh spaces for the non-elite, and within a generation Vitruvius would comment that porticoes like Pompey's were a salve to the eyes and the body.[36] Perhaps the most far-reaching change was Julius Caesar's *lex Julia municipalis*, which set

---

[31] Vitruvius, *De architectura* 1.4, 1.6, 8.3; also Varro, *De re rustica* 1.4.4–5, 12.2.

[32] See Levine (1971); Vasaly (1993) 141–5; Hughes (1994) 65–70; Fagan (2002) 85–100.

[33] See also Frontinus, *De aquae ductu urbis Romae* 9.1; Columella, *De re rustica* 1.5.1–3; de Kleijn (2001) 84–91; Hughes (1994) 161. The naming of the Aqua Tepula for its lukewarm, murky stream suggests that Romans were discerning about its quality.

[34] Grimal (1943) 83–5, 127–32; Gleason (1990) and (1994); Coarelli (1997) 545–53; see also figure 4.1: 5.

[35] Petronius, *Satyrica* 126.12, 127.8, 128.4; Grimal (1943) 84.

[36] Grimal (1943) 128–9; Purcell (2001); Vitruvius, *De architectura* 5.9.5, 5.9.9.

4.4 Portico garden of Pompey, hypothetical reconstruction drawing by L. Cockerham Catalano. Image reproduced with the permission of Kathryn Gleason.

aediles in charge of enforcing prescribed rules for maintenance and cleaning of public spaces, porticoes and streets within the city and a mile beyond. It prohibited standing water in the streets and imposed severe restrictions on daytime traffic, forbidding heavy goods wagons from using the streets except in the late afternoon and after dark. A few trades received exemptions, including, apparently, refuse collection.[37]

Improvements continued apace: after Caesar's assassination the second triumvirate built a public latrine on the site of the act in Pompey's Curia, perhaps the first public latrine in Rome (Cassius Dio 47.19.1). As aedile of consular rank in 33 BC, Marcus Vipsanius Agrippa made much of cleaning out the Tiber and the sewers, famously riding through the latter in a boat (Suetonius, *Augustus* 30.1). He constructed two new aqueducts, the Aqua Julia (33 BC) and the Aqua Virgo (24–19 BC) (figure 4.1: f–g), as well as providing additional public fountains and repairing existing aqueducts thoroughly, using gangs of slaves (see Thornton and Thornton 1989). The Aqua Virgo fed a new bath building on the Campus Martius (25–19 BC), which Agrippa bequeathed to the public free of charge after his death in 12 BC.[38] So grand was it, so lavishly appointed and so well supplied with heated water, that the traditional term *balneae* proved inadequate to describe it, which resulted in the name Thermae Agrippae. Augustus' legislation defined new officers for the upkeep of streets (*stenoparchoi*) and

[37] *Corpus Inscriptionum Latinarum* 1² 593, 20–82; Robinson (1992) 59–82.
[38] Yegül (1992) 133–7 and (2010); Fagan (2002); see figure 4.1: 6.

neighbourhoods (*vicomagistri*), who answered to the aediles.[39] He built yet another aqueduct, the Aqua Alsietina (11–4 BC; figure 4.1: h), repaired existing aqueducts, and established the *curatores aquarum*, or water commissioners, to continue Agrippa's work and to maintain the water supply according to senatorial resolutions in 11 BC and the *lex Quinctia* of 9 BC (Robinson (1992) 86–7).[40]

These rapid changes may reflect evolving attitudes to cleanliness. Perhaps population growth rendered them necessary. But they also coincide with the effective breakdown of the Republic, which makes it likely that politics were at play. I believe that one of the most powerful factors in Republican urbanism was the constitution itself.[41] A legacy of the sixth-century kings, with their monumental building initiatives like the temple of Jupiter Optimus Maximus on the Capitoline, was that Roman politicians understood the force of visual culture (Davies 2006). Interventions in the urban landscape reflected and effected power, and early lawmakers went to some lengths to ensure that no individual could exploit visual culture too effectively for his personal advancement; therein lay a path back to kingship. Thus commissioning a public monument – a temple, for instance, or a basilica – was the sole domain of elected magistrates (chiefly censors, aediles and consuls), who all served alongside at least one colleague in office. The senate had veto power over their projects.[42] Term limits on those offices – a year for most, eighteen months for censors – closely constrained what politicians aspired to accomplish during that time, as did potential charges of tyranny. A temple was feasible (in planning and conception, at least); an urban programme on the scale the late kings apparently aspired to and emperors achieved was not. This, I believe – and not the speed of reconstruction after the Gallic Sack, as Livy would have it – is the single most important factor in the piecemeal growth of the early city.[43] In terms of pollution-fighting measures, there were a few grand and exorbitantly expensive initiatives, noted above. Aqueducts brought life-giving water, but verged precipitously on the kingly; according to Diodorus Siculus, Romans recognized that by naming

---

[39] Cassius Dio 55–8; Robinson (1992) 62–3; Lott (2004).

[40] The *curatores riparum*, who took charge of clearing the Tiber of obstructions, may have been established under Augustus, before being formalized under Tiberius. See Suetonius, *Divus Augustus* 37; Cassius Dio 57.14.7–8; Tacitus, *Annales* 1.76; Aldrete (2007) 198–203.

[41] Davies (2007) 308–11 and (forthcoming). On the constitution, see Lintott (1999a).

[42] Lomas (1997) 27. For a discussion of how this system plays out in temple construction, see Orlin (1997).

[43] Livy 5.55.2–5; Davies (forthcoming). See also DeLaine (2002) 222, who argues that the brevity of annual or five-yearly magistracies in the Republic explains an 'imperative to impress by building quickly'.

4.5 Cloaca Maxima. Photo: Penelope Davies.

his aqueduct for himself Appius Claudius gained a 'deathless monument'.[44] Frontinus specifies that the later Aqua Marcia was a senatorial mandate: the senate commissioned the praetor Quintus Marcius Rex to amplify the city's water supply, but even then deliberations on the subject of its construction took three years, with Marcius eventually prevailing against the advice of the Sibylline Books.[45]

If one's mark on the city had to be limited, enterprises associated exclusively with pollution were a risky proposition to which to tie a political future. At some point in the second century BC, the Cloaca Maxima was barrel vaulted with huge blocks of tufo (figure 4.5). This was a massive engineering feat, which must have improved the face of the inner city by closing off what had probably become a foetid channel; yet it is notable that no name survives to which to attribute the task definitively.[46] It may have fallen in the censorship of M. Porcius Cato and L. Valerius Flaccus in 184, which was famed above all for its severity and austerity: they 'let contracts for works to be undertaken from funds designated for that purpose, . . . cleaning out the sewers wherever it was necessary, and constructing new sewers on the Aventine and elsewhere where they did not yet exist'.[47]

---

[44] Diodorus Siculus 20.36.1; see also MacBain (1980); Humm (1996).
[45] Frontinus, *De aquae ductu urbis Romae* 1.7.
[46] The primary channel was still open in the third century: see Plautus, *Curculio* 476.
[47] Livy 39.44.5–5; see also Plutarch, *Cato Maior* 19.1; Astin (1978).

Moreover, it is notable that the authorities did not establish baths in Rome during the Republic even though Romans increasingly linked bathing with good health, particularly when the Greek physician Asclepiades of Prusa began to practise in Rome at the end of the second century, advocating a regime of diet, exercise and bathing.[48] Perhaps the luxury they represented was deemed an inappropriate use of public funds, or their reputation for housing vice was too overwhelming – both Garrett Fagan's suggestions.[49] As Emily Gowers recognizes, a politician could only address pollution by becoming sullied himself (Gowers (1995) 31). *Plus ça change*: a recent United Nations study concluded that 'the toilet and the latrine, which helped revolutionize public health in New York, London and Paris more than a century ago, are among the most underused tools to combat poverty and disease in the developing world . . . "Issues dealing with human excrement tend not to figure prominently in the programs of political parties contesting elections or the agendas of governments", said Kevin Watkins, the main author of the report. "They're the unwanted guests at the table."'[50]

This changed with the breakdown of the Republic, especially with the dissolution of term limits and collegiality. The constitution provided for the appointment in time of dire emergency of a dictator, who would serve six months or for the duration of the emergency, whichever was shorter (Lintott (1999a) 109–13). The post had not been used for 120 years when Lucius Cornelius Sulla became dictator for at least seven months in 81 BC, with a mandate to restore the Republic.[51] In 77, the Senate granted Pompey special propraetorian authority to fight the rebelling consul M. Aemilius Lepidus, though Pompey had yet to hold senatorial office; when required to disband his army, he sought proconsular *imperium* against Q. Sertorius in Spain and his government in exile – again, without having been senator. In 67, when he was still too young even to hold a consulship, the Gabinian law gave him special constitutional authority throughout the Mediterranean and vast resources to deal with piracy; and the coveted command against Mithridates in 66, usurped from L. Licinius Lucullus, was the *coup de grâce* (see, in general, Seager (1979)). Even a cursory glance at Pompey's role in Roman politics in these years shows that he had no time for constitutional

---

[48] Pliny, *Naturalis historia* 26.7–8; Phillips (1973) 77–84; Yegül (1992) 354; Robinson (1984); Fagan (2002) 85–100.

[49] Fagan (2002) 104–7; also Robinson (1984) 1067.

[50] Dugger (2006). See also Koloski-Ostrow (1996) 84: 'Roman authorities were acquainted with the technology of sewage disposal, but they used it more for good political advantage than for improving health conditions.'

[51] Plutarch, *Sulla* 34.4–5; Appian, *Bella civilia* 1.105; Gruen (1974) 12; Thein (2002) 354–8; Keaveney (2005) 135–6, 142.

niceties or term limits. As for Caesar, his ambitions to supreme authority led up to and were amply rewarded by the grant of dictatorship, first for ten years in 48 BC and then in 45 for life, prefiguring Augustus' sole authority and Principate.[52] Deliberately flouting constitutional term limits on sole authority, and with huge amounts of war booty at their disposal to spend in the public interest, these men could aspire to grander urban interventions than their true Republican forebears. With longer in power, they could embrace pollution-fighting measures alongside other 'untainted' projects, and address the standard of living of the non-elite without neglecting the powerful elite. Previously one of the less glamorous duties of the aedile and censor, maintenance of the infrastructure thereby took on a different hue, as late Republican leaders cast themselves in the role of the city's protector and caretaker – just as Romans came to embrace the notion of a *pater patriae*, or father of the fatherland.[53] With the visible role of providing for the entire city's wellbeing, political support accrued to the 'caretaker'; that support, in turn, helped to mask the equation of his sole authority with tyranny.

Using comparanda from imperial Rome and early modern cities, this essay assesses the role of pollution in Republican city planning and factors that were at play in countering it. Although relative cleanliness may have been part of Roman civic identity, the very restrictions that protected the Republican system may have posed a barrier to greater advances.

---

[52] Weinstock (1971); Meier (1982).

[53] See Cicero, *De officiis* 2.60 for a contemporary assessment of the value of utilitarian projects: 'The expenditure of money is better justified when it is made on walls, shipsheds, harbours, aqueducts and all those works that are useful to the community. Although what is handed out, like cash down, is more pleasing in the moment, still, [public improvements] are more pleasing with posterity.' Cicero refered to Marius as *patrem patriae, parentem . . . huiusce rei publicae possumus dicere* (*Pro Rabirio reo perduellionis* 27). The image was drawn from the Hellenistic characterization of saviour-benefactors as fathers: see Cicero, *Pro Flacco* 60: *Mithridatem deum, . . . patrem, . . . conservatorem Asiae nominabant Graeci*; Curt. 9.3.16: *regem, patrem, dominum Alexandrum*. The title *pater patriae* was first bestowed on Cicero, unofficially, after the Catilinarian conspiracy, and then officially on Caesar. Augustus received the honour in 2 BC. See Cassius Dio 44.4.4; Suetonius, *Divus Iulius* 85; *Augustus, Res Gestae* 35; Suetonius, *Divus Augustus* 58; Brunt and Moore (1967) 80.

JOHN HOPKINS

For over two thousand years, historians have marvelled at the Cloaca Maxima. From Livy and Frontinus to Cassiodorus, and from sixteenth-century guidebooks to twentieth-century classical studies, Rome's largest and oldest sewer has fascinated both scholars of ancient construction techniques and historians of urban life. Most of this literature underscores the Cloaca's architectural significance or its feculence, or both.[1] In the same breath, Livy mentions the filth that ran through it and the engineering prowess of those who first built it, while Pliny praises its engineering precisely because it withstood torrents of filthy flood waters for centuries.[2] Enlightenment critics lauded the Cloaca as the archetype of beneficial public architecture as opposed to vainglorious pomp: deliberately inverting received hierarchies of decorum, the rationalist Francesco Milizia, writing in 1787, conceived his history of Roman buildings 'from the Cloaca Maxima to the Sacristy of St Peter's: from the best to the worst'.[3] More recent scholarship has still highlighted the architectural impressiveness or the polluted function of the drain, and stresses the sewer's role in Roman 'politics of cleanliness', and as an architectural signifier of civic ambition under Rome's early rulers.[4]

Far less attention has been paid to the persisting cultural significance of the Cloaca Maxima for Roman viewers, including the seemingly paradoxical possibility that this unglamorous, workaday drain was not just a prized monument of functionality and engineering but one that was somehow sacrosanct. Scholars have long noted that the Cloaca originated as a

My sincere thanks to the organizers and participants of the 2007 conference for their feedback, and especially to Mark Bradley for all his work on this volume and encouragement on the topic. I am grateful to Penelope Davies and Jeffrey Collins for their feedback on various drafts and to the anonymous readers for their comments.

[1] On the Cloaca Maxima as a marvel of Roman engineering: Lanciani (1890) 95–6; Holland (1961) 349–50; Reimers (1989) 137–8 and (1991) 111–14; Edwards (1996) 106; Bradley (2006); Hopkins (2007) 1–4. On the Cloaca as a sewage system: Platner and Ashby (1929a); Scobie (1986) 418–20; Gowers (1995); Villoresi (2006) 2–26.

[2] Livy 1.56; Pliny, *Naturalis historia* 36.24.103–107. It was praised as late as the sixth century by Cassiodorus: *Variae* 3.30.

[3] I thank Jeffrey Collins for this reference: Milizia (1787) 205. The Cloaca Maxima is discussed in detail also at pp. 17–19.

[4] Bradley (2006); Hopkins (2007) 1–4.

canalized natural stream, and since Romans believed moving water to be holy, some have argued that the waters of the Cloaca maintained this religious aura.[5] In a similar way, the Cloaca's status as one of the city's oldest structures garnered continuous praise from Roman authors, who invested it with an historical prestige that seems to have reinforced its hallowedness.[6] Meanwhile, anthropologist Mary Douglas has complicated modern dichotomies of purity and pollution by studying how non-Western or non-modern peoples either accommodated or actively synthesized the two.[7] Overall, these perspectives have led a handful of scholars to propose that ancient Romans perceived this hidden watercourse in ways dramatically different from how sewers are understood in modern Western cultures: not just as a praiseworthy public utility, but as a living testament to Rome's origins and a monumentalized marker of a hallowed stream. In this chapter, I assess this reinterpretation from an archaeological perspective, suggesting through a close examination of the remains both of the Cloaca itself and of associated monuments that Romans did indeed see something sacred in their sewer.

From stream to sewer: the Cloaca Maxima in context

As the city of Rome and its population grew during the Republic, the need arose for substantial drainage systems, *cloacae*, to keep streets and buildings as free from flood waters and refuse as possible (figure 5.1). Unlike modern septic sewers, which collect every kind of foul material, Rome's ancient urban drains were flushed with natural run-off and excess from fountains, carrying only the occasional debris that collected in the city's streets and perhaps domestic kitchens. They probably did not receive human waste from public or private latrines.[8] In fact, the translation of *cloaca* to mean 'sewer' is imprecise; the Latin word seems rather to have roots tied to the verb *cluere*, which meant 'to cleanse with running water'.[9] Thus, while the waters of the *cloacae* were certainly unclean, they were more similar to a refuse drain than a sewer *per se*. From the early Republic, drains of this sort

---

[5] E.g. Holland (1961) 33, 349–50. On the sanctity of moving water in general: Edlund-Berry (2006).

[6] See below.

[7] Douglas (1966) esp. 161–2. This chapter aims to reassess the Cloaca Maxima with this open mind.

[8] Though it cannot be entirely ruled out. Bodily waste was famously used for laundering clothes and (by way of carts that collected solid waste) for fertilizer: Scobie (1986) 408–18.

[9] See below, n. 62 for further discussion of this.

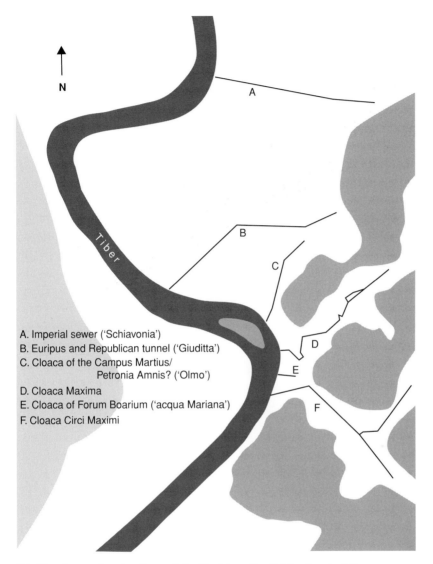

5.1 The *cloacae* of ancient Rome. John Hopkins, after C. Moccheggiani Carpano.

removed filth from around the centre of Rome; the Cloaca Circi Maximi drained the area of the Circus Maximus and later seems to have connected to culverts for the Colosseum and perhaps the Baths of Caracalla.[10] In the Campus Martius, two mid-Republican *cloacae* ran from the future site of the Pantheon and from the north slope of the Capitoline to the Tiber. Under the Empire, a tract was constructed from the Pincian Hill to the Tiber,

[10] Moccheggiani Carpano (1984) 169, 73–6.

5.2 Cloaca Maxima before vaulting. Reconstruction: John Hopkins.

draining the northern Campus.[11] These four systems are just a few of the
many subterranean channels that drained the ancient city, criss-crossing the
valleys and slopes of the hills.

The focus of this essay is the Cloaca *Maxima*, the oldest of the *cloacae*
and by far the most monumental. Its inception seems connected with a
purposeful reclamation of the Forum valley in the seventh century, when

---

[11] Moccheggiani Carpano (1984) 169–70; Bauer (1989) 46.

5.3 Cloaca Maxima. Earliest tract (in cappellaccio) with later concrete vault. Photo: John Hopkins.

some form of canal was employed to collect and direct a freshwater stream that ran through the area.[12] At what time Romans began polluting the stream is uncertain, but for a while at least, it remained a natural stream, not a sewer. Some time in the fifth century the Maxima was monumentalized in stone (figures 5.1–5.3, 5.4: A), and contemporaneously or soon thereafter Romans drained still marshy areas around the temples of Saturn and Castor with the earliest known conduits to feed into it (figure 5.4: B).[13] In the middle

---

[12] Hopkins (2007) 8 and Hopkins (2010) 50–6, appendix; cf. Ammerman (1990) 636.

[13] The date is a correction to Hopkins (2007) 8–11. In recent explorations of the Maxima I was able to confirm that the tops of these earliest walls of the Maxima are higher in elevation than a pavement of the Forum that is commonly dated *c.* 450. This indicates that the remains of the extant stone Cloaca Maxima should date to 450 or later; a problem in absolute dating for this structure, however, is that it is not possible to excavate through the portion of the Forum under which runs the Maxima. It is therefore not possible to get an exact correspondence between the walls of the Maxima in one area and the elevation of the different Forum pavements in that

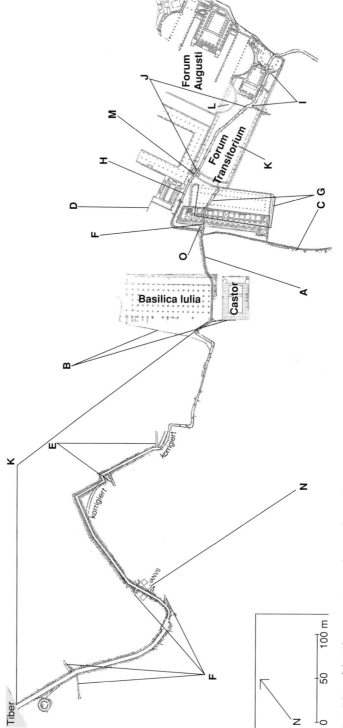

5.4 Map of the Cloaca Maxima. John Hopkins, after H. Bauer.

Republic, they added a large duct alongside the Via Sacra (figure 5.4: C) and from the middle to late Republic new shafts joined these, draining the Tullianum (figure 5.4: D), the slopes of the Palatine and Capitoline hills (figure 5.4: E) and the Velabrum and Forum Boarium (figure 5.4: F).[14] Augustus and Domitian overhauled much of the system, and as each emperor built the imperial fora, he constructed new drains to direct waste to the Maxima.[15] The result of the continuous extensions and repairs to the sewer is a monumental subterranean drainage network exhibiting a patchwork of Roman building techniques.

Louise Adams Holland was the first to suggest that the main conduit of the Maxima channelled sacred waters. For her, the telltale sign was its twisting path through the Forum Romanum, Velabrum and Forum Boarium; indeed, the meandering channel contrasts so sharply with the decidedly straight courses of Rome's other *cloacae* that it would seem at first glance to betray poor engineering (figures 5.1, 5.4).[16] The numerous elbows restrict water flow and cause back-ups, just as swift-moving waters slam into the sharp turns and destabilize walls, rendering such a circuitous course tectonically unsound. Yet the engineering involved in the expansive vaults of the Cloaca, alongside the hydraulic expertise witnessed in other civic and imperial water-management systems, bears witness to Romans' expertise in this line of construction, making it unlikely that they would create, much less maintain, potentially deleterious bends in the course of the Cloaca Maxima without a specific imperative. Rather, it seems the Maxima meandered because it followed a predetermined path that Romans would not or could not change. Holland understood that this was probably a waterway, noting that the stream running through the middle of the Forum in Rome's earliest days would not have disappeared when Romans began inhabiting the valley. She suggested instead that the Maxima was used to channel it, and recent research into the origins of the Forum and the Maxima itself seem to confirm this.[17]

---

same area. One must therefore compare elevations from excavations in different parts of the Forum. In this case, the lowest elevation for the *c.* 450 pavement is nearly equal to the elevation of the top of the cappellaccio walls of the Maxima. Hopkins (2010) appendix. For additions, cf. Platner and Ashby (1929a); Bauer (1989) 48–9; Richardson (1992a); Hopkins (2007).

[14] Picozzi (1975) 3–6; Bauer (1989) 49–51.

[15] Lanciani (1890) 96–8; Ashby (1901) 138; Hülsen (1902) 36–7; Platner and Ashby (1929a); Blake and Van Deman (1947) 60, 159, 341; Picozzi (1975); Moccheggiani Carpano (1984) 165; Bauer (1989) 47–9 and (1993) 288–9.

[16] Holland (1961) 349–50; Cressedi (1984); Bauer (1989) Abb. 19. cf. Lanciani (1897) 29.

[17] See above, n. 14.

Subsequent scholars have largely accepted this view, but without addressing the critical fact that in the second century, one area of the Maxima, which I discuss below, began to cause problems; despite this, Romans maintained its course for centuries. If the channel were nothing but a sewer, and the stream's waters simply a means to flush it, why did Romans keep repairing a problematic and circuitous tunnel when they could have rebuilt one that was much straighter and more functional? Holland argues they did so because the stream it originally channelled was sacred.[18] She refers to Servius and Varro, who state that streams, rivers, brooks and springs were holy, 'in the augural sense "living [moving] water"', and that special regulations had long been in order regarding such 'living water'.[19] Recently, Ingrid Edlund-Berry has recalled the pervasive power of moving water in ancient Roman sources and argues it held a much stronger role in religious practice and consciousness than previously thought.[20] The Maxima was by no means the only sacred stream in the city.[21] Two important examples of living water are the Petronia Amnis in the Campus Martius, which formed the northern boundary of the city, and the Circus brook, which formed the southern. Both required a transgressor's propitiation, and in spite of problems they caused, Romans maintained both of their natural courses.[22] Holland argues that Republican and early imperial Romans revered the waters that the Maxima channelled in precisely the same way they revered the Petronia Amnis and other sacred streams: as living water.[23]

But perhaps it was not religion alone that dictated Romans' reverence of the Cloaca Maxima. Its archaic origin seems also to have led Romans to venerate it as a symbol of their great and ancient past. Among others, Ann Vasaly and Catherine Edwards argue that monuments and places in the city of Rome had an especially strong power to channel memory and evoke history.[24] In Cicero's *De finibus*, Piso remarks, 'Places have so great a power of suggestion that the technical art of memory is with good reason

---

[18] Holland (1961) 33, 349–50.

[19] Servius, *Ad Aeneidem* 2.719; Varro, *De lingua Latina* 5.123. Cf. Holland (1961) 18 n. 39, 19 n. 42, 95–6.

[20] Edlund-Berry (2006) 162–3.   [21] Holland (1961) 8–20, 9–49, appendix C.

[22] Holland (1961) 33 n. 18; Domaszewski (1975) 217 n. 2. Cf. Justinian, *Digesta* 43.13; Tacitus, *Annales* 1.79.

[23] Holland (1961) 6, 19, 21–2; Domaszewski (1975) 217–18. Cf. Cicero, *De natura deorum* 3.52 (though Holland translates *propinquorum fluminum* as 'streams of the neighbourhoods', instead of 'of neighbouring streams', and suggests the Spinon and Nodinus must therefore refer to the streams of the Forum and Circus valleys; cf. Loeb translation); Festus L.146.17, 284, 485; Servius, *Ad Aeneidem* 2.719.

[24] Vasaly (1993) 29–33; Edwards (1996) 2, 29.

based upon them' (*tanta vis admonitionis inest in locis; ut non sine causa ex iis memoriae ducta sit disciplina*).[25] Writing in a city outfitted with the Tabularium, Pompey's Theatre and Portico, Caesar's Forum, a newly monumentalized Capitoline Temple and a multitude of extensive aqueducts, Livy remembers the archaic Cloaca Maxima as a 'monument for which the new magnificence of these days could scarcely produce a match'.[26] Dionysius of Halicarnassus places the earliest manifestation of the Cloaca among Romans' greatest achievements, and writing of a city that had seen Nero's Golden House, Augustus' Forum, early work on the Colosseum and countless other wonders, Pliny praises the power and grandeur of the 700-year-old Cloaca.[27] Edwards suggests these authors and others praise the Maxima because it was built without foreign (especially Greek) inspiration or tectonics.[28]

I suggest that the historical symbolism of the structure also compelled Romans to praise it. The Cloaca was one of the earliest monuments Romans built and by the Empire was one of few from that period that remained and functioned. What is more, by the late Republic, historians tied its waters to the cleansing of soldiers after the Sabine War and the formation of boundaries between tribes in the early city, two of the most momentous historical traditions in the formation of Roman culture.[29] In the late Republic and early Empire, Romans celebrated their origins to an unprecedented degree.[30] Cicero, Livy, Virgil and others looked to Rome's earliest history as a time of virtuous men and great achievements, and Augustus spent much of his career alluding to early Roman history through an extensive propagandistic programme.[31] As a monument that contemporary authors regarded as among the greatest achievements of early Roman architects, and as a structure that still served the city, the Cloaca Maxima would have been a symbol of these origins, and in addition to channelling sacred living water, the structure would have aroused the memory of Rome's earliest history.

---

[25] Cicero, *De finibus* 5.2. Cf. Vasaly (1993) 29–33; Edwards (1996) 29.   [26] Livy 1.56.2.

[27] Dionysius of Halicarnassus 3.67.5; Pliny, *Naturalis historia* 36.104–9.

[28] Edwards (1996) 106.

[29] Pliny, *Naturalis historia* 15.119–120, 122 and cf. Coarelli (1983) 79–89.

[30] Gowing (2005) esp. 18.

[31] For example: Zanker (1990); Galinsky (1992) 457–75 and (1996); Ingholt (1969) 177–87; Kleiner (1978) 753–85 and (2005) 197–233; Martin (1982); Kellum (1985); Holliday (1990) 542–57; Reeder (1992) 265–307; Pollini (1995) 262–82; Favro (1996) and (2005) 234–63; Davies (2000) 1–19, 41–74; Clarke (2005) 264–80; Welch (2005); Rehak and Younger (2006).

Maintaining and marking the Maxima

If the Maxima was hallowed, it stands to reason that its architecture and perhaps its site would reveal traces of that reverence.

The first place to look is the structure itself. That Romans largely maintained its peculiar path is a first indication of its sanctity. In the early second century BC in the Forum, however, they parted from their persistent preservation of its site for the first and only time, when they built the Basilica Fulvia directly on top of the Cloaca Maxima (figure 5.4: G). As the basilica rose and fell during the second and first centuries BC it did an enormous amount of damage to the sewer. Christian Hülsen, Tenney Frank and Bauer note five different masonries crammed together under the *tabernae novae* alone, revealing constant reconstruction.[32] The last of these dates to 34–33 BC, when Augustus or Agrippa completed a major overhaul of the Maxima from the Basilica Aemilia to the foot of the Quirinal.[33] After this, it was diverted around the north side of the Basilica Aemilia, and up the Argiletum to the Suburra (figure 5.4: H).

The date and impetus for this detour have caused intense scholarly debate and merit a fresh look. Heinrich Bauer and Edoardo Tortorici suggest that Domitian built the new tract during the construction of his Forum Transitorium and temple of Minerva.[34] They give no compelling reason for Domitian to have done this, as the Basilica Aemilia did not collapse into the Maxima during his reign or that of his immediate predecessors, and neither Frontinus nor any other author report damage to the system in the late first century AD.[35] Still, they insist the masonry of the new tract exhibits the fingerprints of Domitianic construction. Bauer identifies Domitianic lapis Albanus as the material used for a distinct section built to circumvent the temple of Minerva (figures 5.4: I, 5.5), and Tortorici adds that in the long stretch between the temple of Minerva and the north side of the Basilica Aemilia the channel was vaulted in Domitianic concrete (figures 5.4: J, 5.6).[36] A recent exploration of the canal confirms that lapis Albanus similar in dimensions to that used in the temple of Minerva was in fact the material

---

[32] Ashby (1901) 138; Hülsen (1902) 38–57; Frank (1924b) 66–75; Bauer (1989) 49 and (1993) 288–9.

[33] Bauer (1989) 48–9 and (1993) 288–9; Morselli *et al.* (1989) 47.

[34] Bauer (1977) 310 and (1983) 48; Morselli *et al.* (1989) 47.

[35] The only feasible explanation would be that the colonnade of the new Forum Transitorium or the stalls outside the Templum Pacis were on top of the Maxima, but Bauer's own reconstructions (and those of many others) reveal that these are well to the southeast of the old tract of the Maxima.

[36] Bauer (1997) 310 and (1983) 48; Morselli *et al.* (1989) 47.

5.5 Cloaca Maxima. Domitianic tract east of temple of Minerva in lapis Albanus.
Photo: John Hopkins.

5.6 Cloaca Maxima. Augustan tract under Forum Transitorium in lapis Gabinus with
Domitianic concrete vault. Photo: John Hopkins.

5.7 Cloaca Maxima. Detail of Augustan tufo and Domitianic concrete under Forum Transitorium. Photo: John Hopkins.

of choice for the section by that temple, and the date of the vault seems certain, as it cuts an Augustan pavement at the corner of the Curia and uses Flavian *bipedales*.[37] Yet Rodolfo Lanciani and Claudio Moccheggiani Carpano both note that the Domitianic cocciopesto vault rests upon large cut-stone masonry walls built in the same lapis Gabinus that Augustus used in his Forum (figures 5.6, 5.7), not the lapis Albanus used in the section adjacent to the temple of Minerva.[38] The use of two different building techniques from two different periods for walls and vault creates a conundrum. It seems counterintuitive for Domitian's engineers to bring in workers and an apparatus to build walls in out-of-date lapis Gabinus masonry (especially when lapis Albanus was already on site for construction) and then to contract a separate team of workers with a different expertise to create a concrete vault; it is far more likely, given Domitian's penchant for concrete, that his engineers would have built the entire structure with cocciopesto.[39] What is more, the section of the diversion adjacent to the north and west

[37] I owe many thanks to Luca Antognoli and his crew for taking me through the Cloaca Maxima twice and allowing me to photograph and measure various sections. Morselli *et al.* (1989) 151–70.

[38] Lanciani (1890) 96; Moccheggiani Carpano (1984) 171. The stone exhibits the clear lines of stratification that are prominent in lapis Gabinus and absent in lapis Albanus; cf. Jackson and Marra (2006).

[39] For the use of cocciopesto this early: Lancaster (2005) 58, 65, 211–12.

sides of the Basilica Aemilia is both walled *and* vaulted in lapis Gabinus (figure 5.4: H), and at the point where the vault material changes from concrete to stone, the walls are continuous. Thus, builders used several techniques in the diversion: for the long stretch from the *western* side of the Basilica Aemilia to the temple of Minerva they employed lapis Gabinus walls, but part of this was vaulted in the same stone and part in concrete; then, for the dog-leg around the temple of Minerva, they used lapis Albanus throughout. The medley of construction materials has long baffled scholars, but new evidence provides an explanation.

Largely on the basis of his study of the Forum Transitorium, particularly the turn the Cloaca makes halfway up the open space (figure 5.4: K), Bauer assigns the entire project to Domitian. Bauer argues that engineers should have made the turn at the northern corner of the Basilica Aemilia, but that a massive temple to Janus occupying the west end of the forum forced the tract to remain straight until just past its east facade.[40] No other explanation then existed for the peculiar turn, and most scholars accepted that the diversion of the Maxima was contemporary with or postdated Bauer's putative Domitianic temple of Janus. Excavations in the 1990s revealed that no such temple existed, however, leaving no good explanation for the Maxima's curious curve in the middle of the Forum Transitorium.[41] Ongoing excavations in the Forum of Augustus may provide a new reason. There archaeologists have found a different obstruction: two *exedrae* at the western end of the complex that match the extant *exedrae* at the eastern end (figure 5.4: L).[42] The Maxima turns to the east just before the southern edge of the southwest *exedra*, and it seems likely that this is the structure that forced the change in the direction of the new tract of the Maxima.

If the monument that forced the curve in the new Maxima already existed under Augustus, the diversion itself may have been built long before Domitian's reign, and this would account for the changes in masonry. The Basilica Aemilia collapsed for a third time in 14 BC, when the Forum of Augustus was under way, and it saw yet another reconstruction *c.* AD 22; thus there are two possible occasions for Augustus or Tiberius to have created this new tract.[43] I suggest that Augustus chose to move the sewer around the Basilica Aemilia after the Basilica's third collapse and interruption of the

---

[40] Cf. Bauer (1977) 310 and (1983); Morselli *et al.* (1989) 47, 68.

[41] Morselli *et al.* (1989) 53; Viscogliosi (2009) 202–9.

[42] The excavations are as yet unpublished, but see Favro (2005) 254 for a reconstruction of the western *exedrae*.

[43] Though the Basilica Aemilia's early destruction *c.* 14 BC would predate the Forum of Augustus, Cassius Dio and others report that the Basilica remained under construction for some time. For the 14 BC reconstruction: Cassius Dio. 54.24; for the AD 22 reconstruction: Tacitus, *Annales* III.72.

Maxima's service in 14 BC. The new route ran around the new building, across the Argiletum and curved at the *exedra* of his Forum. Later, while building the Forum Transitorium, Domitian reconstructed the tract's vaulting, giving it the appearance it has today, and diverted only a small section around his temple of Minerva, using lapis Albanus since it was on hand for the construction of the temple and forum. He did not change the lapis Gabinus vault along the north or west of the Basilica Aemilia because his construction zone did not extend over those sections.

At first glance, Augustus' transposition of one section of the Maxima would seem to indicate a decreased concern for the waters it channelled, but closer investigation reveals that this change in course did not go unaddressed. Holland's discussion of the Forum stream focuses on the role that Janus played in mitigating religious rites involved in crossing living water.[44] Romans so often traversed the Forum that it would have been impractical to take the requisite auspices every time one needed to cross the waterway.[45] In order to appease the sacred power of the stream permanently, early Romans, by tradition Numa Pompilius, established a *templum* to Janus at the site of the Cloaca Maxima.[46] Filippo Coarelli has identified a late antique brick structure at the northwest corner of the Basilica Aemilia as the shrine of Janus in the Forum, and, following Holland, he suggests that as long as there had been a shrine to Janus it stood in this spot.[47] Originally, however, there was no shrine at all, since the temple of Janus was a *templum* in the strictest sense, that is, a sanctified area and not a building.[48] Given the proximity of the later sanctuary to the Cloaca Maxima, around which Holland suggests the *templum* was originally sited, Janus' early role as propitiator of the Forum stream seems possible. Holland goes on to suggest that two other temples of Janus cross the Maxima: one in the Forum Transitorium and one in the Velabrum (figures 5.4: M, 5.4: N).[49] It is the one in the Forum Transitorium that is most important for a religious reading of the Maxima and of its continued reverence even as its path was changed in the early Empire.

Excavations in the 1990s revealed the foundations of a *quadrifrons* directly crossing the Cloaca at the juncture of the Forum Romanum, Argiletum and

---

[44] Holland (1961) 26–39, 50–2, 65.

[45] For examples of frequent official needs to cross the forum: Holland (1961) 22 n. 8. One should also consider more common and unceremonious needs of getting from one side of the city to the other, which would be impossible anywhere between the Argiletum and Tiber without either taking auspices or the presence of such mitigation as Holland suggests.

[46] Livy 1.19.2. Cf. Holland (1961) 22–3.   [47] Coarelli (1983) 89–97; Holland (1961) 25.

[48] Holland (1961) 25; Coarelli (1983) 89–97.   [49] Holland (1961) 26–7; Coarelli (1983) 79–85.

Forum of Caesar, and the excavators maintain that this must be an arch of Janus, which Ovid, Statius and others mention.[50] Scholars have long known that the arch existed, but they were unsure of its precise location. These remains may belong to that arch, but in any case ancient sources are clear that somewhere on the Argiletum, very close to the *templum iani* in the Forum, there was a second arch sacred to the god. The proximity of two *templa* on the edge of the Forum Romanum has long confused scholars. Why two so close together, and if they were built to propitiate the sacred waters of the Maxima, why did Romans think two were needed there, but none at all for much longer stretches in other areas?[51] A natural explanation would be that the second temple on the Argiletum – directly above the diversion – was built in response to the transposition of the stream and to mitigate the sacred powers of the waters that flowed through a new channel. This line of reasoning was not possible for Bauer or other scholars of the Maxima, because they thought Domitian created the diversion, long after Ovid had first referred to the sacred arch.[52] But if the diversion was created in response to the collapse of the Basilica Aemilia in 14 BC, this would account not only for the multiple masonry styles, but also for the Augustan date of the second temple of Janus. By this logic, Augustus finally succumbed to the need to move the channel around the Basilica, but in appreciation of its sanctity, he built a *templum iani* over the new branch. The construction of the temple in tandem with the new branch of the Maxima would suggest a continued connection between Janus and the stream as well as the Maxima's continued religious significance through the first century and to some degree until the late empire, when the temples of Janus were refurbished for the last time.[53]

A third temple of Janus is thought to have existed in the Velabrum, and Holland argued that it had a similar function. For centuries the colossal *quadrifrons* by S. Giorgio in Velabro was thought to be an arch dedicated to Janus, but it is now identified as the Arcus Divi Constantini (figure 5.4: N). Still, it seems to have supplanted an earlier structure, and Holland argues that this structure was probably dedicated to Janus and marked the locus where canals from the Forum Boarium enter the sacred stream

---

[50] The *quadrifrons* is undeniably Domitianic, but presumably they mean this is a reconstruction of an original by Augustus. The Augustan pavement in this particular area was destroyed completely by Domitian's Forum Transitorium. See Ovid, *Fasti* 1.258; Bauer (1977) 301 and (1983) 177; Morselli *et al.* (1989) 244–5, figs 132–7, 219–22, 146, tav. 1.

[51] Cf. Holland (1961) 26 n. 65.

[52] Perhaps supplanted by the Domitianic structure. On literary evidence for the *templum iani* in the Argiletum/Forum Transitorium: Ovid, *Fasti* 1.258; Martial 28; Statius, *Silvae* 4.1; Servius, *Ad Aeneidem* 7.604. On these passages cf. Holland (1961) 95, 102.

[53] Cf. Procopius, *De bello Gothico* 1.25.

of the Cloaca Maxima (figure 5.4: E).[54] A connection between the earlier remains and the Maxima is speculative, but textual sources that mention a *templum jani* in the area (and associated with the Maxima) would seem to suggest a widespread connection between shrines of that god and the stream channelled below.[55] Overall, the shrines directly above the Cloaca dedicated to Janus, whose traditional role was to mitigate the powers of sacred water, are the first examples of Romans showing a religious deference to their 'sewer' or at least the waters in it. The first *templum*, in the Forum, reveals an early and longstanding worship of the god and his powers at the site of the Cloaca, reaching as far back as Rome's proto-history and forward into the late Empire, when the shrine was reconstructed for the last time. The second *templum*, along the Argiletum, explains how Augustus could move the stream and still maintain its sacred nature, and the third, if indeed related to Janus, was a manifestation of the propitiation of its waters downstream, as it assumed the scourges of the surrounding valley.

Less than five metres from the late antique shrine of Janus in the Forum and immediately adjacent to the Cloaca Maxima, the temple of Venus Cloacina seems to have functioned as yet another sacred marker of the stream (figure 5.4: O). The extant structure dates only as far back as Sulla's dictatorship, but textual sources mention its use throughout the Republic and as early as the archaic period.[56] A *denarius* struck in 39 BC shows the temple's circular base interrupted by a short rectangular stair on the north and with a balustrade on top that encircles two sculptures, identified by Pliny as statues of Venus.[57] The sanctuary's proximity to the Cloaca is a clear physical connection between the two structures, and the homonymous and linguistic ties between Venus' epithet, Cloacina, and the word *cloaca* reveals an undeniable association.

Carel Van Essen was the first to question *why* Romans would associate the two.[58] He notes that the purifying and fertile qualities associated with water were commonly associated with Venus and that Venus Cloacina was a goddess of both fertility and purification whose powers were tied to the

---

[54] On the attribution of the arch in the Velabrum: Coarelli (1968) and (1988) 8, 12; Cressedi (1984) 276; Pensabene and Panella (1994–5) 29; on the Republican finds under the arch and their identification: Holland (1961) 38–9. Cf. Richter (1889) pl. 37 and Cressedi (1984) 276, fig. 3.

[55] Holland (1961) 38–9.

[56] For full citations of work on Venus Cloacina through the early 1980s: Vaglieri (1900) 61–2 and (1903) 97–9; Coarelli (1983) 84 nn. 19, 20 on the literary tradition; cf. Pliny, *Naturalis historia* 35.119–22; Livy 3.48.5; and later, St Augustine, *De civitate dei* 4.8; Lactantius, *Divinae institutiones* 1.20.11.

[57] Grueber (1910) 573, 7–8; Pliny, *Naturalis historia* 15.119.   [58] Van Essen (1956).

use of sacred moving water. Of course, in the Empire when the Maxima was covered and soiled, its waters would not have been used at the shrine, but the implication is that originally the Forum stream was used in purification rites at the temple. Pliny and Livy would seem to confirm exactly that. They state that after the Sabine War, soldiers purified themselves of the stain of battle at Venus' shrine and with the waters of the Forum stream.[59] As with other textual sources, the testimony should be used with caution when trying to imagine the early city; nevertheless, it offers a valuable insight into the Cloaca and its sanctity because it reveals that in the late Republic and early Empire, a religious connection between the shrine of the goddess and the waters of the stream housed in the Maxima persisted in Roman lore. The association may have continued into the Empire, as the goddess retained her capacity as a purification deity for brides who visited her to cleanse themselves and ensure fertility.[60]

Linguistic study only reinforces the connection. In his work, Van Essen recalls ties between *cloaca*/Cloacina and the Latin *cluere*, to cleanse, and in further study, Pontus Reimers notes that the essential meaning of the word *cluere* suggests running water like that in a stream.[61] Given the historical tradition linking the shrine with the moving water in the stream, the linguistic connection between the two and their conspicuous homonymous and proximal relationship, it is hard to imagine that Romans would not connect the structures in their minds and that the sanctity of the shrine and of waters associated with it would not reflect on the adjacent Maxima and its stream.

While these sacred spaces were distinct from the Maxima's own architecture, the last two markers of the great sewer were part of the structure itself and its sacred, historic function. Scholars have long believed that the Bocca della Verità is in fact an ancient drain cover for the Maxima (figure 5.8). In 1715, Crescimbeni suggested that the expense lavished on the fine sculpting and Phrygian marble, and the representation of a deity on the stone, did not fit the proposed function.[62] Yet through the twentieth century and into the twenty-first, scholars have defended the identification.[63] The worn holes at the eyes, nostrils and mouth of the marble disk seem clearly to have

---

[59] Pliny, *Naturalis historia* 15.119–120, 122 and cf. Coarelli (1983) 85.

[60] Note Pliny's discussion of her function as well as St Augustine, *De civitate dei* 4.8; Minucius Felix, *Octavius* 25.8; Lactantius, *Divinae institutiones* 1.20.11.

[61] Van Essen (1956) 139–44; Pliny, *Naturalis historia* 15.122. Cf. Reimers (1989)137–8, n. 8.

[62] Crescimbeni (1715).

[63] Barry (2011) 7–37. Barry gives an extensive discussion of possible locations, and does not assign any particular spot; curiously, he does not suggest the Cloaca Maxima as a possibility.

5.8 Bocca della Verità, here proposed as a drain cover for Cloaca Maxima with relief of Oceanus. Photo: John Hopkins.

facilitated drainage, as does its slight concavity. What is more, its size is roughly the same as a travertine drain cover found in a tributary to the Cloaca near the Round Temple in the Forum Boarium, and the uniform weathering of the sculpture suggests it was washed evenly and heavily by water draining into it.[64]

Lanciani identifies the deity on the disk as Oceanus, with lobster claws protruding from his thick head of hair.[65] In the Empire, when the drain cover was almost certainly created, Oceanus had come to be synonymous with the power of the emperor, Caesar being the first to conquer a land across Ocean: Britain.[66] But the iconography here seems much more likely to suggest Ocean's primordial sacred powers, which mosaics and *paterae* throughout the empire recalled, and which can be traced back through

---

[64] Nash (1968) 261. This is based on my own measurement of the Bocca della Verità and personal correspondence with Mark Bradley.

[65] Lanciani (1891) 56. Most recently this identification has been confirmed by Fabio Barry: Barry (2011).

[66] Bajard (1998) 177–84; Williams (1999); Santoro Bianchi (2001) 88–92.

5.9 Mouth of Cloaca Maxima. Photo: John Hopkins.

Roman traditions to Hesiod and Homer.[67] In this mythology, Oceanus represented the most venerated of waters; Romans believed his powers were more ancient than Neptune's and his waters the only ones sacrilegious to navigate.[68] What is more, he was both the source of the world's water and its final destination, the perpetuator, and one theory suggests the face on the drain cover was meant to recall the eternal, cyclical power of the stream that passed below.[69] Though his image in this sculpture may only have recalled that water ran underfoot, his ties to primordial sacred streams such as the one channelled through the Maxima suggest more: that he marked the religious or mythological character of this water. For now, the Bocca della Verità is the only surviving example of such an opulent drain cover, and it may have been the only marker of its kind; regardless, Romans lavished great expense on an expertly sculpted marble relief whose purpose was to drain refuse into the Cloaca, and it is therefore yet another example of Romans highlighting the pathway of their sewer.[70]

The last marker of the Maxima is perhaps the most directly tied to it. Just south of the Pons Aemilius, Romans aggrandized the mouth of the Cloaca in the first century BC with a triple arch embedded in the Tiber wall (figures 5.9, 5.10: A). The arched facade was a distinct initiative; sometime in

[67] Voute (1972); Cahn (1981) 33; Bajard (1998) 185–6; Williams (1999) 311. Cf. Ambühl (1894) 1154; cf. Hesiod, *Theogonia* 133.

[68] Ambühl (1894) 1154; Bajard (1998) 188, 91; Williams (1999) 311–12.

[69] Barry (2011).

[70] And it seems they put much effort into the symbolism, with Phrygian marble, meant to evoke water and a stock of other iconographic and stylistic idiosyncrasies: cf. Barry (2011).

5.10 Mouths of three *cloacae* on the Tiber. Photo: R. Lanciani, after Cressedi (1984).

the middle Republic, the Tiber embankment was built out of tufo giallo della Via Tiberina, and, according to Marion Blake and Lanciani, who examined the arch and wall during the modern Tiber reclamation project, the large lapis Gabinus arches were added at the mouth of the Maxima in the first century BC.[71] Though six other *cloacae* emptied into the Tiber, the mouth of the Maxima is the only one that evidences such elaboration. To the south, an outlet from the Forum Boarium, now covered by the embankment, was finished with an unmarked opening in plain masonry (figure 5.10: B). The mouth of the Cloaca Circi Maximi 20 metres south of this small *cloaca* was elaborated with a single arch, perhaps indicating its archaic heritage (figure 5.10: C), but the dimensions and arched embellishment of the Maxima's mouth seem to have been unmatched (figure 5.10: A).[72] Such embellishment would not have facilitated drainage and would have served only to mark the Maxima's entrance into the Tiber with a monumental image. Even if this facade was by no means as resplendent as the triumphal arches or other major monuments in the city, it was nonetheless exceptional and technologically superfluous. Had Romans seen the Maxima as nothing but an unsavoury waste chute, surely they would have done everything to hide

[71] Lanciani (1897) 62–3; Blake and Van Deman (1947) 125, 159; Cressedi (1984) 273–4, figs 5–8.
[72] Cressedi (1984) 273–4, figs 5–8.

its issue into the Tiber. Instead, they isolated this lone outlet and called attention to its opening with the triple-coursed archway.

Sacred pollution

For nearly five hundred years, as the city centre rose and fell on top of it, Romans maintained the Maxima's original course. Its persistent reconstruction under the Basilica Aemilia and its snaking path through the Velabrum and Forum Boarium suggest a reverence for the religious stream it originally channelled and for the power of the structure to recall and promote the earliest history of Rome. Religious ties to the origins of the temples of Janus and Venus Cloacina further reveal the sanctity of the Maxima, and the historical tradition of their origins in the regal period must have compounded the Cloaca's ties to Rome's most ancient past. As reminders of the subterranean sacred stream below, the temples and other markers also recall a dramatic change in the Maxima's image during the Republic: its transition from visible canal to invisible underground drain. Sometime in the middle Republic, probably in the second century BC if not before, Romans vaulted the entire length of the Maxima, thrusting the channel into a new obscurity.[73] Over the subsequent centuries, though, they by no means forgot or neglected the drain; instead they continued to refurbish it and build new shrines to propitiate the hidden waters. In place of a simple travertine drain cover, the marble disk of Oceanus signified the sacred primordial nature of the stream waters that ran below, and at the mouth of this most ancient stream, an arched facade proclaimed its terminus. Each of these monuments reveals that Romans went to great lengths and expense to highlight the path of their primary sewer, and, whether for religious reasons, historical pride or some combination of both, a re-analysis of the construction and marking of the Cloaca Maxima indicates that Romans believed it to be a hallowed monument.

In this way, the Maxima is markedly different from modern sewers, and when contemplating its significance in antiquity, one must seek to recover

---

[73] The date is controversial, but a reference in Plautus' *Curculio* would seem to indicate that it was open to the sky at least until the second century. Plautus, *Curculio* 475–6: *in foro infimo boni homines atque dites ambulant / in medio propter canalem, ibi ostentatores meri* ('in the lower forum walk the good, wealthy men; in the middle, near the canal, there [one finds] the utterly boastful ones'). Also, it was probably not covered solely because of the pollutants therein, but also because the pavement of the Forum had risen so high that the deep channel was either too dangerous, or was simply now low enough to allow for vaulting and a continuous pavement above it.

its distinct and enduring place in Roman culture. Rome's main sewer was not merely a waste receptacle, nor was its impressive vaulted construction of sole importance. The attention paid to the monument, its waters, its path and its outlet reveal a broader sense of respect, appreciation and veneration. Thus, like many polluted aspects of ancient and non-Western culture, the Cloaca Maxima held multivalent meanings for Romans who lived and built around it. Its waters were undoubtedly polluted, soiled; they were not hygienic and were apparently foul enough to demand a cover in the middle Republic. But by the same token, they remained inherently tied both to the natural sacred stream that continues to flush the drain even today, and to the vaunted history of earliest Rome. Romans' reverence for this ancient channel seems therefore to have led them to conflate sacred and polluted in the Maxima, and, however repulsive it was, the structure's sanctity and historicity demanded that Romans venerate it.

6 | Crime and punishment on the Capitoline Hill

MARK BRADLEY

'Only when *you*, as I have long been urging, get out, Catiline, will this dreadful sewage be drained out, *out* of the city' (*sin tu, quod te iam dudum hortor, exieris, exhaurietur ex urbe tuorum comitum magna et perniciosa sentina*).[1] In 63 BC Cicero, Rome's greatest orator, squashed the conspiracy of Catiline with a masterful display of rhetorical invective. This was no easy job: Catiline was a leading Roman noble and a popular politician, and Cicero needed to 'out' Catiline as a criminal mastermind and eliminate him from the city. Cicero calls him every name under the sun: a gladiator, a wild beast, a parricide – metaphors that placed Catiline at the very edges of Roman identity. But no metaphor offered a more potent means of segregation than that of disease, dirt and sewage. 'Leave and cleanse the city!' (*purga urbem!*), Cicero ordered Catiline in the first of his four damning speeches. 'Rome will be free from fear only if there is a wall between you and us.' Cicero leaves us in no doubt as to how he approaches his role as consul concerning this 'matter out of place': *exire ex urbe iubet consul hostem*, 'the consul orders the public enemy to get out, out of the city'. This disease, he goes on, must be expelled and isolated in its entirety. The second of the four speeches that Cicero delivers against Catiline is in part a celebration of success: Catiline is gone, departed, escaped, exposed and expelled – although Cicero is professionally committed to the idea that there is more purging to be done. For the moment, he proudly announces that the city seems to rejoice because it has spewed forth (*euomuerit*) such a great disease and cast it outdoors. He imagines a future city purged of all Catiline's cronies: 'O fortunate Republic, if it manages to evacuate this sewage from the city.' With Catiline alone 'drained away', the state already seems relieved and refreshed. The metaphor leaves little to the imagination. It is the job of his consulship, as Cicero puts it a few sections later, to remove *all* the city's unclean (*impuri impudicique*).

As the previous chapter has shown, describing Catiline as sewage was more than just a playground insult: dirt existed to be cleaned up and put back in its place, and Cicero's metaphor appealed to a fundamental Roman

[1] Cicero, *In Catilinam* 1.12–13.

concern with purity and pollution, with order and disorder. The Cloaca Maxima, Rome's ancient and monumental sewer system, as well as being a mechanism for expelling hundreds of tonnes of human waste from the city every day was also a popular focus for social and political symbolism, and it is no accident that Cicero's effective and incisive rhetoric compared the elimination of the city's criminals to the evacuation of sewage or the eradication of disease.[2] And, as recent scholars have argued, metaphor is more than just a rhetorical or linguistic phenomenon: it is an essential part of cognitive human thought, a way of organizing and making sense of the world.[3]

This chapter, then, maintains the same focus as previous chapters on Rome's urban topography and dirt removal, but shifts the emphasis on to the role of description, representation and rhetoric in the ways that the city's inhabitants made sense of their landscape.[4] It does this by concentrating on the representation and control of one particular type of pollution, a theme which will become critical to approaches to the subject in later chapters of this volume: the identification and elimination of the 'criminal'. How were Rome's most notorious criminals identified? How were they punished, and where did this take place? How were their bodies disposed of? And how were these processes described?[5] This chapter will concentrate on the monumental complex of Rome's Capitoline Hill, the site of the city's execution chamber and the Tarpeian rock, as well as the home of the city's foremost deities and the focus for that most central of Rome's ritual ceremonies, the triumph. Instead of focusing on movement in and out of the city, as Hopkins did with his study of the sewers and as I began this

---

[2] I have discussed some of these themes in Bradley (2006). Cf. Bauman (1996) 35–49 on Cicero's approach to punishment. These same metaphors are evident in early Christian invective: see Leyerle (2009) on the use of the language of waste and excrement by the fourth-century Antiochan priest John Chrysostom to mobilize disgust towards extravagant and consumptive lifestyles; Leyerle argues that popular familiarity with sewer imagery in the ancient city fed this discourse.

[3] See in particular Lakoff and Turner (1980); cf. Kövecses (2002). For a recent attempt to explore these ideas in Ciceronian discourse, see Sjöblad (2009).

[4] On this theme in the early Empire, see Nicolet (1991); cf. Vasaly (1993) on topography in Ciceronian oratory; Edwards (1996) on textual approaches to the city.

[5] The fullest study of capital punishment and its significance in late Republican and early imperial Rome (as well as some key differences between the two periods) is David (1984); see also Baroin (2010) on the significance of various types of damage to the bodies of Roman magistrates. Cf. Kyle (1998), who explores similar questions but focuses principally on gladiatorial games (see ch. 7 more generally on criminal punishment). See also Mustakallio (1994) on capital penalties in early historiography; more generally, see Robinson (2007) esp. 179–97. See also Charlier (2009) on violent deaths in antiquity. Bauman (1996), esp. 1–20, discusses Roman views on the nature and function of criminal punishment.

chapter, it will concentrate on how Roman writers engaged with movement upward and downward within this landscape, and particularly the routes taken by the city's criminals.[6]

From a very early stage in the city's history, the Romans transformed their proverbial seven hills – *montes*, as they often described them – into central hubs of social, political and religious activity.[7] Every hill told a story: you could not go up one or down one without thinking which Romans had done so before you and why. Varro, among others, speculates on the primordial relationship between the name *Septimontium* or 'seven mountains' traditionally given to the site of Rome, and the ancient festival of the same name celebrated on 11 December. The hills were a geological template for Roman identity. The Palatine Hill was the stronghold of the patricians in the early Republic, the home of the wealthy and, under the Principate, the site of imperial palaces. The Aventine Hill was the refuge of the plebeians during their strikes in the early Republic, and the site of a host of plebeian sanctuaries. And from these hills, Pliny the Elder traced seven streams that flowed into one channel of the Cloaca Maxima. The topography of central Rome was a rich playing field for creative symbolism.

The Mons Capitolinus was Rome's most imposing hill and the city's main fortress. From here, the extraordinary sixth-century BC temple of Jupiter Optimus Maximus looked down over the Roman Forum (see figure 6.1). Cicero describes the Capitoline Hill as the headpiece of the city, the topographical equivalent of the human head where Plato located reason.[8] The hill was two-pronged; on one side the *Capitolium*, on the other the *arx* or citadel. In the Roman triumph, the *Capitolium* was the climax of the triumphant general's ascent, the spot where he would make his sacrifice to Jupiter after he had been literally – and socially – elevated in the ceremonial procession to the top. According to some interpretations, he would become Jupiter for a day, climbing to the top of the social pecking order at the same time as climbing to the top of the Capitoline. Not all in the triumph were so lucky: this was the moment when the enemy leader, after being paraded through the streets, would be cast down into the prison

---

[6] The significance of the ejection of criminals outside the *pomerium* is evident from the representation of or legal rationale for exile, or the practice of crucifixion outside the city walls: see Claasen (1999) on Roman exile; Kyle (1998) 53 on crucifixion; cf. 214 on exposure of prodigious babies.

[7] For a major new study of the cultural history of the hills of Rome, see Vout (forthcoming).

[8] Cicero, *Tusculanae disputationes* 1.10.20. On the city as a living organic system with head, limbs and guts, and the history of this idea, see Gowers (1995). On the Capitoline Hill, see Rodocanachi (1906); Giannelli (1993).

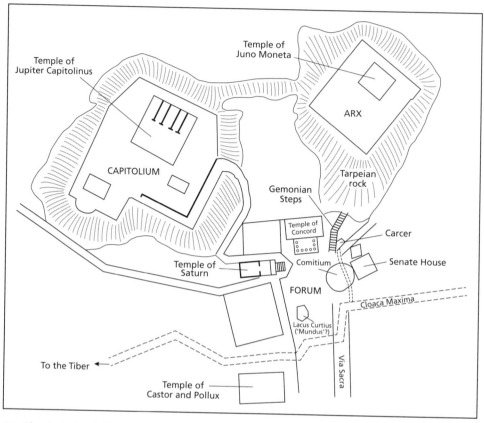

6.1 The Capitoline Hill complex in the mid-first century BC. Illustration: Mark Bradley.

on the hill's slopes and executed.[9] Here was the site of the Tarpeian rock, down from which Rome's most conspicuous criminals were thrown to their death. Immediately to the north was the *pomerium*, the traditional sacred boundary of the city which marked inside and outside, and immediately to the south was the Forum, the epicentre of Roman civic life. The Capitoline Hill also overlooked, as well as the Senate house and *comitium* where judgements were passed, the ancient Carcer Tullianum, now the Mamertine Prison, where criminals were strangled and their bodies then tossed on a further journey down the Gemonian Steps to be mangled by the mob in the

---

[9] See Kyle (1998) 217–18. For an excellent discussion of the fate of triumphal captives, see Beard (2007) ch. 4, esp. 128–32 on the execution. More recently, see Östenberg (2009), esp. ch. 3 on the spectacle of the triumphal captives. Josephus' description of the triumph of Vespasian and Titus in AD 70 (*Bellum Judaicum* 7.153–5) offers a vivid account of the ritual execution of the Jewish leader Simon in the Carcer, which had to be completed before the concluding sacrifice to Jupiter on the Capitolium.

Forum, or possibly even flushed into a channel of the Cloaca Maxima to be pumped into the Tiber and out of the city.[10]

It is little surprise, then, that this evocative landscape fed the imagination of a number of ancient writers. The Capitoline, with its steep slopes and sharp drops, its sanctuaries and its prison, could be seen as a scaled-down image of the world in the heart of Rome, with the gods and the enlightened at its top, and condemned outcasts at its bottom. Movement up and down this landscape could be made to express one's position within this hierarchy.

The Tarpeian rock

This inquiry will begin at the top of the Tarpeian rock, a steep cliff on the Capitoline Hill from which convicted criminals were thrown to their death. Its precise location is debated; the modern Hill has been changed beyond recognition by landslides and erosion, as well as by numerous post-classical building projects. It seems most likely, however, to have been the part of the Hill that overlooks the Forum, and – as F. Coarelli argued – part of a monumental complex that included the Carcer and the Gemonian Steps (see figure 6.1).[11] It was not unusual, particularly in verse, to classify the Capitoline Hill as a whole as *Tarpeius rupes* or *Tarpeia saxa*, particularly when one wished to emphasize its foreboding aspect; the Latin epic poet Silius Italicus uses this expression no fewer than twenty times, and pictures Jupiter thundering menacingly on its peaks.[12] Varro tells us that *mons Tarpeius* was

[10] See David (1984) for an exploration of various aspects of the topographical relationship between these landscape features; see esp. 134: 'le chemin de la mort manifestait publiquement l'expulsion du monde des vivants'. On the use of the Tiber for the convenient disposal of the dead, and the symbolic uses of water for purgation, see Kyle (1998) 213–41: 'water was a classic way to dispose of polluted objects, prodigies, and unwanted creatures – beings rejected from or never accepted into the community' (214); Kyle makes an important point about the river's position outside the formal *pomerium* of the city. See also Kyle (1998) 218 for the Gemonian Steps. On the traditional punishment for parricide (the criminal tied up in a sack with various vicious creatures), see Kyle (1998) 216–17; Robinson (2007) 40–6. See Cicero, *Pro Sestio* 77 on the sewers 'stuffed' with the bodies of citizens following the public violence of 57 BC; Gowers (1993) 14–16 on the sewers as political 'waste-disposal units'; cf. Gowers (1995) 28; Kyle (1998) 221.

[11] Coarelli (1985) 80–7. See Wiseman (1999) for an up-to-date synopsis, with fig. 114 for the four major hypothetical locations for the *saxum*. The most credible area of the modern landscape (the southeast corner of the Capitoline, and the tourist 'rock') now seems very unlikely. Further on its location, see Dionysius of Halicarnassus, *Antiquitates Romanae* 7.35.5, cf. 8.78.5; Cassius Dio 58.15.2–3. Cf. Wiseman (1990) on the topography of the central Forum area.

[12] Cf. Propertius 4.1.7 on early Rome: *Tarpeiusque pater nuda de rupe tonabat*. Virgil, *Aeneid* 8.347ff. (Evander's description of *Tarpeia sedes*, a *locus* imbued with *dira religio* and marked by

an older name for the whole hill, until a human head was discovered while digging the foundations of the temple of Jupiter, establishing the 'Capitoline' as the figurative headpiece of the city (and the capital of the world).[13] Fastened to its base, according to the Lucan scholia, was a wooden construction called the *robur* which caught the bodies of the victims; from here they may have been tipped on to the Gemonian Steps, where they were exposed or dragged down into the Forum below.[14] The route taken by the fallen criminal was etched into the landscape. So it is that the early Roman hero Manlius and the 'Tarpeian citadel' are engraved strategically at the top (*in summo*) of Aeneas' shield (Virgil, *Aeneid* 8.652), far away (*procul hinc*, 666) from Tartarus and the punishments of Catiline. It is also the Tarpeian rock that, according to Propertius, was under threat from the foul mosquito nets (*foeda conopia*) of Cleopatra – a striking antithesis between Rome's purgative rock and the polluting material luxuries of the East. At least in the literary imagination, the Tarpeian rock was a powerful symbol of permanent Roman *seueritas*.[15]

Those considered out of place in the Roman system were dis-placed from the rock; punishment *e saxo* was enshrined in the *XII Tabulae*, and this was the fate met by a long historical catalogue of tyrants, false witnesses, treacherous allies, men guilty of incest, magicians, and slaves who had betrayed their masters – all of them the sorts of criminals who had earned the most conspicuous type of public execution.[16] One of its earliest recorded

---

a terrifying *saxum*); Juvenal 13.78 (an oath taken *per Tarpeia fulmina*). On the use of this synecdoche in Silius Italicus, see for example 1.541 (the Tarpeian rock described as habitable by the gods, *superis habitabile saxum*); 2.33 (*Tarpeios... scopulos praeruptaque saxa*); 5.109 (personification of a trembling *Tarpeia rupes*); 5.82 (Corvinus swears by *Tarpeia saxa*); 5.164 (Hannibal's goal of setting up camp on the Tarpeian rock); 12.609 (all the gods in battle array on Rome's seven citadels, with Jupiter on the Tarpeian heights, *Tarpeio sublime uertice*). Cf. Lucan 3.154 (*rupes Tarpeia sonat*); Claudian 15.30, 21.214 (captives mounting the rock, *Tarpeias pressis subeant ceruicibus arces*), 28.45, 28.375 (the rock as a vantage point for Jupiter, *Tarpeia spectanda Ioui*); Statius, *Silvae* 3.4.105 (*Tarpeia... templa*).

[13] Varro, *De lingua Latina* 5.41. This event is recorded also at Pliny, *Naturalis historia* 28.15.

[14] Scholia, *Lucan* B 2.125. Lucan describes the dreadful image of the *robur* spattered with blood in civil war: *saeua tribunicio maduerunt robora tabo*. On the corpse's onward journey down the Gemonian Steps, see Valerius Maximus 6.3.1d, *de robore praecipitati*. On various interpretations of the *robur*, see David (1984) 144–5; cf. 133 on the Gemonian Steps.

[15] Propertius 3.11.45. See Martial 9.1.5, where the *Tarpeia saxa summi patris* stand as a yardstick for the durability of the Flavian *gens*.

[16] Tyrants: see Dionysius of Halicarnassus, *Antiquitates Romanae* 8.78.5, and below pp. 109–10 on Manlius Capitolinus; false witnesses: Aulus Gellius 20.1.53 (*e saxo Tarpeio deiceretur*); cf. Plautus, *Curculio* 268–9; allies: Livy 25.7.11–13 on treacherous allies in the Second Punic War who, after receiving a popular beating in the *Comitium*, were tossed off the rock (*deducti in comitium uirgisque adprobante populo caesi de saxo deiciuntur*); incest: Tacitus, *Annales* 6.19.1 on a rich Spaniard found guilty of incest with his daughter in AD 33 and flung from the rock

victims, Spurius Cassius (consul 494 BC), was accused of tyranny and cast off the rock, quite properly as far as Dionysius was concerned (*epichorios... hē kolasis*, 8.78.5), in full view of the citizens and down into the Forum. Nor was Tarpeian rock execution a feature of the dim and distant Roman past: it continued well into the Empire.[17] And the prepositions and prefixes used by the texts describing this punishment are telling: the criminals are decisively cast *down* (*de*, *kata*, *ex*) and eliminated. This was legalized, landscaped waste disposal, down and out.[18]

Roman writers were of course as interested in the examples where the system went wrong as in those where it functioned properly. Valerius Maximus describes the supporters of the Republican demagogues the Gracchi hurled down (*praecipitati*) from the rock.[19] The conservative *seueritas* of the whole system, in fact, was open to ridicule: Horace satirizes political ambition as a desire to be the man who throws citizens down from the rock, or hands them to the prison executioner.[20] Cassius Dio catalogues the political victims of Tiberius, Caligula and Claudius who plunged unfairly to their deaths from the rock, *de saxo*.[21] One of the rock's most famous mistakes was Manlius Capitolinus, the saviour of the Gallic Wars and defender of the citadel in the early Republic. The man perched on top of Aeneas' shield was, after accusations of tyranny in 385 BC, thrown from the very rock which he had famously rescued and which he had made his home. Livy's account is dramatic: although Manlius was considered 'heavenly' (*caelestes*) for his heroic actions and shared his surname 'Capitolinus' with Jupiter, and like Jupiter had his home on the top of the Capitoline Hill, he was cast into the

---

(*saxo Tarpeio deicitur*); magicians: Tacitus, *Annales* 2.32.3 on the dangerous *magus* L. Pitanius under Tiberius; slaves: Valerius Maximus 6.5.7 on Sulla *iustissimus*, who had a slave who had betrayed his master tossed off the rock (*praecipitari protinus saxo Tarpeio*); cf. Aulus Gellius 11.18.8 on slaves caught thieving subject to *e saxo praecipitari*. See also David (1984) 136–7 (Republican rock victims); 168–9 (imperial rock victims). On the myth of the rock's eponymous founder Tarpeia, see Plutarch, *Romulus* 18.1 (where the connection between Tarpeia and the Tarpeian rock is made incidental); cf. Propertius 4.4, esp. 95–6; Varro, *De lingua Latina* 5.41 on the etymology of the Capitoline: the *rupes* is a *monumentum* of Tarpeia's name.

17 See Kyle (1998) 219–20.

18 Cf. David (1984) 135 on the idea of execution from a cliff-edge as part of a broad Mediterranean pattern of 'consécration aux dieux'.

19 Or, as one manuscript has it, *de robore* (the wooden base of the rock from which bodies were tossed down the Gemonian Steps). Following the systematic abuse of the corpses of Gaius Gracchus and (according to Plutarch, *Gaius Gracchus* 17.5) 3,000 of his followers, and their ejection into the Tiber, the consul L. Opimius restored the temple of Concord as part of a lustration (Appian, *Bella civilia* 1.7.27); it must be significant that this temple stood next to the Carcer and at the heart of the execution complex.

20 Horace, *Satirae* 1.6.39: *deicere de saxo ciuis aut tradere Cadmo*.

21 On Tiberius, Dio 57.22.5; on Caligula, Dio 59.18.3; on Claudius, 60.18.4.

dark abyss (*tenebrae*) of the prison at the mercy of the executioner, and then thrown down (*deiecerunt*) by the tribunes from the rock.[22] Livy, who had a little earlier (6.17.4) described the Gauls scaling the Tarpeian rock, adds that many thought the Capitoline Hill was violated (*uiolatum*) by the blood of its saviour. Manlius' 'reversal of fortune', his steep fall from power, was – at least for one Roman historian – evocatively expressed by the Capitoline landscape.[23]

Controversia 1.3 ('*Incesta de Saxo*') is part of an imaginative rhetorical debate put together by the Elder Seneca in the early decades of the first century AD. It explores the strange case of an unchaste woman who, condemned to die, appealed to Vesta and then survived the fall from the rock.[24] The question was, should she pay the penalty again? This was one of the Elder Seneca's many creative and idealized *Controversiae*, exploring the edges of Roman cultural values and the central role of rhetoric in determining courses of action when the system goes radically wrong.

It is witness to the enduring potency of pollution and purity in contemporary thought that much of the *Controversia*'s argument concentrates on the contaminated status of the unchaste woman post-fall: she is categorically *fallen*, and Seneca's introduction to the debate hammers home this fundamental point: *de saxo, deiciatur, deiceretur de saxo, deiecta uixit*. The first prosecutor begins with the idea that the appeal to Vesta was itself an act of pollution (*uiolare*), itself a rejection of any claim to modesty (*uiolasset inter altaria . . . uiolare supplicio*, 1). This was, he points out, an additional offence (*alterum incestum*) and a reinforcement of her condemnation (*damnata est quia incesta erat, deiecta est quia damnata erat*). She had already been contaminated by the hands of the executioner (*carnificis manu incesta*), and the rock itself contaminates those who stand upon it (*loco incestarum stetisti*). He goes on: she is both unchaste (*incesta*) *and* condemned (*damnata*) and cast down (*deiecta*). The casting down (*deicienda*) must be repeated again

---

[22] Livy 6.17.5: *et quem prope caelestem, cognomine certe Capitolino Ioui parem fecerint eum pati uinctum in carcere, in tenebris obnoxiam carnificis arbitrio ducere animam?*; cf. Livy 6.20.12: *locusque idem in uno homine et eximiae gloriae monumentum et poenae ultimae fuit*. See also Livy, *Periochae* 58; Pliny, *Naturalis historia* 7.143 for Metellus Macedonicus' close shave with the rock; cf. Velleius Paterculus 2.4.2 on a political victim of the Sullan dictatorship.

[23] Cf. Livy 7.10.3: *Gallorum agmen ex rupe Tarpeia deiecit*. Plutarch, *Camillus* 36.7 is somewhat more approving of the punishment, although he draws attention to the power of the landscape to sway the trial; cf. Florus 1.17.8. On the interplay of rhetoric and topography in accounts of Manlius' trial, see Wiseman (1979).

[24] Survival was not unknown: see Dio 4.17.8 on Roman clemency in sparing survivors of the fall. The declaimers even make out that Seneca's *incesta* could be a Vestal (although this is never specified). See Winterbottom (1974) 88.

and again until the purpose of her casting down has been achieved. The verb is repeated seven times in the first prosecutor's short speech.

Another speaker, Arellius Fuscus, quotes the law that condemned criminals to the rock: the rock is a sheer drop of immense height, a grim sight; its deadly function therefore was quite unambiguous.[25] Arellius continues with his own even more dramatic description, reminiscent of epic accounts of Hades: *stat moles abscisa in profundum* – the crag stands cut away into the depths, spiky with rocks everywhere designed to crush and mangle the fallen bodies. He continues with morbid flair, as if the rock were some living organism: its sides bristle with projecting crags, *inhorrent scopulis enascentibus latera*. The site was chosen so that the condemned should not be cast down – *deiciantur* – more than once. The fall was a one-way ticket.

Another speaker offers a different take on the situation: the woman is so polluted that she is cast down by the upper gods and even rejected by the lower ones (*a superis deiecta, ab inferis non recepta*). One would require a very special rock for such a criminal. Another speaker repeats the same themes: the woman had the audacity to get back up after her fall and go and pollute with her very touch the temple of Vesta, *quidquid secundum deos sanctissimum est contactu suo polluit* (4). Being consigned to the rock is being consigned to the rock – a permanent sentence. But, the speaker goes on (not without some humour), she bounced back to the temple (*in templum resiluit*). Another speaker remarks that the idea of acquitting this woman now is tantamount to digging up a condemned priestess from the underworld (*ab inferis eruatis sacerdotem*). Yet another imagines her bouncing around the landscape, tossed (*iactata*) between the gods above and the gods below.[26] She has lived again, the speaker argues, only for the purpose of receiving a new penalty. With its upper world and its lower world, Rome's topography of crime and punishment is envisaged in this imaginative rhetorical exercise as a kind of cameo picture where heaven and hell intersected with Roman life.[27]

---

[25] The MS at 1.3.3 is corrupt. For details, see Winterbottom (1974) 92.

[26] This speaker makes the point that the executioner had broken the fillet of Vesta carried by the woman, symbolically severing her connection to the upper gods (from this point she is imagined to be a Vestal Virgin).

[27] Cestius Pius (a narrator, who appears to be flexible in his stance) then suggests that it is normal for somebody facing the Capitoline's heights (*altitudo montis*) to pray to the gods; it is *horrenda* even for people looking down in safety, and even the executioner (*carnifex*) recoils as he pushes his victims. Other, less successful, spins include that the woman's body had been hardened by drugs (*ueneficiis corpus induruit*, 11) and so bounced off the rocks unharmed (*saxa reuerberet inultum*); that the priestess's clothes billowed out and acted as a parachute; or that she was a priestess good at leaping (*uirgo desultrix*), and had taken thorough lessons in falling.

The Carcer Tullianum

According to Lucretius, alongside the 'terrible fall downwards from the rock' (*horribilis de saxo iactu' deorsum*), the Carcer was the traditional alternative site of punishment for Roman transgressors.[28] This stood just north of the Forum on the slopes of the Capitoline Hill, sharing a wall with the temple of Concordia, and in an area of the city that contained many of the oldest and most ritually important sites and monuments (*media urbe, inminens foro*, Livy 1.33.8), such as the *comitium* and the Curia.[29] It appears to have been built alongside the Gemonian Steps down which condemned criminals were ritually flung, and was connected by a drainage channel to the Cloaca Maxima which ran beneath the Forum. It was set within a stretch of early tufo quarries (*lautumiae*) on the site of a natural spring, although one Roman orator jokes that there is nothing 'clean' (*lautus*) about the prison.[30] The prison was ancient, traditionally founded by the king Ancus Marcius to deter crime, and its lower part, the 'Tullianum' or execution chamber, was said (probably because of its name) to have been added by Servius Tullius.[31] In Pliny's account of early time-keeping, the movement of the sun from the Maenian column down to the prison marked out the day's *suprema hora*.[32]

The Carcer of course has little in common with the giant detention centres of modern Rome: the Regina Coeli in Trastevere, the prison at Rebibbia or the military prison at Forte Boccea.[33] Today's remains consist of just two rooms (the Tullianum being the lower of the two) preserved in a sixteenth-century Christian church, although recent excavations have revealed several adjoining rooms. Rome's ancient prison – the central prison serving a population bordering on a million people – was a monument of miniature proportions, and (like comparable sites around the ancient Mediterranean) normally a place of execution rather than a place of detention.[34] The

---

[28] Lucretius, *De rerum natura* 3.1016.

[29] See David (1984) 131–2 on the *comitium*; cf. 134–5 on this as the site where public beatings took place prior to an execution.

[30] The rhetorician Asilius Sabinus, cited by Seneca, *Controversiae* 9.4.20. Cf. Livy 32.26.17: *carceris lautumiarum*; Varro, *De lingua Latina* 5.151. Further on the spring and the drain, see Kyle (1998) 218.

[31] Varro, *De lingua Latina* 5.151; Festus 490L. At 482L, Festus connects '*Tullianum*' with '*tullius*' (jet of water); a reference, possibly, to the famous spring at the site. See also Propertius 4.4.3ff.; cf. Richardson (1992b) for an imaginative conjecture on the relationship between spring and prison.

[32] Pliny, *Naturalis historia* 7.212.   [33] See Robinson (2007) 113–14.

[34] For the dimensions, see Coarelli (1993c). The upper, and larger, of the two surviving rooms was approx. 3.6 by 5 metres. The lower room is slightly more than 2 metres high. Cf. Dio 57.50 on Metellus (consul 60 BC) imprisoned by Pompey's tribune; the consul made the absurd

execution chamber itself, the Tullianum, appears to have been accessed only through an opening in the floor of the upper chamber. After criminals had been thrown down into the Tullianum, they were garrotted and their bodies – at least if they were politically significant enough – were traditionally flung down the Gemonian Steps into the Forum. So, for example, Valerius Maximus, in a graphic account of old-fashioned Roman *seueritas*, recalls the punishment in 236 BC of Marcus Claudius, who had instigated a shameful treaty (*pax turpis*) with the Corsi, an enemy tribe (6.3.3): Claudius' treaty was nullified, his freedom (*libertas*) stripped from him, his life (*spiritus*) extinguished – and, Valerius continues in a flurry of retribution, his body *befouled* with the insult of prison and the abhorred stigma of the Gemonian Steps (*corpus contumelia carceris et detestanda Gemoniarum scalarum nota foedauit*).[35] In addition, as noted above (p. 107), there was a specially constructed drainage channel that connected the Carcer to the Cloaca Maxima, which ran beneath the Forum and out into the Tiber. The prison, then, was a plughole from which Rome's down-and-out could be cast down and out.

Sallust, describing the execution of Catiline's five fellow conspirators, paints a vivid picture of the Prison:

> est in carcere locus, quod Tullianum appellatur, ubi paululum ascenderis ad laeuam, circiter duodecim pedes humi depressus. eum muniunt undique parietes atque insuper camera lapideis fornicibus iuncta: sed incultu, tenebris, odore foeda atque terribilis eius facies est. in eum locum postquam demissus est Lentulus, uindices rerum capitalium, quibus praeceptum erat, laqueo gulam fregere.

> There is in the prison a place, which is called the Tullianum, when you have climbed a little way to the left, about twelve feet below the surface of the ground.[36] Walls enclose it on all sides, and above a vault formed by stone arches: but its appearance is foul and terrible in its uncivilized state, darkness and stench. After Lentulus was let down into this place, the executioners to whom the order had been given broke his neck with a rope. (Sallust, *Bellum Catilinae* 56.3–5)

Sallust introduces his description of the Carcer, in which the Catilinarian conspirators met their end, with a classic epic formula – *est in carcere*

---

request that the Senate assemble there, even going as far as suggesting that a hole be knocked through the wall of the cell to allow them entrance. See David (1984) 139–40 on 'prison' sites at Sparta, Corinth and elsewhere in Italy.

[35] Cf. Plutarch, *Marius* 12.3 on the execution of Jugurtha in the Carcer after Marius' triumph, where he paid the proper penalty for his transgressions (*eichen axia dikē tōn asebēmatōn*).

[36] See McGushin (1980) 275 on the likely means of entrance to the Tullianum in Sallust's day. One MS reads *descenderis* rather than *ascenderis*, but the movement implied by Sallust is unclear.

locus – which endowed the passage with literary *gravitas*.[37] This description in fact picks up Cato's description of the underworld just a few sections earlier (52.13), which is also *foeda* (foul) and *inculta* (uncivilized), and set out like a fortress. And the commentator Heinze has noted that this particular description echoes another description of the underworld given by Sallust's contemporary Lucretius.[38] Even though it was not really underground, Sallust describes the Carcer as 'set down into the ground' (*humi depressus*) and Lentulus was cast *down* (*demissus*) into it.

The stereotypically hellish environment of the Carcer is wonderfully captured by a *declamatio* of Calpurnius Flaccus describing the unusual case of a convicted parricide who, anticipating the legal stipulation that such a criminal must be kept in custody for a year before facing punishment, pleads for a year in the public Carcer rather than a year in his father's custody (he appears to have murdered his stepmother). In order to convince the court that the Carcer is punishment enough, he vividly imagines the long wait for execution:

> uideo carcerem publicum, saxis ingentibus structum, angustis foraminibus tenuem lucis umbram recipientem. in hunc rei abiecti robur Tullianumque prospiciunt, et, quotiens iacentes ferrati postis stridor excitat, exanimantur, et alienum supplicium [ex]spectando suum discunt. sonant uerbera, cibus recusantibus spurca manu carnificis ingeritur. sedet ianitor inexorabili pectore, qui matre flente siccos teneat oculos. inluuies corpus exasperat, manum catenae premunt.

> I can visualize the state prison, constructed of huge stone blocks, receiving through the narrow chinks just a faint semblance of light. Culprits cast away (*abiecti*) into this prison look forward to the oak post (*robur*) and the execution chamber (*Tullianum*), and, whenever the creaking of the iron-bound door stirs the sprawling prisoners, they lose their breath and by watching somebody else's punishment they learn about their own. Whips crack, food is delivered by the filthy hands of the executioner to men who then refuse it. The hard-hearted doorkeeper sits by, a man whose eyes would remain dry even when his mother was weeping. Filth roughens our bodies and chains grip our hands tight. (Calpurnius Flaccus, *Declamatio* 4[39])

---

[37] Cf. Livy 21.54.1 (*erat in medio riuus*); Virgil, *Aeneid* 2.21 (with Austin); echoing *Iliad* 2.811; *Odyssey* 3.293. Other examples in Bömer (1957) 131.

[38] See Heinze (1897) 191–2 on Lucretius, *De rerum natura* 3.1014–23 (specifically its influence on Sallust's *incultu... facies est*). Lucretius 3.1017 lists the tortures that Romans lined up for criminals: *uerbera carnifices robur pix lammina taedae*; the *robur* probably refers to the Tullianum, picking up the *carcer* in line 1016. For a topographical discussion of the Tullianum in Sallust's account, see Frank (1924a). On the description by Diodorus Siculus 31.9.7 of the prison at Alba Fucens in which Persius was detained, also a possible model for Sallust's account of the Tullianum, see David (1984) 140.

[39] See Sussmann (1994) 103–7 for a detailed commentary on this *declamatio*.

Part of the defendant's dramatic plea is an indulgent description of city-dungeon hellishness: torture, torments, groans and chains (*cruciatus, gemitus, tormenta, uincula*), shady and polluted with befouled hands (*spurca manus*) and dirt (*inluuies*) – all a rhetorical reminder that there is no better place for Rome's fallen. The long wait for execution is made to sound like unburied souls waiting for passage across the Styx: the *carnifex* (literally the 'butcher') and the doorkeeper (*ianitor*) are the prison's answer to Cerberus and Charon.[40] The passage is also reminiscent of Virgil's description of tortured souls in *Aeneid* book 6, an episode that reminded Servius, when he was writing his commentaries on the *Aeneid*, of the city's prison.[41] Prison ekphrasis stressed counter-civilization: as Calpurnius Flaccus' prosecution goes on, here it was impossible to show any trace of shame (*ibi te patiar includi ubi non erubescas?*). Indeed, such was the ignominy of the Carcer that suicide was a preferred route to death.[42] This was Rome's lower world, a place of sadness, fear and dirt.[43]

Decline and fall

The downfall of the praetorian prefect Sejanus in AD 31 following Tiberius' discovery of his treachery has captivated the imagination of writers and historians, ancient and modern alike. Perhaps due to his distance from the events in question, the Roman historian Cassius Dio, writing (in Greek) at the start of the third century, missed no opportunity to play on the rich symbolism of the landscape as he described the lead-up to Sejanus'

---

[40] On the liminality of the *carnifex* and his rejection from the community, see David (1984) 144.

[41] Servius, *Ad Aeneidem* 6.573: *et est secutus ordinem iuris antiqui: nam post habitam quaestionem in Tullianum ad ultimum supplicium mittebantur.*

[42] Most notably the young Flaccus, son of Gaius Gracchus' supporter, who dashed out his brains on the stone portal of the prison, rather than entering the place (*protinusque inliso capite in postem lapideum ianuae carceris effusoque cerebro expirauit*, Velleius Paterculus 2.7.2); similarly, Herennius Siculus, another supporter, who followed suit to avoid a public punishment (*spiritum posuit, uno gradu a publico supplicio manuque carnificis citerior*, Valerius Maximus 9.12.6).

[43] It was ideologically a place of no return (in spite of numerous political pardons and escapes); thus Varro, *De lingua Latina* 5.151 forcing an etymology: *carcer a coercendo, quod exire prohibentur*. Only the guilty would be thrown into the dungeons; any form of escape would mean that, somewhere along the line, the system had gone wrong. Cf. Livy 29.22.10 on the criminal Pleminius, whose plans to escape (*effringendi carceris fugiendique haberet occasionem*) were thwarted by the Senate and who was forthwith thrown down into the Tullianum (*delegatum in Tullianum ex senatus consulto*); cf. 34.44.8 on the same episode, *Pleminius in inferiorem demissus carcerem est necatusque*. On the idea of the prison as a site for impending death (by strangulation, starvation or suicide, with a catalogue of examples), see David (1984) esp. 142–3; cf. 145 'un lieu souterrain qui n'appartenait pas au monde des vivants'; 155 on the terrifying uncertainty and ambiguity surrounding detention.

fall (58.5.6ff.).[44] He begins by telling how, among the many portents of Sejanus' impending doom, the prefect's bodyguards, as they were accompanying Sejanus after a sacrifice down from the Capitol to the Forum (*eis tēn agoran katēei*), were separated from their master and diverted along the path leading to the prison. Subsequently they tumbled head over heels down the Gemonian Steps (*kai kata tōn anabasmōn kath'ōn hoi dikaoioumenoi erriptounto kationtes ōlisthon kai katepeson*). Later, Sejanus' auspices elicited a raucous flock of crows which forthwith flew and perched on the prison. The process of (literally) going *down* in the world (Dio's alliterative and cacophonic repetition of *kata* leaves little doubt about the direction of Sejanus' career at this point) was fraught with topographical hazards in the heart of Rome; the landscape itself was designed to exterminate and expel those who did not 'fit in'.

Then, when he was finally convicted in the Senate house, Sejanus was dragged *down* into the prison (*katesuron*, 58.11.1[45]), a movement matched by the hurling down, beating down and dragging down of his images (*kateballon kai katekopton kai katesuron*, 58.11.3); later that day, he was condemned to death (*katepsēphisato*, 58.11.4) and after he was strangled his body was thrown down the Gemonian Steps (*kata te tōn anabasmōn*), where it was abused by the mob for three days and then thrown into the Tiber to be flushed for good out of the city.[46] In addition, all those convicted of having connections with Sejanus were similarly disposed of (58.15.12): the accused were crowded together in the prison, and those who had been condemned (*hoi katauēphisthentes*) paid the penalty there or were hurled down from the Capitol by the tribunes or the consuls (*apo tou Kapitōliou . . . katekrēmnizonto*), their bodies subsequently cast out (*erripteto*) into the Forum and later cast into the river.[47]

This isolated example of capital punishment ekphrasis demonstrates how writers could parade the machinery of dirt control and elimination by toying symbolically with the lie of the land. As this chapter has demonstrated, this is not a discourse that Dio has made up himself: the topography of the city of Rome, as befitted the capital of the civilized world (at least as its own inhabitants perceived it), was charged with a kind of symbolic electricity

---

[44] See Kyle (1998) 221–2 on this passage.

[45] Although he would have had to climb the Capitol a little in order the reach the Carcer from the Senate house.

[46] Cf. Velleius Paterculus 2.6.7 on the bodies of the Gracchi thrown into the Tiber (*corpus, ita Gai mira crudelitate uictorum in Tiberim deiectum est*).

[47] On the question of the 'correct' topography and etiquette for Roman execution, see Seneca, *Controversiae* 9.2. Cf. Tacitus, *Annales* 6.19 on Tiberius' horrific treatment of the corpses of Sejanus' followers and their disposal in the Tiber.

that invited creative Roman thinkers to discover cosmic meanings in its landscape and monuments. So it is, for example, that Tacitus' account of atrocities in the city during the civil wars of AD 69 includes an elaborate account of Vitellius' unruly soldiers storming the Capitol, forcing their way upwards against the landscape itself – the Hundred Steps, the Asylum and the Tarpeian rock – and eventually setting fire to Jupiter's temple.[48] This was, Tacitus concludes, a crime of the saddest and foulest sort (*facinus luctuosissimum foedissimumque*): not even the Gauls had been able to violate it (*temerare*), but it was destroyed by the madness of emperors (*furor principum*).[49] However, Vitellius paid the penalty for this insult to the city: a few sections later (84–5), when Vespasian's troops entered the city, he was captured, his hands were tied behind his back, and he was led with tattered robes (*laniata ueste*) – a foul spectacle (*foedum spectaculum*) – to the Gemonian Steps.[50] Tacitus' eloquent account conjures up a vivid picture: the crowd hurling abuse, none weeping, and the degradation of Vitellius' end extinguishing any pity (*deformitas exitus misericordiam abstulerat*). Before he was killed at the Steps and his body mangled by the mob (*uulgus prauitate insectabatur*), he was forced to raise his head (*erigere os*) to look at his own statues falling around him (*cadentis statuas suas*), at the rostra and the spot nearby where the previous emperor Galba had been slain, and finally at the Gemonian Steps where the body of Vespasian's brother Flavius Sabinus had recently been exposed. The landscape around carried the scars of Vitellius' principate and was now afforded the opportunity for retribution.

Suetonius describes Vitellius' end in similar terms, adding that he was pelted with dung and mud (*stercore et caeno*) and his grotesque personal appearance was derided: as well as being an incendiary and a glutton (*incendiarium et patinarium*), he was monstrously tall (*enormis proceritas*), lame and had a face red from wine (*facies rubida*) and an obese belly (*uenter obesus*).[51] According to Suetonius, he was butchered (*excarnificatus* – a rare and graphic verb) on the Gemonian Steps, tortured slowly with tiny, lingering wounds (*minutissimis ictibus*), and after he was dead his body was dragged by a hook into the Tiber (*unco tractus in Tiberim*). The whole of this ritual spectacle – the geography, the insults, the execution – explicitly paraded the marginalization, rejection and disposal of 'matter out of place':

---

[48] Tacitus, *Historiae* 3.71–2.   [49] Cf. Statius, *Silvae* 5.196: *Tarpeio de monte facem.*

[50] Cf. Perkins (1990), who suggests that *foedum spectaculum* was an ironic reference to Vitellius' earlier taste for spectacles; this may be true, but misses an important point about the physical, visible role of the polluted body in public executions.

[51] Suetonius, *Vitellius* 17–18. On Vitellius' obesity, see Bradley (2011) 2.

here were pollution and purification at their most vivid.[52] It is witness to the enduring traditions of execution and public humiliation that in April 1945 the body of Benito Mussolini was also kicked, mutilated, abused and spat upon by the crowd in the Piazza dei Quindici Martiri (where fifteen anti-Fascists had recently been executed) in Milan, and then hung upside down on a meat-hook from the roof of a gas station, where it continued to be ridiculed and stoned.[53] Sejanus, Vitellius, Mussolini: the criminalized leaders of ancient and modern Italy formed an intricate geographic relationship with the landscape around them, their bodies symbolically exposed, abused and eliminated as part of the very fabric of the urban environment.

The ideas explored in this chapter about landscape and crime, then, were not restricted to the Capitoline Hill complex, nor were they limited to the period of the late Republic and early Empire.[54] One of Rome's oldest criminals was the mythical cattle-thief Cacus, in some accounts (Livy, Dionysius of Halicarnassus) a thieving shepherd, in others (Virgil) a fully fledged monster; all agree that he was righteously clubbed to death by Hercules after trying to make off with Hercules' cattle.[55] Cacus' story was indelibly grafted on to the Roman landscape: he lived in a cave on the Palatine Hill (or in some accounts the Aventine), an abode which, in Dionysius' version at least, the indignant Hercules perceived to be well suited to the harbouring

---

[52] See also Scriptores Historiae Augustae, *Commodus* 18.3–6 for a particularly vivid senatorial declaration of pollution and corpse disposal regarding the body of Commodus. In a way reminiscent of a formal curse, this repeatedly describes him as enemy of the state, parricide and gladiator (*hostis patriae, parricida, gladiator*) and prescribes that his body be mangled in the *spoliarium* (a place for depositing the corpses of gladiators) and dragged by a hook through the dust to the Tiber, and the memory of this 'foul gladiator' (*impuri gladiatoris*) be utterly wiped away. Cf. Scriptores Historiae Augustae, *Heliogabalus* 17.1–3 on Elagabalus, who was killed in a latrine; his body was dragged through the streets and around the Circus, thrust into a sewer and then out into the Tiber. See David (1984), esp. 167–75 on the highly charged symbolism of the Capitoline Hill complex as a site for public humiliation and corpse-mutilation during the Principate, in a period when popular participation in political decision-making was limited.

[53] On the complex story of Mussolini's body *post mortem*, see Foot (1999).

[54] One could pursue this inquiry by exploring the Spartan exposure of infirm babies on Mount Taygetos, or the treatment of bodies in Dionysiac rituals.

[55] Livy 1.7.5–12; Dionysius of Halicarnassus, *Antiquitates Romanae* 1.39.2ff. (he twice refers to Cacus' crimes as *kakourga*); Virgil, *Aeneid* 8.193–272, transferring Cacus' traditional Palatine home to the Aventine, which lay outside the traditional *pomerium* and may have been chosen to dissociate Augustus' Palatine home from the site of the crime, as well as to suggest links with Remus, who had his home on the Aventine. This suggestively connects Cacus to a welter of historical or contemporary Roman figures implicated in disorder: Remus, Antony, Turnus. Cacus the criminal appears to have derived from an earlier myth which placed Cacus–Caca as a chthonic fire deity associated with the Palatine Hill, and himself the victim of a cattle theft: see Fordyce (1977) 224; cf. Small (1982) on Cacus and Marsyas. The *scalae Caci* on the Palatine Hill probably earned their name in the pre-criminal phase of the myth.

of criminals (*kakourgōn hupodochais eutheton*) and therefore demolished, burying Cacus' body beneath it.[56] Afterwards, Hercules cleansed himself of the murder in the river (*hagnisas de tōi potamōi ton phonon*) before setting up an altar to Jupiter the Discoverer. This story, one of the city's founding myths, was an aetiology for the Roman worship of Hercules, and it is characteristic of Roman values and principles – at least those of the late first century BC – that it involves the elimination of a prototype criminal and the cleansing of the city, as well as being intimately tied up with aspects of Rome's physical landscape.[57]

Conclusion

The high points and the low points of the Roman landscape offered a geographic template for articulating the high points and low points of Roman life and Roman history. The very lie of the land, and the monuments that the Romans installed in the land, evoked a system of cultural hierarchies. Movement up and down this landscape could be made to express one's position within this hierarchy: Seneca's tautology of the unchaste maiden's descent down (*de*) from the Tarpeian rock or Dio's striking account of Sejanus' *down*fall (*kata*), to take just two examples, envisage the Capitoline Hill as a topographical mechanism for the disposal of the city's criminals. For some, the Roman urban landscape could even be visualized as a scaled-down image of the universe. The fourth-century poet Claudian imagined the heights of Rome as the abodes of the gods: the emperor – the 'resident god' (*deus habitans*) – commanding high jurisdiction (*summum ius*) on the Palatine with the palace raising its head above the Forum that lay at its feet; Jupiter on the Capitoline, cloud-capped statues and temples supported by massive ridges towering into the heavens – and below these, the Giants, doomed to eternal punishment, hanging from the Tarpeian rock (*iuuat infra tecta Tonantis / cernere Tarpeia pendentes rupe Gigantas*, 44–5). In a literary

---

[56] Virgil's account offers a particularly graphic ekphrasis of the foreboding landscape: the crags suspended on rocks (*rupes suspensa saxis*, 190ff.), the scattered boulders (*disiectae moles*), the desolation of Cacus' home in the mountain, and the great destruction (*ingens ruina*) dealt out by the collapsed stones. The mountain hid a deep cave that stretched down into a vast chamber (*hic spelunca fuit uasto summota recessu*); inaccessible to the sun's rays, there dwelt the terrible appearance (*dira facies*) of the half-human (*semihomo*) Cacus.

[57] In Livy's account (1.7.5–12), Hercules is accused by the other shepherds of murder (*manifesta caedes*), until the king Evander hears about 'the crime and the reasons for the crime' (*facinus facinorisque causa*) and sees that Hercules is more *augustus* than a normal man, at which Evander immediately sets up an altar to him.

blend of nature and art, Claudian drives home the message: no other city, he proudly proclaims, could fitly be the home of the world's rulers.[58]

This image of the city of Rome as a monumental interface on earth between heaven and hell is a common theme: as well as the Tarpeian rock, a fall from which could be imagined as a descent into hell, and the Carcer, a liminal glimpse of Hades for the city's criminals, the chthonic rituals surrounding the disposal of bodies in the gladiatorial arena (another site for Rome's social 'outcasts' and a significantly liminal space) are well known.[59] Furthermore, the Mundus – an early monument normally situated on the Palatine but identified by some in the Forum – was sometimes interpreted as a gateway to the underworld.[60] In the floor of the Mundus was an opening that led to a shaft or chamber sacred to the *Di Manes* (the Ancestral Spirits), opened only on 24 August, 5 October and 8 November, which were unlucky days in the Roman calendar. The Mundus marked the centre of the city, and therefore the centre of the universe, described by Varro, Plutarch and Servius as the point at which the planes of the cosmos met.[61] As with many early monuments, the educated *literati* of imperial Rome reached little agreement over the significance of the Mundus; what they could agree on, however, was the need to imbue this obscure little archaic shrine in the imperial metropolis with microcosmic significance. It would, of course, be a mistake to think that these grand ideas were passing through the minds of every Roman wandering around central Rome, but they were recurring themes

---

[58] Claudian, *De sexto consulatu Honorii Augusti* 39ff.

[59] See for example Kyle (1998) 155–8: the term *noxii* was used to describe convicts condemned to the arena. See also 128–33 on Roman ideas about damnation beyond death, and 242–64 on Christian ideas about damnation and the afterlife.

[60] Festus (142M) calls it the *mundus Cereris: Cereris qui mundus appellatur, qui ter in anno solet patere; viiii Kal. Sept. et iii Non. Octobr. Et vi Id. Novembr. qui uel enim dictus est quod terra mouetur.* He later quotes Cato, who described it as a domed structure large enough for a man to enter (*mundo nomen impositum est ab eo mundo, qui supra nos est: forma enim eius est, ut ex is qui intrauere cognoscere potui, adsimilis illae,* 157M). There appears to have been an earlier form of the Mundus on the Palatine Hill: an underground chamber was discovered beneath the Domus Augustana in 1914; see Platner and Ashby (1929b), who also suggest that the *lapis niger* was an altar to the gods of the underworld. The *umbilicus Romae*, a cylindrical, brick-faced core standing north of the imperial rostra hemicycle, and close by the Milliarium Aureum, may have represented the centre point of the city and the empire (so modelled on the Greek *omphalos*), and may have been connected with the *comitium* 'Mundus', but the surviving monument is Severan or post-Severan. See Warde-Fowler (1912), who suggests that the association of the Mundus with underworld rites was a Greek/Etruscan innovation.

[61] Varro, *De lingua Latina* 7.6, cited in Macrobius 1.16.18: *mundus cum patet, deorum tristium atque inferum quasi ianua patet*; Plutarch, *Romulus* 11 on the *mundus* situated in or around the imperial *comitium* (Plutarch describes the *mundus* as the compass point (*to kentron*) around which the *pomerium* was drawn; Servius, *Ad Aeneidem* 3.134: *quidam aras superorum deorum uolunt esse, medioximorum, id est marinorum focos, inferorum mundos*; cf. Ovid, *Fasti* 4.821ff.

which cut across several different genres of literature, and serve to remind us of how cultures can derive meaning from their landscapes. It is perhaps telling that the city of Sheffield is sometimes described as the topographical twin city of Rome in the United Kingdom; although its own seven hills are considerably larger than those of Rome, Sheffield's hilly landscape has seldom been described as a locus for the meeting of heaven and hell.[62] The landscape itself only provides the scaffolding for topographical meaning; ancient Rome provided the voices, the history, the myths and the rituals that gave it substance.

Invasion and expulsion, rise and fall, erection and destruction: pollution and propriety were important indices in the topography of the city of Rome. Even in antiquity, this city was built on ancient myths, told and retold, about the establishment of moral codes, the preservation of divine favour, and the removal of elements that threatened that favour. Its traditions and rituals, evocative features of Roman historiography, rhetoric and epic, were fundamental to how the city represented, policed and eliminated its most threatening criminals – whether parricides, would-be tyrants, treacherous slaves, bad emperors or unchaste Vestals (further, see Schultz, this volume). That these processes could be likened to the evacuation of sewage, the curing of disease or even consignment to Hades, and that the physical infrastructure responsible for these processes could be perceived as a source of contamination, indicate necessarily and incontrovertibly that pollution and purity were central factors in the way the city represented and policed crime from a very early stage in its history.

---

[62] On Sheffield ('the ugliest town in the Old World') and its attested seven hills like Rome, see George Orwell's 'The road to Wigan Pier', in Orwell and Angus (1968) 191; discussed at http://mdfs.net/Docs/Sheffield/Hills (accessed July 2011).

7 | On the burial of unchaste Vestal Virgins

CELIA E. SCHULTZ

For an investigation into Roman methods of handling ritual impurity, the priesthood of Vesta provides an excellent case study. Not only were the priestesses responsible for performing numerous rites that cleansed the city and people of pollution,[1] but they were also capable of bringing pollution upon the Roman state by failing to maintain their own sacred status. The Vestals purified Rome, and Rome's purifiers had to be pure and perfect themselves. Candidates for the priesthood, all young girls between the ages of six and ten, had to be free from any physical defect, including speech impediment or hearing loss. The new priestess had to be *patrima et matrima*, a common ritual requirement in Roman religion – not just for Vestals – that meant both of the candidate's parents had to be alive.[2] The most important aspect of a Vestal's purity was, of course, her virginity, which she was required to maintain for the thirty years of her service to the Roman people. As was argued in a seminal article by Mary Beard (1980), who freely acknowledges her debt to Mary Douglas,[3] a Vestal's liminal existence between girlhood and fully realized adult womanhood was the crucial marker of her sacredness.

Some of what follows has already appeared in Schultz (2010), which is intended as a companion to this chapter. Those portions of the earlier argument reprinted here appear with the permission of the *Journal of the American Academy of Religion*. Abbreviations used within this chapter are as follows: *OLD = Oxford Latin dictionary*, ed. P. G. W. Glare. Oxford (1982); *RE = Paulys Real-Encyclopädie der classischen Altertumswissenschaft*, eds G. Wissowa, W. Kroll, K. Mittelhaus, and K. Ziegler. Stuttgart (1894–1978); *MRR =* Broughton, T. R. S. *The magistrates of the Roman Republic I–III.* Atlanta: Scholars Press (1951–86); *SBA =* Shackleton Bailey, D. R. (ed.) *Cicero's Letters to Atticus* (7 volumes). Cambridge University Press (1965–71).

[1] A strong argument for purification being the primary concern of the college has been made by Wildfang (2006) 1–36.

[2] For an explanation of the requirement, using the variant forms *patrimes et matrimes*, see the relevant entries (82L and 113L) in W. M. Lindsay's 1913 edition in the Bibliotheca Teubneriana series of Paulus' eighth-century redaction of Festus' *Glossaria Latina*, a late second-century dictionary, which is itself a redaction of an earlier dictionary compiled by the Augustan-age scholar Verrius Flaccus. The status was also required of young men and women who took part in observances such as weddings (as attendants, Festus, *Glossaria Latina* 282L), public expiatory rituals (e.g. Livy 37.3.6), and the opening of public games (Cicero, *De haruspicum responso* 23). The *locus classicus* for the requirements for candidates for Vesta's priesthood is Aulus Gellius 1.12.1–8.

[3] In the 'affectionate critique' in Beard (1995) 167 especially: 'What Mary Douglas had done for the pangolin and for the prohibitions of Leviticus, I could do for the Roman Vestals.' Though

The purificatory function of the priestesses, whether acting in the seclu-sion of the goddess's temple or out in public view, is perhaps most clearly illustrated by a number of ritual obligations that required the priestesses to use water – for the Romans the purifying substance *par excellence* (Edlund-Berry (2006) 162–80) – of the purest, most unpolluted kind. Two important such tasks were the washing of sacred items within the temple and the prepa-ration of materials needed for other purification rites. The priestesses were required to transport water from sacred springs back to Vesta's temple in special containers that could not be set down (further on this, see Fantham, this volume). Thus the water could never become still before it was used, nor could it come into contact with profane earth.[4]

So central were the Vestals to the stability of the *pax deorum* that any failure to perform their duties or observe the restrictions placed upon them had serious repercussions for the state, and so was dealt with in a heavy-handed manner. The Vestal who allowed the goddess's fire to go out was beaten.[5] The Vestal who broke her vow of chastity was buried alive. Once accusations of sexual transgression surfaced, the accused priestess was immediately removed from her duties, even before her case could be investigated – so important was it that she perform them only in a pure state.[6] She was then tried by the pontifical college.[7] If convicted, the polluted priestess was carried through the city in a funeral-like procession.[8] She was dressed in funereal garb, and the covers of the litter on which she rode were tied down so that no one could see her or hear her cries. She was attended by grieving friends and relatives, while the people looked on in mournful silence. When the procession arrived at the Campus Sceleratus[9] just inside the Colline Gate, the priestess descended a ladder into an underground chamber that had been hollowed out of a small ridge. The bystanders averted their eyes

---

some aspects of her initial analysis have been set aside, the basic argument of Beard (1980) is now widely accepted.

[4] Servius, *Ad Aeneidem* 11.339; Festus, *Glossaria Latina* 152L; Wildfang (2006) 10–11.

[5] Dionysius of Halicarnassus 2.67.3; Paulus ex Festus, *Glossaria Latina* 94L. We do not know what punishment, if any, was meted out to other public priests or priestesses for dereliction of duty.

[6] Livy 8.15.7–8, Dionysius of Halicarnassus 2.68.3–5, Plutarch, *Numa* 10.4.

[7] The manner in which accusations of Vestal promiscuity were handled is unique within Roman judicial procedure. It was not dealt with by criminal law, nor is there a parallel process for other priests (so far as we know) under religious law. The issue of the *pontifex maximus*'s jurisdiction in cases of this sort has been addressed by Cornell (1981) and Lovisi (1998).

[8] What follows is a composite picture of the ceremony drawn from the accounts provided by Dionysius of Halicarnassus 2.67.4 and Plutarch, *Numa* 10.4–7.

[9] In addition to the sources in the previous note, see Festus, *Glossaria Latina* 448L; Livy 8.15.8; Servius, *Ad Aeneidem* 11.206. The Campus Sceleratus has not been identified conclusively, but it should probably be identified with the area just south of the gate (Coarelli (1993a)).

as she climbed down into the cell, where she would find certain ritual items set out for her: some oil, water and milk; a bit of bread; a burning lamp; a bed. Then the ladder was pulled up and the opening sealed so that no trace remained.

The punishment of unchaste Vestals attracted considerable attention from ancient writers, and continues to appeal. A perennial favorite topic for scholarly investigation, the issue has been addressed at least twice in the last few years: in a monograph by R. L. Wildfang (2006) and in a significant article by H. Parker (2004).[10] Though these studies take very different approaches to the Vestals in general,[11] and to more specific questions of the significance of their duties and the metaphorical meaning of their virginity, there is agreement on the nature of the punishment inflicted on Vestals convicted of violating their vow of chastity. Wildfang and Parker both understand the live interment as a form of sacrifice, similar to, but not exactly identical with, the live burial of pairs of Gauls and Greeks that the Romans performed on three occasions during the Republic (and perhaps once in the early Empire).[12] An opposing point of view has been put forward most recently by Gradel (2002) 237, who argues that the punishment visited upon the Vestals was 'simply a form of execution', and the rite of burying Gauls and Greeks was not a sacrifice.

In what follows, I propose an interpretation of the ancient sources that does not line up with either of the positions I have just sketched out. The live interment of a Vestal was not just another form of execution: the priestess's status, the form of her punishment, the body responsible for adjudicating her case and the location of her burial all militate against this interpretation. Nor was the punishment of an unchaste Vestal a sacrifice. Though similar in some important ways to the only unequivocal case of human sacrifice among the Romans, that is, the live burial of Gauls and Greeks, the punishment of the Vestals was not identical to it in the minds

[10]   Another recent monograph on the Vestals by Martini (2004) puts great importance on the issue of Vestal *incestum*, but does not touch on the question of whether or not the live interment of convicted priestesses was a sacrifice. Mekacher (2006) focuses on material evidence for the priesthood and is not interested in the issue of Vestal *incestum*.

[11]   Wildfang (2006) considers the Vestals as an historical phenomenon. Parker (2004) takes an anthropological approach.

[12]   An oblique reference to human sacrifice in the imperial period comes from the Elder Pliny, *Naturalis historia* 28.12: *boario vero in foro Graecum Graecamque defossos aut aliarum gentium cum quibus tum res esset, etiam nostra aetas vidit.* Though Pliny does not mention Gauls, it is generally assumed he is speaking of the same rite that was observed in the Republic. For doubt about the truthfulness of Pliny's statement, see Gradel (2002) 236–7, who argues that it is an unintentional repetition of a sentence in the author's source – a result of Pliny's hasty method of composition.

of the Romans. A close examination of the language our sources use to talk about these two instances of live interment indicates that the ancients had more than one intellectual category for rites that required a human death. This is quite different from modern, popular usage that commonly describes any religiously motivated killing of a human being as a 'sacrifice'.

Fortunately, many historians of religion and anthropologists make a distinction, which will be followed here, between human sacrifice and a more broadly defined category of ritual murder that fits very well with distinctions drawn in the ancient sources.[13] In its widest sense, ritual murder is a rite performed repeatedly in specific circumstances in a prescribed fashion that requires the killing of a human being. Ritual murder is not necessarily part of the regular worship of a particular deity or group of deities, though it might be. Human sacrifice, on the other hand, is a special case of ritual murder. It is, of course, killing in a prescribed manner performed in a specific circumstance, but human sacrifice must also be part of the worship of the gods. It must have a recipient: a sacrifice must be offered to some being or other. In sum, all human sacrifice is ritual murder, but not all ritual murder is human sacrifice. The live interment of unchaste Vestals is certainly ritual murder, but it is not human sacrifice.

In her extended treatment of the priesthood of Vesta, Wildfang (2006) 58–9 argues that the live interment of a Vestal was understood by the Romans as a sacrifice; more specifically, the former priestess became an expiatory offering to the goddess she used to serve. Wildfang's interpretation is based on the form of the ceremony, which involves the same method of killing as the ritual murder of Gauls and Greeks. Wildfang sees the choice of a live burial in an underground chamber (*cubiculum subterraneum*; Pliny, *Epistulae* 4.11.9) as a reflection of Vesta's close association with the earth.[14]

A close analysis of the language of our ancient sources, however, indicates that the Romans did not see things this way. No ancient author ever refers to the priestesses as *hostiae* or *victimae*, nor are the Vestals described as offerings to Vesta or any other god. More significant is the fact that the punishment itself is never discussed in the terms of Roman sacrifice. Descriptions of the burial are most often terse expressions like *viva defossa est, terra obrutae sunt* or *zōsa katoruttetai*.[15] Where our sources use one of the many Latin words for 'kill', the chosen term never has any religious overtones, for example

---

[13] The discussion by Hughes ((1991) 1–3) is very helpful in clarifying the distinction. Hughes himself, however, defers to popular custom, using 'sacrifice' to describe actions that do not meet the technical definition he has just laid out.

[14] E.g. Ovid, *Fasti* 6.267 and 457–60; Dionysius of Halicarnassus 2.66.3.

[15] E.g. Livy, *Periochae* 2 and 14; Plutarch, *Numa* 10.4; Jerome, *Chronica* a. 216 a. C.

necare (Adams (1973) 280–90). No ancient author uses Greek *thuein* or Latin *immolare* or *sacrificare*, the words most commonly applied to the ritual slaughter of animals.

The absence of sacrificial vocabulary in descriptions of the punishment of Vestals stands in sharp contrast to accounts of another form of live interment practised at Rome. Under the direction of the Sibylline Books, the Romans buried alive pairs of Gauls and Greeks, one man and one woman each, on as many as three occasions during the course of the Republic, around 228 BC, in 216[16] and in 114/113.[17] It is certain that the latter two instances coincided with the only two convictions of Vestal Virgins for unchastity in the period between the Second Carthaginian War and the end of the Republic; it is possible that the earliest burial of Gauls and Greeks also followed the discovery of an unchaste Vestal.[18] The pattern of events in 216 and 114/113 suggests that the latter ritual murders were prompted by the priestesses' convictions. Yet despite their apparent similarity and temporal proximity, the two types of live interment had different significances, as is made clear by Livy's description of the events of 216 in reaction to Roman losses at Cannae and Canusium (22.57.2–6):

> Territi etiam super tantas clades cum ceteris prodigiis, tum quod duae Vestales eo anno, Opimia atque Floronia, stupri compertae et altera sub terra, uti mos est, ad portam Collinam necata fuerat, altera sibimet ipsa mortem consciverat; L. Cantilius scriba pontificius, quos nunc minores pontifices appellant, qui cum Floronia stuprum fecerat, a pontifice maximo eo usque virgis in comitio caesus erat ut inter verbera exspiraret. Hoc nefas cum inter tot, ut fit, clades in prodigium versum esset, decemviri libros adire iussi sunt et Q. Fabius Pictor Delphos ad oraculum missus est sciscitatum quibus precibus suppliciisque deos possent placare et quaenam futura finis tantis cladibus foret. Interim ex fatalibus libris sacrificia aliquot extraordinaria facta, inter quae Gallus et Galla, Graecus et Graeca in foro boario sub terram vivi demissi sunt in

---

[16] Livy 22.57.2–6; Plutarch, *Fabius* 18.3; Cassius Hemina fr. 32 Peter.

[17] In addition to Plutarch, *Quaestiones Romanae* 83, discussed below, the most important sources for the events of 114 and 113 are Livy, *Periochae* 63; Fenestella *apud* Macrobius, *Saturnalia* 1.10.5; Asconius on *Pro Milone* 45C–6C; Dio 26, frag. 87; Obsequens 37; Orosius 5.15.20–2.

[18] The earliest known burial of Gauls and Greeks is consistently dated in the sources to 228 BC: Fraschetti (1981) 65 n. 33; Eckstein (1982) 76–7, with additional sources. Uncertainty about its connection with Vestal unchastity derives from inconsistency among the sources (Livy, *Periochae* 20; Dionysius of Halicarnassus 2.69.1–3; Valerius Maximus 8.1, abs. 5; Pliny, *Naturalis historia* 28.12; St Augustine, *De civitate dei* 10.16, among others) about the name of the priestess, the date of her trial, and even whether she was actually convicted of *incestum*. The most detailed argument for a link between the burial of 228 and a Vestal scandal was put forward by Cichorius ((1922) 7–21), whose position has been widely accepted, with Eckstein (1982) an exception.

locum saxo consaeptum, iam ante hostiis humanis, minime Romano sacro, imbutum.

These terrible disasters aside, the Romans were also alarmed by a number of prodigies, especially the conviction that year of two Vestals, Opimia and Floronia, on charges of sexual misconduct. One had received the traditional penalty of being buried alive near the Porta Collina, and the other had committed suicide. The man guilty of the misconduct with Floronia, Lucius Cantilius, was the secretary of a pontiff—one of those officers that are these days called minor pontiffs—and he was so badly beaten in the Comitium, at the orders of the pontifex maximus, that he expired under the lash. Occurring as it did along with all the other misfortunes, this piece of sacrilege was naturally interpreted as a portent, and the decemvirs were therefore instructed to consult the Books. Quintus Fabius Pictor was also sent to the oracle at Delphi to ask what prayers and acts of supplication they could employ to appease the gods, and further enquire when the Romans would see an end to their great disasters. Meanwhile a number of outlandish sacrifices were conducted on instructions from the Books of Fate. These included a Gallic man and woman, and a Greek man and woman, being interred alive in the Forum Boarium, in a spot enclosed with stones which had already been the scene of this very un-Roman practice of human sacrifice.[19]

Here the killing of the Vestal who did not pre-empt punishment by committing suicide is described in neutral terms (*necata fuerat*). The sacrificial nature of the second interment, that is, the burial of the Gauls and Greeks, is clear from the identification of the rite itself as one of the *sacrificia extraordinaria* ordered by the Sibylline Books, and of the underground room where they were buried as a place that had already seen human victims (*hostiis humanis*).

Plutarch makes a similar distinction in the eighty-third of his *Quaestiones Romanae*, where he wonders why, upon hearing that a foreign tribe called the Bletonesii[20] had sacrificed a man, the Romans initially planned to punish the leaders of the tribe, but when it was learned that the sacrifice had been done according to custom (*epei de nomōi*) the Romans let the leaders go free, only forbidding them to perform such a rite in the future. Plutarch finds this remarkable since the Romans themselves had buried alive pairs of Gauls and Greeks in the Forum Boarium not too many years before. Plutarch here probably refers to the outlawing of human sacrifice in 97 BC, recorded by the Elder Pliny (*Naturalis Historia* 30.3.12). Thus the burial of Gauls and Greeks

[19] Trans. by J. C. Yardley (Oxford World's Classics, 2006).
[20] *Aliunde ignoti.* Most scholars follow Cichorius (1922) 9–10 in identifying the tribe with the Bletisa in Spain.

would be the sacrifice of 114/113. Following the format he has established for his *Quaestiones Graecae* and *Romanae*, Plutarch then offers a number of possible answers to the question he raises, each answer itself phrased as a question. The last of these asks whether the reason is that the Romans thought that human sacrifice (here he uses sacrificial vocabulary: *to men theois thuein anthrōpous*) performed because of tradition and custom (*ethei kai nomōi*) was wrong, yet acceptable if ordered by the Sibylline Books. He goes on to recount the revelation of three unchaste Vestals, whom he says were punished (*ekolasthēsan*). Disturbed by these events, Plutarch says, the Romans consulted the Sibylline Books, which in turn ordered the burial of Gauls and Greeks.

While it may appear at first glance that Plutarch provides two instances of human sacrifice, one of which was sparked by the other, on closer inspection it becomes clear that the story of the Vestals intervenes only as background to the burial of the Gauls and Greeks. Plutarch's introductory question had proposed that the role of the Sibylline Books made human sacrifice acceptable, and the Sibylline Books were only involved with the second interment. Plutarch does not discuss the burial of the Vestals as a sacrifice.

So, why not sacrifice a Vestal Virgin? There were, no doubt, several factors that prevented the Romans from offering an unchaste priestess as a sacrifice to the gods. At the most basic level, however, the very nature of the priestess's transgression would have made it impossible no matter what other concerns arose. The sources emphasize the convicted priestesses' impure, polluted state, frequently describing them with phrases like *in corruptione deprehensa*[21] and *tas diaphthareisas*.[22] As mentioned above, a priestess's purity was so central to her duties that she was not permitted to perform them if it was under any suspicion. The charge against the accused was the *crimen incesti*. Though the Latin is closely related to English 'incest', it is not synonymous with it. An individual who is *incestus* is in the state of not being *castus*, of being ritually unclean, of not properly observing ritual requirements,[23] not just sexual abstinence but, in other cults, dietary restrictions and the like. A priestess who was guilty of *incestum* was no

---

[21] Jerome, *Chronica* a. 276 a. C.

[22] Dionysius of Halicarnassus 3.67.3. Cf. Plutarch, *Quaestiones Romanae* 83 = *Moralia* 284B.

[23] *OLD*, s.v. For examples of *castus* referring to dietary restrictions in the cult of Jupiter, see Schultz (2006) 55–7. The most famous case of *incestum* was P. Clodius' infiltration of the Bona Dea's December ritual, which was supposed to be attended only by women, in 62 BC (Cicero, *Epistulae ad Atticum* 1.13.3 and 1.16.9–10 = *SBA* 13.3 and 16.9–10; Scholia Bobiensia on Cicero p. 89.18–28 Stangl; Valerius Maximus 8.5.5; Plutarch, *Cicero* 28.1–29.7). See also the discussion in Lennon, this volume, p. 51.

longer ritually pure, which surely would have made her an unacceptable offering.

As a point of comparison, animal sacrificial victims had to meet specific criteria at all points in order to be suitable gifts for the gods. Different deities had preferences beyond simply the species of animal to which they were partial, requiring victims that were male or female; mature or juvenile; intact or castrated; pregnant or not; black, white or red.[24] The parallel between the necessary perfection of a priest and the necessary perfection of a sacrificial victim is even made explicit in *Controversia* 4.2 of the Elder Seneca. The issue at hand in this text is the removal from his priesthood of L. Caecilius Metellus who, as *pontifex maximus*, was blinded after rescuing sacred items out of the burning temple of Vesta.[25] He was not supposed to have entered the temple's inner sanctum and, although he acted in the goddess's interest, he was blinded as a punishment for seeing what he should not. The argument in favour of taking Metellus's priesthood away points out that the perfection demanded for offerings to the gods must certainly extend to their priests: *sacerdos non integri corporis quasi mali ominis res vitanda est. hoc etiam in victimis notatur, quanto magis in sacerdotibus.* Whether pollution came upon an individual for good reasons or bad, offerings to and the attendants of the divine needed to be perfect.

If the live interment of unchaste Vestals was not a sacrifice in the opinion of ancient authors, the question remains as to how they understood it. Unfortunately, none of our sources directly addresses the issue, so we are left to look for parallels with other Roman practices. We have already considered the rite to which it bears the closest relationship – the sacrifice of Gauls and Greeks – and have seen that the ancients did not conceive of these two types of ritual murder as identical.

The punishment of unchaste Vestals has less obvious similarities to a third type of ritual murder practised by the Romans: the disposal of hermaphrodites (on which, see MacBain (1982) 127–35). On fourteen occasions between 209 and 92 BC,[26] androgyne infants and children – the oldest

---

[24] The most important sources are collected by Wissowa (1912) 412–16.

[25] Pliny, *Naturalis historia* 7.141; Cicero, *De senectute* 30. MRR I.218, s.a. 243; E. Groag, *RE* III.1203–4, s.v. 'Caecilius (72)'.

[26] An additional two instances, in 104 and 99 BC, may have been passed over by Obsequens in his epitome of Livy's history. Though neither Obsequens 43 nor 46 mentions hermaphrodites, both entries preserve a record of the traditional expiation of androgyne prodigies: donations made to Ceres and Proserpina and a wandering chorus of twenty-seven *virgines*. See MacBain (1982) 131–2. No hermaphrodites are reported after 92 BC. Instances where individuals were reported to have sudden, miraculous sex changes are not included here, though such reports may refer to the same phenomenon.

is identified as being sixteen (Livy 31.12.6) – were included among the prodigies reported to the Roman Senate. It appears that a routine method for handling such discoveries was developed, probably after the successful expiation of 207.[27] The children were drowned by the *haruspices* in the sea, though in one instance a hermaphrodite discovered in the area around Ferentinum was disposed of in a river.[28] The Romans then observed a regular set of expiatory rituals, most importantly offerings made to Ceres and Proserpina by matrons of the city and the procession of a chorus of twenty-seven virgins.

Both hermaphrodites and unchaste Vestals presented the Romans with a contamination of sexual categories: the former combined two sexes in a single body, the latter combined virginity and violation. In addition, the presence of both types of polluted individuals was taken as an indication of dire things to come for the Roman people as a whole.[29] In contrast, the Gauls and Greeks appear to have been free of any ritual taint, and their burial was intended to remedy a rupture in the *pax deorum*. Finally, neither the drowning of the hermaphrodites nor the live interment of Vestals is described by any ancient source as a sacrifice, whereas the burials of the heterogeneous pairs are explicitly identified in this way.

In the disposal of hermaphrodites and unchaste Vestals, we see efforts by the community to remove the individual from the profane world in a way that minimized the responsibility of the community itself for the death. Plutarch attributes this motivation to the Romans explicitly with regard to the burial of Vestals.[30] That the drowning of hermaphrodites was meant to remove a polluting presence from the earth is emphasized in Livy's report of the first such instance in 207 BC: *id vero haruspices ex Etruria acciti foedum ac turpe prodigium dicere: extorrem agro Romano, procul terrae contactu, alto*

---

[27] The one hermaphrodite reported prior to 207 BC was apparently not singled out for special treatment (Livy 27.11.4–6).

[28] Obsequens 27 presumably refers to the nearby Sacco river.

[29] It is certain that the hermaphrodites were taken as *prodigia*. See, for example, Livy 27.37.5–7. That an unchaste Vestal was also considered a *prodigium* was argued by Wissowa (1923–4) 203. Several scholars have taken up this position as well, most recently Parker (2004) 586. The language of Livy's description of the events of 216, however, indicates that the unchaste Vestals were not prodigies in their own right, but that the revelation of their transgressions caused excessive anxiety in a population already frightened by disastrous military defeats and numerous supernatural occurrences. Wildfang (2006) 56 is most probably correct in her assertion that human actions are not, in and of themselves, prodigies. An individual's actions can disrupt the *pax deorum*, and this rupture is then indicated by the appearance of *prodigia*, but the actions themselves are not *prodigia*.

[30] Plutarch, *Numa* 10.5 and *Quaestiones Romanae* 96 = *Moralia* 286F–287A.

mergendum. vivum in arcam condidere provectumque in mare proiecerunt (27.37.6).[31]

This passage from Livy raises another point of similarity between the rituals for unchaste Vestals and hermaphroditic children: both treat their objects as if they were dead, even before the actual act of murder. As we have seen, convicted priestesses were dressed in funeral clothing and led through the city in a procession very much like the aristocratic funeral procession to which the Vestal's status would have entitled her had she lived to a ripe old age. The children, none of whom seems to have come from wealthy or locally prominent families, were put into *arcae*, coffins or containers in which the bodies of the poor were carried (*OLD*, s.v. (2)). By enclosing the children in this way, the Romans kept the sea, as well as the land, free of contamination, if we may extrapolate from Cicero's description of another Roman practice: the placing of those who murdered a close relative or patron (parricides) into a sack before drowning them:[32]

> Insui voluerunt in culleum vivos atque ita in flumen deici. O singularem sapientiam, iudices! Nonne videntur hunc hominem ex rerum natura sustulisse et eripuisse cui repente caelum, solem, aquam terramque ademerint ut, qui eum necasset unde ipse natus esset, careret eis rebus omnibus ex quibus omnia nata esse dicuntur? Noluerunt feris corpus obicere ne bestiis quoque quae tantum scelus attigissent immanioribus uteremur; non sic nudos in flumen deicere ne, cum delati essent in mare, ipsum polluerent quo cetera quae violata sunt expiari putantur; denique nihil tam vile neque tam volgare est cuius partem ullam reliquerint. Etenim quid tam est commune quam spiritus vivis, terra mortuis, mare fluctuantibus, litus eiectis? Ita vivunt, dum possunt, ut ducere animam de caelo non queant, ita moriuntur ut eorum ossa terra non tangat, ita iactantur fluctibus ut numquam abluantur, ita postremo eiciuntur ut ne ad saxa quidem mortui conquiescant.

> They therefore stipulated that parricides should be sewn up in a sack while still alive and thrown into a river. What remarkable wisdom they showed, gentlemen! Do they not seem to have cut the parricide off and separated him from the whole realm of nature, depriving him at a stroke of sky, sun, water and earth – and thus ensuring that he who had killed the man who gave

---

[31] 'Soothsayers brought in from Etruria declared it to be a foul and loathsome prodigy, that the child should be taken from Roman territory, kept from all contact with the ground, and sent to the bottom of the sea. It was then placed, alive, in a box, taken out to sea, and thrown in' (trans. J. C. Yardley).

[32] A detailed investigation of the range of murders that fell under the heading of *parricidium* is offered by Lassen (1992). The offender was put into a sack (*culleus*) with a dog, a rooster, a snake and a monkey (*Digesta*, 48.9.9). See Briquel (1980) for an extended discussion.

him life should himself be denied the elements from which, it is said, all life derives? They did not want his body to be exposed to wild animals, in case the animals should turn more savage after coming into contact with such a monstrosity. Nor did they want to throw him naked into a river, for fear that his body, carried down to the sea, might pollute that very element by which all other defilements are thought to be purified. In short, there is nothing so cheap or so commonly available that they allowed parricides to share in it. For what is so free as air to the living, earth to the dead, the sea to those tossed by the waves, or the land to those cast ashore? Yet these men live, while they can, without being able to draw breath from the open air; they die without earth touching their bones; they are tossed by the waves without ever being cleansed; and in the end they are cast ashore without being granted, even on the rocks, a resting-place in death.[33] (*Pro Roscio Amerino* 70–2)

Although the Vestals were not placed in any sort of container before being buried, the Romans seem to have made efforts to limit their contact with the world in the interval between conviction and burial. Plutarch stresses that the women were neither seen nor heard on their way to the Campus Sceleratus (*Numa* 10.6–7). They were heavily veiled and were carried through the streets of the city in a litter covered so that their cries were muffled. With the exception of weeping friends and relatives, the crowd attending the procession remained silent and did not watch as the priestess entered the underground chamber.[34]

The choice of live interment over drowning for the Vestals remains unexplained. The difference in method should not be taken as an indication that the punishment meted out to a convicted priestess was just another form of secular execution. While some offences against the gods, particularly those that involved the whole community such as the desecration of a public temple, were punishable under Roman criminal law, more personal offences were outside its jurisdiction (Cornell (1981) 29). When criminal law did demand capital punishment, the Romans preferred strangulation, beating or precipitation (the technical term for throwing someone off a cliff) for dispatching secular criminals.[35] As far as we know, for the Romans live burial was only practised as ritual murder. The very method by which

---

[33] Trans. by D. H. Berry (Oxford World's Classics, 2000).

[34] On the transportive nature of Roman purification rituals in general, see Lennon, this volume, pp. 55–6.

[35] For more on punishment types, see Bradley's chapter 6 in this volume, especially pp. 110–11 where he discusses the case of a woman, not explicitly identified as a Vestal, who survived precipitation after conviction on a charge of *incestum*.

they died marks the unchaste Vestals as still existing within the sacred realm.

The sacred nature of live interment is further underlined by the fact that the Vestals and the Gallic and Greek victims were buried within the city walls. Generally, burials were restricted to areas outside the walls,[36] though there was a tradition of prominent individuals and their relatives buried within the city in the early Republic (Cicero, *De legibus* 2.58). Into the imperial period, burial within the city continued to be an honour given to Vestals, and to emperors as well since, as Servius reports, they were above the law (*quia legibus non tenentur*). Servius says this unusual status pertained even to those Vestals who had committed transgressions since they, too, were buried *in urbe*.[37]

These live interments, all within the *pomerium*, the sacred boundary of the city, share certain features with another religious habit the Romans practised within sacred boundaries, that is, the burial of defunct cult materials. Legal sources tell us that once an item was consecrated to a god (or gods) it became subject to divine law, that is, it belonged to the divine and could not be returned to the realm of the profane.[38] A result of this eternal sacrality was that obsolete votive offerings, organic sacrificial material, and even architectural elements from a temple that were polluted through accidental or intentional violence could not be disposed of as if they were regular refuse. Instead, these items were routinely buried within sanctuary precincts.[39] Like a broken statuette or bones left over from a burned sacrifice, an unchaste Vestal might be seen, in a sense, as a decommissioned sacred entity, as something polluted but still sacred: not profane, but not ritually perfect (and therefore not ritually usable) either.

It is even possible that the Vestals were, in fact, officially decommissioned before burial. A fragment of the Elder Cato's speech *De auguribus* may preserve a reference to the *exauguratio* of unchaste Vestals:

---

[36] The interdiction is included in the XII *Tabulae* (Cicero, *De legibus* 2.58); Bodel (1994) 33–5.

[37] Servius, *Ad Aeneidem* 11.206. Lovisi (1998) 728–9 highlights the distinction Servius makes between *habent sepulchra* and *in Campo Scelerato obruebantur*, and raises the question of whether the Romans perceived the Campus Sceleratus as a sort of cemetery. For an argument that Servius is wrong, that the intramural burial only applied to unchaste Vestals, see Mekacher (2006) 32.

[38] Gaius, *Institutiones* 2.2–5; Watson (1992) 55–7.

[39] For a catalogue of the kinds of items disposed of in votive deposits, see Bouma (1996) III.132–49. The typological analyses of votive deposits of Fenelli (1975) and Comella (1981) are limited to man-made votive material and do not take into account sacrificial remains. Glinister (2000) offers an excellent discussion of the ritual burial of architectural items.

> Probrum viriginis Vestalis ut capite puniretur, vir qui eam incestavisset ver-
> beribus necaretur, lex fixa in atrio Libertatis cum multis alis legibus incendio
> consumpta est, ut ait M. Cato in ea oratione quae de auguribus inscribitur.
> adicit quoque virgines Vestales sacerdotio exaugurat[as] . . .

The text presented here is that of H. Jordan's 1860 Teubner edition of the
extant fragments of Cato (LXVII: *De auguribus*).[40] *Exaugurare* is a very rare,
technical word for an augural action, the opposite of *inaugurare*, by which
something was removed from its full sacral status. The verb and its substan-
tive appear in only a handful of contexts. The best-known *exauguratio* is the
repurposing of sanctuaries of other gods on the Capitoline Hill to make way
for the temple to Jupiter Optimus Maximus during the reign of Tarquinius
Superbus.[41] Servius twice uses the term to refer to a rite intended to bring
about the destruction of a city by persuading the gods to abandon it, and
explains that the ritual is identical to the rite of *inauguratio* by which cities
were founded (*eodem ritu quo conditae*).[42]

   Jordan's reconstruction of the last word in the fragment quoted above
draws significant support from a comparable passage in Gellius, where he
discusses honours bestowed on the Vestal Taracia, who, in the mythical
early period of Rome's history, bestowed the Campus Martius upon the
Roman people. In addition to numerous other distinctions, upon reaching the age of forty, Taracia was permitted to leave the priesthood of her
own accord and marry (*ius ei potestasque exaugurandi atque nubendi facta
est munificentiae et beneficii gratia*, 7.7.4).[43] As one might expect, given
the permanence of sacrality under Roman religious law, exauguration does
not appear to have completely removed the Vestal from the sacred sphere.
Several ancient sources emphasize the convicted priestess's sanctified state,
such as Servius' assertion (on *Aeneid* 11.206) that convicted priestesses were
buried within the city because they were not bound by law, and the Younger
Pliny's description of the Vestal Cornelia, condemned by Domitian in AD 83,
as she brushed off assistance offered to her by the executioner as she
descended into the underground chamber, lest she be contaminated by his

---

[40] A less sensible, but not negligible, reading of the last sentence is offered by W. M. Lindsay in his
1913 Teubner edition of Festus, *Glossaria Latina* 277L where our fragment is preserved: *adicit
quoque virgines Vestales sacerdotio ex augurali . . .*

[41] Cato, *Origines* 1.23 Jordan = Festus, *Glossaria Latina* 160L; Livy 1.55.2–3.

[42] Servius, *Ad Aeneidem* 4.212, also 2.351.

[43] Cf. Scriptores Historiae Augustae, *Marcus Aurelius Antoninus (Caracalla)* 4.4.

foul touch: *foedum contactum quasi plane a casto puroque corpore novissima sanctitate reiecit* (*Epistulae* 4.11.9).[44]

In conclusion, ritual murder, broadly defined, was not unfamiliar to the Romans. Indeed, Roman society felt the need for it more than twenty times in the 150 years between 230 and 80 BC. Though there are many important similarities among the methods used for hermaphrodites, convicted Vestals and the pairs of Gauls and Greeks, there are equally important differences, the most significant of which is that these three rituals did not serve a single purpose. The only instances identified by the ancients as sacrifice are the live burials of Gauls and Greeks. It is also the case that the Gallic and Greek victims are never discussed as aberrations or perversions of the natural order, unlike the other two groups considered here. In contrast, the vast majority of ritual murders were intended to remove from the Roman community, in a way that left no one to blame, individuals tainted by an irreparable pollution: the mixing of sexual categories. The very presence of hermaphroditic children and Vestals no longer virgins tainted the Roman state as a whole. Such polluting presences needed to be removed permanently and bloodlessly so that the contagion they harboured could spread no further.

---

[44] Cf. Plutarch, *Numa* 10.2 and *Quaestiones Romana* 96 = *Moralia* 286F–287A. The contagion of the executioner's touch is also stressed in Seneca, *Controversiae* 1.3, on which see Bradley, this volume, pp. 110–11.

PART II

Modernity

8 | Fra Girolamo Savonarola and the aesthetics of pollution in fifteenth-century Rome

ALESSIO ASSONITIS

Facts regarding Fra Girolamo Savonarola's early life and training can be gleaned from the quasi-hagiographic accounts written in the first half of the sixteenth century by Placido Cinozzi (*c.* 1503), Benedetto Luschino (1518–23), the anonymous author of the *Vita latina* (*c.* 1527), Giovanfrancesco Pico della Mirandola (1530) and the so-called 'Pseudo-Burlamacchi' (post-1530).[1] Also valuable are the *en passant* references that the Frate, as he was commonly known, made about his youth in his sermons, letters, treatises and poems – the majority of which have been worked into scholarly biographies published in more recent times.[2] Ancient and modern biographers alike addressed the various degrees of impact that Michele Savonarola and his ideas had on his grandson Girolamo.[3] A devout and learned physician, Michele carefully monitored Girolamo's education, introducing his grandchild to the study of Latin and instilling in him those precepts of moral integrity that he fervently believed in and attempted to observe.[4] Having resided in the patriarchal *domus* in Ferrara through his adolescence, Girolamo was not immune to Michele's fulminations against contemporary culture and, at the same time, was well aware of the power and esteem that Michele enjoyed as personal physician of Niccolò and Lionello d'Este. This rigid intellectual grounding – humanist in all effects, but not

---

I would like to thank Stefano Dall'Aglio and Sheila Barker for their help and suggestions. The manuscript sources I have used in this chapter are abbreviated as follows: Biblioteca Nazionale Centrale, Florence, *Conventi Soppressi* (BNCF, *Conv. Sopp.*); Biblioteca Medicea Laurenziana, Florence, *San Marco* (BML, *San Marco*); Biblioteca Apostolica Vaticana, Rome, *Ottoboniano Latino* (BAV, *Ottob. Lat.*); Biblioteca Angelica, Rome (BA); Archivio Generalizio dell'Ordine dei Predicatori, Rome, *Serie XIV* (AGOP, XIV).

1 See Cinozzi (1898) 2; Luschino (2002) 16–24, 297–303; BNCF, *Conv. Sopp.*, J.VII.28, 'Vita Beati Hieronymi, martiris, doctoris, virginis ac prophetae eximii', fols. 1v–3v; Pico della Mirandola (1998) 7–11; Ginori Conti (1937) 3–10.

2 Villari (1930) I.1–22; Schnitzer (1931) I.1–28; Ridolfi (1997) 3–11, 271–5, 397–8; Cordero (1986) 3–25.

3 On Michele Savonarola, see Segarizzi (1900); Liotta (1963); Samaritani (1976).

4 'Vita Beati Hieronymi, martiris, doctoris, virginis ac prophetae eximii', fols. 1r–1v; Pico della Mirandola (1998) 7–9; Ginori Conti (1937) 4–7; Luschino (2002) 19. On Girolamo Savonarola's Latin, see Donnini (1999) 75–120.

subject to fashionable philosophical speculations – proved fundamental for Girolamo's perception of the incongruities of his times.

Michele's treatises on ethics and religious practice were ponderous scholarly works which displayed a thorough knowledge of the Old Testament and of Patristic and Thomistic thought (Samaritani (1976) 44–8). Moral rectitude and the reform of customs were the topics tackled most. Michele often lamented that courts had become cesspools of corruption. In the dedicatory letter of the *De vera republica et digna saeculari militia*, he urged his patron and patient, Lionello d'Este, to safeguard the morality of the clergy and the welfare of the destitute, orphans and widows[5] – the same social categories to which Michele provided free cures (Pico della Mirandola (1998) 8). In another treatise, he extolled apostolic poverty and exhorted his readers to observe civil self-constraint.[6] However, many measures to correct moral laxity proved to be inapplicable and were criticized for their lack of *realpolitik*. While his medical works were appreciated in his time – particularly his treatises on urine, fevers, gout, plague, diet and pregnancy – his writings on moral philosophy, the majority of which still remain in manuscript form, were naïvely argued. Michele was also attracted by various forms of superstition and believed in odd medieval Church legends, both practices which Girolamo would later condemn (Schnitzer (1931) 1.8–9; Cordero (1986) 6–7). Perhaps unable to express fully with the pen his reforming zeal and sense of piety, Michele tried to join the Hospitallers of St John of Jerusalem, just two years before Girolamo's birth (Samartitani (1976) 90; Luschino (2002) 298).

The extent to which Michele's religious fervour shaped Girolamo's awareness of the 'great misery of the world' and the 'iniquity of mankind' is subject to debate.[7] As a child, Girolamo did not manifest any particular vocational proclivity.[8] His religious formation was certainly regimented, but it did not prepare him in any special way for priesthood. Soon after his grandfather's death, Savonarola's father, Niccolò, pushed him towards the study of natural sciences and philosophy.[9] These were perhaps incentives for Girolamo to follow in Michele's footsteps in the medical trade: a career that his younger brother Alberto would later undertake.[10] However,

---

[5] Schnitzer (1931) 1.4; Segarizzi (1900) 40–4, 74–5; Samaritani (1976) 62–4.

[6] BAV, *Ottob. Lat.* 1667, 'De sapiente et insipiente'. On this text and other moral treatises written by Michele Savonarola, see Segarizzi (1900) 78–81 and Samaritani (1976) 306ff.

[7] Savonarola used these very words in a letter written to his father dated 25 April 1475 (Savonarola (1959c) 3).

[8] Cinozzi (1898) 3; Ginori Conti (1937) 6; Villari (1930) 1.3–5; Schnitzer (1931): 1.12, 26.

[9] On Niccolò Savonarola, see Cordero (1986) 4, 9, 10, 12–13, 24.

[10] Luschino (2002) 19; Schnitzer (1931) 1.10–11.

Girolamo's interest in curing bodies immediately turned into an urge to heal souls.[11] In courtly and elegant Ferrara, Girolamo first experienced distrust for ecclesiastical authorities, lamenting the discrepancy between their interior commitment and exterior conduct (Savonarola (1955b) I.262–3; Pico della Mirandola (1998) 10). He detected the ambiguous role they played in courts as well as the unchecked infiltration of pagan models in Christian spheres (Savonarola (1955b) I.78). The interference of the sacred with the profane (and vice versa), along with an unrelenting desire for ostentation, were common under Borso d'Este, who was duke of Ferrara during much of Savonarola's youth. Borso often promoted public ceremonies and processions that flaunted his quasi-divine status.[12] Memorable were the pagan *entrata* on the Po that he organized for Pius II Piccolomini in 1460, which the young Girolamo may have witnessed (Villari (1930) I.11–12), and Borso's lavish procession towards Rome, where he received his ducal investiture from Paul II Barbo in 1471.[13]

Aside from basic knowledge of medical notions, Girolamo inherited from Michele a sense of pious concern for the poor as well as a methodical approach to combating vice, which the Dominican friar envisioned as a contagion: a contagion whose origin had precise geo-political co-ordinates. This chapter intends to chart the trajectory of Girolamo's critical assessment of Rome and the contemporary Church as it emerges from his prolific production, most of which was published during his lifetime or shortly thereafter. Throughout his sermons, treatises, letters and poetic writings, Savonarola often referred to 'Rome' and the 'Church' interchangeably, in a quasi-synecdochal manner. This was more than just a rhetorical device: the humanist city – with its ancient monuments and holy sites – came to be identified with the very institution which was chosen by God to uphold and safeguard the purity of the Christian faith. Ironically, the language employed by Savonarola to chastise papal Rome was very similar, in some cases identical, to that employed in the proceedings at the Lateran Council of 649 to condemn heresy and heretics.[14] Savonarola condemned the rampant secularization, laxity and opulence of ecclesiastical authorities, claiming that their filth and dirt were contaminating the purity of the Church. Terms

[11] Savonarola (1959c) 8; Ginori Conti (1937) 8.
[12] Gundersheimer (1988) 57–72, 68–69; Molteni (1995) 37.
[13] Celani (1890) 361–450, esp. 372–3; Chiappini (1967) 131; Gundersheimer (1988) 67–8.
[14] As discussed by Katy Cubitt in her paper 'The jet-black spiderwebs of heresy: pollution and the language of heresy in seventh- and eighth-century Rome' at the conference 'Pollution and Propriety: Dirt, Disease and Hygiene in Rome from Antiquity to Modernity', British School at Rome, 21–2 June 2007.

such as *maculare*, *immondizia* and *sporcizia* are not rare in his sermons and writings. In the eyes of the Frate, the See of Peter had become a see of pollution, with the Church itself the pollutant. In a sort of domino effect, prelates tainted the lower clergy with moral turpitude, and these, in turn, fouled the populace whom they served. Strictly connected with hygiene and pollution, another recurrent and powerful metaphor that Savonarola used to evoke the corrupting powers of Rome was the decay of the body. Images of putrefying corpses, plague-stricken cadavers, poisoned organs, pallid complexions, purging processes, food poisoning and nauseating miasmas populated many of his sermons. He diagnosed the causes of Rome's moral malaise and, as I shall examine in the final part of this chapter, he identified treatments for the purification and healing of those affected by it.

Girolamo's earliest invectives against Rome and the Church date back to the Ferrarese period. The Petrarchian canzoni *De ruina mundi* (1472) and *De ruina ecclesiae* (1475) (Savonarola (1968) 3–9), as well as the letter written to his father explaining the reasons for taking the Dominican habit (25 April 1475), voiced his concern for the corruption of ecclesiastical institutions and the devastating repercussions it had on society.[15] These juvenilia also display the first glimmerings of his reforming concerns, which would be programmatically revealed in his Florentine sermons some twenty years later. The theme of Babylonian Rome – a city which, in his words, fostered abjection, rape, thievery, fornication, blasphemy and idolatry – is tightly interwoven in the complex fabric of his millenarian message. Following the tradition of medieval eschatologists and reformers, he declared that the very place that was chosen to set the example of true Christian living and elected by God to be the See of Jubilees was completely saturated with vice.[16]

Up until his execution in Piazza della Signoria on 23 May 1498, Savonarola pursued a project of radical reform whose aim was to restore spiritual purity and moral rectitude in the Church. From the beginning of his pastoral tenure, he recognized that the first step of this ambitious healing could be expedited if first experienced within the safety of convent walls; only afterwards would his friars deliver this *reformatio* to the world at large. To his dismay, various forms of corruption had yet penetrated his Order, especially the Observant branch.[17] Not even San Marco, the convent founded by Saint Antoninus Pierozzi, had been spared. In the early 1490s, Savonarola began assembling a religious community committed to living by the strict tenets

---

[15] Savonarola (1959c) 3–6. On the propagation of this 'poison', see Savonarola (1959b) II.17.

[16] On Savonarola's conflicting rapport with Rome, the papal court and the Curia, see De Maio (1969); Modigliani (1998); Miglio (2001).

[17] On the Dominican Observant tradition and Savonarola, see Polizzotto (2009).

of St Dominic's pauperism (Verde 1983). The popularity of these friars grew rapidly in Florence, particularly in the upper levels of society.[18] With their help and that of an uncorrupted and incorruptible youth brigade, known simply as the *fanciulli*, Savonarola began to curb the city's decadent mores.[19] The 1497 and 1498 bonfires – in which ancient and contemporary works of art burned alongside cosmetics, musical instruments, mirrors and lascivious literature – were forms of communal purification that incarnated this spirit of reform.[20] However, the Frate soon realized that while Florence could possibly be transformed into a perfect Christian city, the Church – not the institution but its leadership and its see – had reached a point of no return. His hopes of seeing it willingly yield to the *simplicitas* of primitive Christianity had been deluded.[21] Ecclesiastical authorities and the clergy at large were guilty of leading fellow Christians to perdition: as such, they had to be reformed at once. This change was only going to take place through divine punishment – which Savonarola had prophesied – along with the imminent arrival of the Pastor Angelicus.[22]

The contempt for Rome and the Church – respectively, the physical epicentre and source of corruption – is a recurring *topos* in the Frate's prolific production. Numerous are his remarks on the extreme disjunction between this city's production of excess and ostentation of luxury, on the one hand, and the Mendicant experience, on the other. According to the Frate, Rome was not merely tainted by the immoral actions of the clergy, but had also become evil incarnate.[23] For this very reason, Savonarola demonstrated no intention of establishing a convent there, even at a time when his reforming project was gaining significant momentum within his own Order.[24] Instead, he discouraged priests from visiting Rome lest they be harmed by the pestilential venom in its air (Savonarola (1955b) ɪɪ.18). Those who did go, he insisted, were irrevocably scorched (Savonarola (1955a) ɪɪ.59–60).

---

[18]   On the followers of Savonarola, known as the *Piagnoni*, see Polizzotto (1994) and Dall'Aglio (2010) 69–147.

[19]   On the *fanciulli*, see Ginori Conti (1937) 119; Schnitzer (1910) 5, 95, 105–6, 112; Cerretani (1994) 232. For their role and that of their leaders, Fra Domenico da Pescia and later Pietro Bernardino, see Trexler (1974) 200–64; Polizzotto (1994) 38–41, 123–7; Niccoli (1996) 279–88.

[20]   On Savonarola's bonfires, see Ciappelli (1997a) 214–33, (1997b) and (1999).

[21]   On Savonarola's notion of *simplicitas*, see Savonarola (1959b).

[22]   The most important work on Savonarola's prophecy still remains Weinstein (1970). Aside from the main biographers listed above in n. 1, see also Vasoli (1962) and Garfagnini (1988).

[23]   The references to Rome as a crossroads for prostitution, sodomy and incest are especially copious in the sermons on Ezekiel (Savonarola 1955a) delivered in the months in between Advent 1496 and Lent 1497.

[24]   The first convent in Rome associated with Savonarola and the Piagnoni, San Silvestro a Monte Cavallo, was established in 1507 (Assonitis 2003).

Romans were accused of soiling God's tabernacle with filth (Savonarola (1898) 37). He also warned that as soon as the stench of its sodomy, rape and prostitution reached the heavens, Rome's lavish palaces would be laid waste by God's wrath (Savonarola (1955a) 2.61–2). Not even the 'Rock of Peter', aptly renamed 'Rock of Satan', would be spared (Savonarola (1955a) 2.57). Simone Filipepi, brother of Sandro Botticelli, who also espoused the Frate's cause, supported such predictions – and their millenarian implications – by confabulating exaggerated accounts of catastrophes in the Vatican area. He reported that in 1501 large portions of Castel Sant'Angelo and the Passetto, as well as the statue of the Archangel Gabriel, had been destroyed. Furthermore, a burst of wind blew scaffolding off the papal palace, injuring Pope Alexander VI, and a three-day flood of the Tiber brought diseases and famine to the city (Filipepi (1898) 468–9). Similar portentous events occurring in the Vatican area were described in Benedetto Luschino's *Vulnera diligentis* (Luschino (2002) 275–6). In late 1512, the Camaldolese reformer Pietro Quirini, who manifested more than just sympathies towards the Savonarolan project of reform, harangued in his *Libellus ad Leonem X* on the condition of the Vatican: a degrading brothel, whose turpitude mirrored that of the Church and its clergy (Mittarelli and Costadoni (1755–73) ix.cols 706–7).

The Frate died having never set foot in Rome. He would receive updates on the pope's transgressions from the Florentine bankers and merchants *romanam Curiam sequentes* that had espoused his project of reform.[25] When, in July of 1495, Pope Alexander VI summoned Savonarola to Rome in order to respond to the accusations of hostility to the papal Curia and dissemination of sectarian propaganda, the Frate knew very well that a trap was being set. He diplomatically answered that he had always wanted to visit the city, specifically the relics of Sts Peter and Paul, but was unwilling to leave Florence due to his poor health (Savonarola (1959c) 71–4). In truth, the Frate had been moderately sceptical about the miraculous powers of relics and icons, cognizant of the placebo effect that these beliefs had on people. The thaumaturgical and apotropaic nature of these objects was in conflict with his fulminations against superstition. In fact, he firmly maintained that an image or statue could never incarnate the divine but could only represent it, and he insisted that Christians should not be fettered by the chains of idolatry (Savonarola (1961) 495–505). An exception was made for the Madonna dell'Impruneta. He fully understood its ancient and almost anthropological connection with the Florentines and so, compelled

---

[25] Bullard (1976) 51–71; Schnitzer (1931) II.276; Assonitis (2003) 243–8.

by popular fervour, he invoked its suppilicatory powers and exploited the icon as a conduit for collective prayer.[26] His assessment of early Christian relics in Rome was unquestionably more caustic. He warned pilgrims and Romans that the veneration of the bodies of Peter and Paul – as well as the relics of other martyrs – was not a means of salvation (Savonarola (1955a) I.224). In fact, he would refer to relics as gold and silver idols (Savonarola (1971–2) II.207). Savonarola compared Rome's fate with that of Jerusalem's, a city which housed holier sites – such as the Holy Sepulchre and the True Cross – and which nonetheless had suffered God's wrath (Savonarola (1962) I.266–7). He assured prelates and Romans that God was soon going to send hordes of barbarians to destroy its sanctuaries and churches, markers of pride rather than humility (Savonarola (1971–2) II.84–109).

The Frate had limited specific interest in attacking Roman antiquities since he considered these to be only one of the many elements constituting the evil of the pagans. He was deeply disturbed by the way modern poets imitated and glorified the fraudulent ancients, who in turn had only fostered superstition and corrupted generations of youths with their stories. More pressing was his concern for preachers who, in order to compete with philosophers and poets in the audience, would take the pulpit only to parade their classical and Neo-Platonic learning – which he termed the 'most pernicious plague' (Savonarola (1982) 256–7). Many of them had so wholeheartedly embraced this trend that Savonarola described their intellect as being caged in the 'jails of the ancients' (Savonarola (1982) 250). His pronouncements on antiquities and the antiquarian tradition were clear: ancient monuments represented the impermanence of paganism; papal ones stood as majestic markers of immoderation and pagan revival. The day after his first bonfire (1497), he compared himself to Pope Gregory the Great, who did not hesitate to smash pagan statues and burn volumes of Livy (Savonarola (1955a) I.147). Savonarola condemned the custom of adorning buildings with mythological themes and ordered that statues of Minerva and Hercules be replaced by images of Christ and the saints (Savonarola (1955a) I.208–9). The Pantheon, he claimed, was noteworthy only in so far as it housed the Christian cross (Savonarola (1971–2) III.71).

His scepticism towards the antiquarian tradition was passed on to his most dedicated follower, Fra Zanobi Acciaiuoli, who was later appointed librarian of the Vatican by Leo X de' Medici.[27] Acciaiuoli worked on the

---

[26] BA 2102, 'De legatione, praedicatione, et supplicio fr. Hieronymi Savonarola Ordinis Praedicatorum', fol. 223v; Filipepi (1898) 465.
[27] Redigonda (1960) 93–4; Morisi Guerra (1991) 89–109; Nardi (1991) 9–64; Verde and Giaconi (1992) 197–200; Assonitis (2003) 267–72.

scholarly editions of Eusebius of Caesarea's *Opusculum contra Hieroclem* (Venice, 1502) and Theodoret of Cyrrhus' *De curatione graecarum affectionum* (Paris, 1519). The works of these two anti-pagan apologists of late antiquity dealt with the purity of Christian faith in the light of the recrudescence of pagan contaminations in the fourth and sixth centuries. Both works were re-introduced to underscore the impossibility of conciliation between Christian spirituality and pagan materialism. Finally, in both instances the abuse of pagan culture was considered to be a venom or contagion, for which Acciaiuoli provided, with his translations, valid antidotes.

Acciaiuoli's most notable work was the 1518 *Oratio in laudem urbis Romae* (Assonitis (2005) 55–63). Around the same time that Raphael perorated in his famous letter to Leo X with the excavation and preservation of ancient ruins ('the cadaver . . . of what once was the queen of the world'), Acciaiuoli championed the *urbs nova et christiana*, deeming it superior to the *urbs antiqua*.[28] He scorned all those architects and sculptors who pioneered the exploration of the caverns and ruins of the city and who heralded their discoveries as models for contemporary art.[29] His reference here to Nero's Domus Aurea and artists such as Pinturicchio and Raphael is not unintentional. Acciaiuoli insisted that the glorious past of pagan Rome is best conveyed by the immense archaeological carcass scattered around the sixteenth-century city.[30] Despite Savonarola's chastising pronouncements, Acciaiuoli insisted that one should visit relics that endowed this city with holiness, including the Column of Flagellation, the Scala Sancta, the Sudarium and the Titulum Crucis as well as the relics of saints.[31] In his concluding statement, Acciaiuoli went so far as to envision Rome as the city of redemption.

From the pulpits of San Marco and, later, Santa Maria del Fiore, Savonarola began in 1490 to reveal his fiery dissent, often in a blunt and coarse style which to many refined ears in the milieu of Lorenzo de' Medici sounded both barbarous and obsolete (Verde 1988). The popular philo-Medicean Augustinian preacher Fra Mariano da Genazzano criticized him for purposely neglecting classical tradition and philosophical discourse in his sermons.[32] Partly in response to this and similar accusations, Savonarola denounced pagan sophistications in contemporary literature and oratory in his 1491 tract entitled *Apologeticus de ratione poeticae artis*.[33] At the

---

[28] On Raphael's letter, its dating and composition, including Baldassare Castiglione's involvement, see Di Teodoro (1994); Shearman (2003) i.500–45.

[29] Acciaiuoli (1518) fol. C4v.   [30] Acciaiuoli (1518) fol. A3v.

[31] Acciaiuoli (1518) fol. C3r.   [32] Perini (1917); Cherubelli (1940).

[33] Savonarola (1982) 211–72, 395–400. Though written and distributed in 1491, the *Apologeticus* came out in 1492.

same time, the Frate's rhetoric gradually began to distance itself from that of itinerant eschatologists who roamed the Italian territory, announcing catastrophic events purely for shock value. He did this by channelling his message of salvation and chastisement via medical language and sensory imagery,[34] some of which he borrowed from his grandfather Michele.[35]

On the fourth Sunday of Lent (13 March) of 1496, Savonarola delivered a famous sermon filled with corporeal metaphors that served to illustrate ecclesiastical corruption better (Savonarola (1971–2) ii.219–20). Italy, he insisted, had been suffering a fatal illness. Its contagion originated from the head (Rome), which delighted in eating putrefying food (ambition, pride and luxury), despite the ultimatum of the doctor (God) whose blade was about to put an end to the suffering. On more than one occasion, Savonarola scanned the limbs of the Christian city to verify the pervasiveness of the disease (Savonarola (1989) 253), insisting that it progressively contaminated the rest of the body: namely, the prelates and lower clergy, who then corrupted the populace they served (Savonarola (1962) i.302). The Church had begun dying from a terminal illness the moment it departed from the *simplicitas* of primitive Christianity (Savonarola (1955a) ii.234–5), whose adepts enjoyed impeccable health.[36] He commonly invoked the gruesome and clashing image of the decaying bodies of prelates, adorned in silk, inside their lavish tombs, adorned with images of Mary (Savonarola (1955a) ii.184). Commenting on a passage in Revelation (22:11), he described in detail the effects of excess and hubris on the human organs: overfed stomachs bursting open; kidneys collapsing due to luxury (Savonarola (1959a) 137). His remedy was penitence and contrition (Savonarola (1955a) ii.152). Given the patient's reluctance to regain his own health (literally, in refusing the medicine), Savonarola began to isolate this contagion and initiate a project of reform that served as a cure (Savonarola (1971–2) ii.220). This ultimately proved ineffective, and the Frate warned of the direct intervention of *Christus Medicus*, whose medication was going to be inexorably severe. He claimed that the Church – its entire body, due to the interpenetration of all its parts – had to undergo a complete palingenesis in order to restore its purity (Savonarola (1955a) ii.171).

---

[34] Terms such as *pestilenzia, febbre, tepido, infirmità, veleno, remedio, guarire, curare, medicare, purgare* and *ammalare* appear often in his works, especially in his sermons.

[35] On the use of fever and tasting metaphors and their similarities with Michele Savonarola's work, see Donnini (1999) 108.

[36] In a sermon delivered on 20 March 1497, Savonarola embarked on a medical diagnosis of the early Christian communities, with particular reference to their balanced humours and clear complexions (Savonarola (1955a) ii.162–78).

Savonarola understood the great potential of the imagery of disease for instilling in his brethren a perception of imminent ruin. His attempts to somatize papal Rome and make manifest the pathology of its moral corruption served to build a solid foundation for his reforming and millenarian discourse. Furthermore, medical and nosological metaphors acquired particular significance in the Florentines' collective concerns about the emergence of the French pox in Florence in 1496 and the 1497 plague, which Savonarola had prophesied as one of God's scourges.[37] His remedies for these physical afflictions were not medical but moral. In an open letter written on 15 July 1497 to his brothers at San Marco, he addressed the issue of the treatment of plague (Savonarola (1959c) 165–7). As *medico spirituale* of his convent, he embarked on a systematic guiding of moral behaviour during such dire times. Though confident in the medical practice ('given by God in order to provide comfort and health to the human body'), he insisted that plague, being divinely sent to punish the decadence of Rome, could only be cured with humility.

Savonarola's answer to Rome's miasma was embedded in St Dominic's message of poverty. Like Dominic, he set out to isolate and form his friar-preachers in pious and simple communities.[38] These spiritual havens were to function also as microcosms in antithesis to the worldly corruption widespread in courts and monasteries. Just as the lofty Medicean villas at Poggio a Caiano and Careggi served as optimal fora for humanist discourse, Savonarola envisioned his religious space as an instrument of salvation in which the rigor of the Rule was fully lived out. Though successful in assembling a cluster of pre-existing convents in which the Dominican Observance was experienced (aside from the mother convent, San Marco, other Piagnone bastions included San Domenico in Fiesole, San Domenico in Pistoia and San Romano at Lucca), the construction of a brand new, self-sufficient convent turned out to be his major unrealized project.[39]

Few aspects of San Marco's architecture met the appropriate conditions for the austere and modest life envisioned by Savonarola. According to the late sixteenth-century Dominican historian Fra Serafino Razzi, the convent had always appeared too magnificent and sumptuous for the friars' standards.[40] Albeit sober, Michelozzo's classicizing architecture (especially in the library and St Antoninus' cloister) constituted not only a marker of

---

[37] Landucci (1883) 132; Arrizabalaga *et al.* (1997) 44.
[38] On St Dominic's convent, see Meersseman (1946); Lippini (1990).
[39] On the convents that espoused Savonarola's reform, see Creytens (1970); Di Agresti (1980); Verde (1994).
[40] BML, *San Marco* 873, 'Cronica della Provincia Romana dell'Ordine de' frati predicatori', fo. 6v.

excess but also the Medicean seal of humanist hegemony. Stylistic issues aside, the pervasiveness of lay infiltration in the convent concerned the Frate. The Medici had disseminated signs of their presence throughout the church in the form of coats of arms, a practice that Savonarola denounced from the pulpit.[41] Even more evident were the references to this family in Fra Angelico's San Marco altarpiece, painted 1438–40, which stood behind the main altar of the church.[42] The figures of Sts John the Evangelist, Cosmas and Lawrence, clustered to the left of Mary, allude to Giovanni di Bicci de' Medici and his two sons Cosimo and Lorenzo, while the Medicean *palle* prominently framed the lavish rug in the foreground.[43] A more disturbing feature infiltrated the cloistral sections of the convent. With Pierozzi's blessing, Cosimo de' Medici had a personal cell built at the end of the lay brothers' dormitory, for his sporadic spiritual retreats.[44]

The convent that Savonarola had in mind was, in essence, a vehicle to re-establish a direct, physical tie with the early fathers of the Church, mediated by the apostolic poverty of St Dominic. Accounts of this structure, to be erected in the area of Montecavo near Careggi, appear in the *Vita latina*, in 'Pseudo-Burlamacchi' (here cited) and in Serafino Razzi's *Cronica*.[45]

> He wanted the said monastery to be famous for its simplicity and not for its precious stones, therefore he did not want marble or sandstone for it, or other similar stones of worldly value; but [he wanted it to be] humble and low to the ground, so that it would not extend too much in height, with small cells divided by partitions made of wooden boards or knotted straw and covered with plaster. He wanted the door jambs, thresholds and window frames made of wood with wooden closures, without iron hardware or locks, so that they would be open even to thieves. [He wanted] the convent and the church to be simple, not vaulted; with columns made of wood or baked bricks; with no curious images inside the church, but simple and pious ones, without any vanities. [He wanted] paraments, linen or woollen garments, to be in accordance with the customs of our parish of San Domenico, and the

---

41 AGOP, xiv, lib. Z, 'Notizie de' soggetti, e cose più memorabili del Convento di San Marco di Firenze dell'Ordine de Predicatori raccolte dal P.L.F. Serafino Maria Loddi del medesimo Ordine. Anno MDCCXXVI', fo. 19. The Frate complained that escutcheons and other insignias appeared on liturgical paraments and in lieu of the Crucifix and condemned them as ostensible marks of external subjugation and secular propaganda (Savonarola (1971–2) ii.26).

42 'Notizie de' soggetti, e cose più memorabili del Convento di San Marco di Firenze', fols. 18–19. See also Improta (1989–90) 106–7; Hood (1993).

43 'Notizie de' soggetti, e cose più memorabili del Convento di San Marco di Firenze', fol. 19.

44 'Notizie de' soggetti, e cose più memorabili del Convento di San Marco di Firenze', fol. 26.

45 'Vita Beati Hieronymi, martiris, doctoris, virginis ac prophetae eximii', fols. 10v–11r; Ginori Conti (1937) 51–4; 'Cronica della Provincia Romana', fol. 6v.

chalices and other liturgical objects to be without excess, so that everything
might exude an odour of devotion and simplicity.[46]

The construction of a new, larger convent became necessary also due to
the extraordinary number of young Florentines who, ignited by the Frate's
reforming zeal, decided to enter San Marco.[47] The issue of space and expan-
sion was addressed by Savonarola in a sermon delivered on 6 April 1496,
in which he proclaimed that his friars were going to increase exponentially
(Savonarola (1971–2) II, fos 357–8). The project for a new religious home
had received the *placet* of Alexander VI, obtained by the Frate's close follower Fra Francesco Salviati while sojourning in Rome in spring 1494.[48]
Savonarola's plan both reflected and, at the same time, sought to inspire
the pursuit of chastity, poverty and obedience. It was a site that was safely
sequestered from the dangers of the 'Roman plague'. More than an urban
convent, Montecavo was to resemble a hermitage or *hospitium*, much like
the rural Santa Maria Maddalena in Pian di Mugnone or Santa Maria a

---

[46] The passages on Montecavo in the *Vita latina* ('Vita Beati Hieronymi, martiris, doctoris,
virginis ac prophetae eximii') and 'Pseudo-Burlamacchi' (Ginori Conti (1937) 51) are
almost identical and are part of a larger section entitled '*Quomodo voluit fabricare novum
conventu*[m]' and '*Come volle hedificare un nuovo convento*' respectively. I choose to cite
'Pseudo-Burlamacchi' since it had a much wider distribution in Piagnoni circles: 'Et voleva che
detto monasterio fusse famoso in ogni semplicità et non in pietre pretiose, perché e' non voleva
che vi fussi marmi o macigni, o altre simile pietre dal mondo stimate, ma humile et basso a
terra, acciochè non si estendessi molto in alto, et piccole celle, li tramezzi delle quali fussino o
d'asse o di canne intessute et le camere intonicate; li stipiti cardinali et soglie degli usci et
finestre voleva che fussino di legno con serrami di legno, senza ferramenti, o chiave, in modo
che alli ladri fussino patenti; e' chiostri et chiesa semplici, sanza volte; le colonne di legno, o
mattoni cotti; in chiesa non figure curiose, ma semplice et devote, senza alcuna vanità. E
paramenti, lani o lini, secondo il costume della parrocchia nostra di Santo Domenico, e' calici
et ogn'altra cosa necesseria al culto divino senza superfluità, in modo che ogni cosa gittassi
odore di devotione et semplicità.' Razzi's passage ('Cronica della Provincia Romana', fol. 6v)
broadly summarizes the content of the two texts, adding little new information: 'Fatto per tanto
priore di San Marco il padre Savonarola è venuto pensiero di fare nuova riforma, dissegnarono
da prima di fabricare un nuovo convento alla loro simile volontà convenevole e fu loro offerto
un sito nella villa di Careggi due miglia incirca, fuori Firenze, sopra di uno ameno colle, et
altissimo per la vita contemplativa, et savevano di già trovata da alcuni cittadini i danari per
detta fabrica. Ma opponendosi dei più vecchi padri del convento, con dire di cotale ritiramento
e ristringimento di vita sarebbe stato causa della morte dei loro figliuoli per le molte astinenze
et austerità, che qui fatte si sarebbero, distolsero i prefati cittadini da quelle imprese.'

[47] San Marco's *Liber vestitionum* recorded how in 1491 (the year of Savonarola's first priorship)
only two novices were admitted at the convent, while in 1493, sixteen were. On the novices
entering San Marco (and San Dominico at Fiesole) see Verde (1983). On the social background
of the novices, see Arrighi (1996). On the lack of physical space in Savonarola's convent, see
Dall'Aglio (2002) 123.

[48] Letter of Fra Francesco Salviati from Rome of 1 May 1494 (Gherardi (1887) 67).

Lecceto, both of which functioned as San Marco's asylums for old or conva-lescing brothers and a temporary hostel for travelling friars.[49] It served as a vehicle of rebirth for the entering novice and as an appropriate abode for the tired friar returning for spiritual nourishment after a hard sojourn on the road. Here, scholarly friars would also pursue their studies while teaching theology and sacred letters. The choice of building materials (wood, straw, plaster and bricks), the sizes of cells, the unvaulted ceilings of the church, and the sober iconographic programme were attempts to revive primitive Dominican architectural precepts.[50] In a letter written to the prioress of San Domenico at Pisa, Savonarola had once complained that the Mendi-cant orders no longer lived in 'poor cells, without any superfluities', but erected 'palaces with marble columns' and 'rooms adequately suitable for nobles'.[51] Exhorting contemporary clergy to forgo possessions and not imi-tate Roman mores, he reminded them how the early Church Fathers lived frugal existences.[52]

This severe expression of conventual pauperism never left the drafting table.[53] The elder members of Savonarola's congregation were staunchly opposed to the convent's construction, while the families of the novices considered the restrictions imposed on friars too austere (Ginori Conti (1937) 53). Building plans may have been abandoned, or possibly post-poned, due to the rapid succession of both local and international crises, namely the expulsion of the Medici and Charles VIII's entry into Florence (Ridolfi (1997) 56–7), the *Flagellum Dei* who would purge the corrupt Roman papacy. Nevertheless, the construction of a 'Savonarolan' convent formed an essential component of the Frate's reforming project. There is no doubt, in fact, that his friars' observance of sobriety and piety *intra moenia* would have served as a model of purity for the Christian world *extra moe-nia*. Most importantly, since Savonarola did not leave a treatise on artistic theory, the account by 'Pseudo-Burlamacchi' provides the programmatic aesthetic stage on which Savonarola's artistic commentary, opinions and caveats, expounded throughout his sermons and treatises, are played out. As such, the extant account of Montecavo represents a starting point for a

[49] Falletti (1988); Burke (2000).
[50] On the legislation involving Dominican architecture, see Meersseman (1946); Montagnes (1974); Sundt (1987); Lippini (1990) 10–35.
[51] This letter is dated 10 September 1493 (Savonarola (1959c) 42–51, esp. 44.)
[52] Some years later, he compared the luxurious palaces of contemporary clergy with the modest living of the early Church Fathers (Savonarola (1962) I.143).
[53] On the planned convent at Montecavo, see Bruschi (1966) 1–7; Assonitis (2003) 225–30; Polizzotto (2009) 183.

theory of an 'anti-Renaissance', as well as a blueprint for a possible means of salvation.

According to Savonarola, Rome was the foremost site of pollution. It was not merely 'matter out of place', but also, given its position within the context of the Christian world, 'matter out of place' which was not marginalized but flourished triumphantly at the centre (Douglas (1996) 2, 35–6, 53–5). Its *sporcizia* and *immondizia* penetrated deep into the fabric of religious and lay society. To resist this disorder, Savonarola constructed purity systems of his own, predicated on the *paupertas* and *simplicitas* of early Christianity as interpreted by the Rule and early constitutions of his Order. His first step was to reform the convent of San Marco and other Dominican convents in central Italy. Second, he planned to build one *ex novo*, in the Florentine countryside, which would fully incarnate the purity of poverty. However, the de-urbanization of this religious community would have transformed the coenobitic nature of the friar-preachers into a congregation of religious recluses.[54] Although Savonarola knew very well that sequestering his friars in rural Montecavo hardly meant an absolute protection against the contagion of sin, he was also aware that relocating them to the periphery would establish a healthy distance from the very dirt that he sought to wipe out.

---

[54] Ginori Conti (1937) 53–4; Di Agresti (1980) 17–18.

9 | Purging filth: plague and responses to it in Rome, 1656–7

DAVID GENTILCORE

Introduction: avoiding plague

For early modern Europeans plague was associated with corruption: filthy and foul living conditions, in particular 'bad air'. This corruption also had a pronounced spiritual and moral dimension. Sacred and profane causation formed an impenetrable tangle. A booklet written in Rome in the 1620s lists the causes of plague. 'The first cause which brings on plague', it states, 'is the activity of corrupt air, which because of the bad qualities it then has, corrupts human bodies, attracting this air through the wrists, breath and pores.'[1] Worst of all were 'vapours corrupted and mixed with water, raised up by putrid and stinking things', occasioned, for instance, by the opening of caves or the digging up of burial grounds, by dying trees (especially walnut and fig) or by the breath of poisonous animals.

The second cause of plague was the patient's own 'disposition'. 'The bodies most prone to this putrefaction are those which practise coitus and which go into the baths.' Sexual intercourse was out because it debilitated the body; baths were out because they heated the body's humours and opened the pores, making the body vulnerable. Contagion was the third cause of plague, in the sense of 'contact and company' with infected bodies. The ambivalence between plague causation in terms of corrupt air (miasmas), expressed as the first cause above, and a notion of 'seeds of contagion' carried by the air, expressed here, is typical of the contemporary understanding of plague. The fourth cause was residing in close proximity to corrupt air. Since 'air is in continual contact with us and leaves its properties on the human body, if the air is corrupt it infuses its corruption into human bodies'.

I would like to express my thanks to Mark Bradley and Sheila Barker. The abbreviations used in this chapter are as follows: Archivio di Stato, Florence, *Archivio Mediceo del Principato* (ASF, *Mediceo*); Archivio di Stato, Rome, *Fondo Università* (ASR, *Università*); Biblioteca Nazionale Centrale, Rome, *Fondo Gesuitico* (BNCF, *Gesuitico*); National Library of Scotland, *Crawford Collection* (NLS, *Crawford*).

[1] BNCF, *Gesuitico* 516, 'Miscellanea medica: no. 1, Remedi preservativi tanto spirituali quanto corporali contro la peste', fols. 18v–21v. All translations are my own unless indicated otherwise.

In this case 'matter out of place', in Mary Douglas' famous formulation, was itself the cause of disease. Putting matter back into its 'proper' place was the solution. In her introduction to *Purity and danger*, Douglas noted that 'eliminating [dirt] is not a negative movement, but a positive effort to organise the environment' (Douglas (1966) 2). And indeed in early modern Europe the elimination of 'corruption' – whether of people or things – was done with a crusading zeal. It meant re-establishing order out of disorder (a dichotomy which was a favourite early modern way of viewing the state of the world). And, just as plague causation was multifaceted, so too the response to plague took various forms. This chapter will analyse responses to plague in mid-seventeenth-century Rome in terms of the creation and enforcing of boundaries, comparing theory and practice. Along the way, we shall consider a handwritten booklet, a health office and a few charlatans.

Let us start by saying something about the booklet from which the just-quoted medical advice is taken. It is called simply 'Remedi preservativi tanto spirituali quanto corporali contro la peste, particolarmente in ordine all'utile comune' ('Preservative remedies against the plague, spiritual as well as corporeal, in particular as regards the common good'). It is undated, but is presumably from the mid- to late 1620s.[2] Who wrote it and why? The anonymous author wrote it in a clear italic hand, in places almost printed, perhaps for the sake of clarity. The advice collected and presented in the booklet is quite basic, providing an elementary summation of contemporary knowledge about plague. There are one or two learned references, for example to Hippocrates, cited in Latin – not done to impress, but quite naturally. The booklet's author was probably a simple churchman, perhaps a Jesuit, since the booklet is part of the Jesuit holding (the *Fondo Gesuitico*) in Rome's Biblioteca Nazionale. It may be directed at parish priests, since there is some specific advice for them. Our author may also have been responsible for nicely binding it together with various other manuscript booklets and printed pamphlets to form a collection, which offers practical advice on how to avoid getting plague or on the public health response.

The booklet begins with the standard acknowledgement that plague is sent by God as a punishment for sins, before turning to the means necessary to preserve the city, Rome, from plague. Just as the plague here was quite real, so the responses were both practical and material – quite a striking contrast with the spiritual and moral solutions advocated by Fra Savonarola

---

[2] This is so because it refers to a prayer used by 'St' Carlo Borromeo against plague (the archbishop of Milan, canonized in 1622), but not to the plague epidemics which struck much of northern and central Italy in 1630.

over a century earlier (see Assonitis, this volume). The key is ridding the city of pollution. The best way to keep plague away is to live somewhere that is 'very healthy indeed'. The sewers must be 'purged of their filth and rubbish', as the ancient Romans did, and this must be done especially during the summer months, when smells carry and the air is easily corrupted. The streets must be kept clean: no refuse, animal skins, tainted wines or even dishwater, since this 'can easily spoil and turn the stomach of passers-by'. The health officials must make sure that animals are not kept in the city, especially pigs and wild animals (evidently regarded as the least clean). The streets should be cleaned and the rubbish dumped into the river.

The city was also to be 'cleansed' of the wrong kinds of people, for the greater good. There was no equality in the face of disease. The familiar list begins with lepers and the victims of other 'skin sores' (read: syphilitics), 'the sight of whom causes horror in those who see them', and who must be kept far away since they are possible sources of infection and because they might 'disturb the moderate gaiety which should be fostered at such times'. Prostitutes and go-betweens must be banished, 'as the cause of corporal and spiritual plague'. Beggars and mendicants qualify for removal because 'by getting drunk and committing acts of depravity they introduce or foster the plague'. The city's Jews were not to be expelled; but they were reminded to keep the rooms, streets and courtyards of the ghetto clean. A health official should be sent to inspect the ghetto once a week.

The best preservative, of course, was to live in a safe 'dwelling place'. 'This place must be high,' the advice went, 'open, so that one can easily see the sky, not too hot or cold, not subject to fogs and far from marshland', which render a place 'pestilential'. Human contact should be limited. 'The remedy is to avoid public conversations, stay on one's own, avoid winds [which can carry the contagion], the infected place and, most of all, plague victims.' On the one hand, the advice is ruthlessly selfish, uncharitable in the extreme. It is difficult to reconcile with the broader religious spirit and the notion of the common good which imbue the booklet. On the other hand, it suggests the kinds of draconian measures that were believed to be the only effective ones in preparing for and, even more so, responding to plague.

'Of those people who because they live filthily put the public health at risk'

So much for the theory; what about the practice? How did Rome react when plague struck the city in 1656–7? In order to answer these questions, I would like to focus on two themes centring on the control of plague within the city. The first is the control of people and goods believed to spread infection, especially in terms of the need to rid the city of 'filth', in all its guises. As we shall see, people bore the brunt of this policy: not so much 'matter out of place' as 'people out of place'. The second theme concerns the ambivalent position of one of the groups supposedly targeted by the health officials: medical charlatans and other itinerant practitioners. We shall explore their activities in Rome during the plague epidemic, in the context of the ongoing search for effective remedies. And we shall survey the reactions of public officials to new medicines proposed in time of plague, notably the alchemical ones so much in vogue during the seventeenth century.

Responsible for dealing with plague and maintaining order in Rome was the city's Congregation of Health. One of the first measures passed by the newly re-instituted *Congregazione della Sanità per liberare la città di Roma dal contagio* ('Congregation of Health for the liberation of the city of Rome from contagion'), set up by Pope Alexander VII and consisting solely of cardinals, was to prohibit 'all gatherings of people of whatever sort whether temporal or spiritual' to limit the spread of plague.[3] This law was soon extended to include all sorts of regular activities which would bring people together. Even the workings of legal tribunals like the Rota were suspended by order of the Congregation.[4]

This legislation is an indication of the hard line the Roman authorities were prepared to follow in responding to an epidemic which had already struck with full force in Naples. The Roman officials benefited from seeing the catastrophe already unfolding further south. The prelate Girolamo Gastaldi was appointed by the pope to be commissioner-general of pest-houses, a role which included the task of reading all the reports and accounts of plague cases in towns between Naples and Rome, as the epidemic made its inexorable way northwards towards the Papal States.[5] From the very first cases in Rome the Congregation was ready to act: closing or placing under strict surveillance all the gates of the city; instituting a regime of health passes for the regulated movement of people and goods, to keep people from suspected areas far away; ensuring the inspection of all goods and

---

[3] NLS, *Crawford* B.16(118), edict of 17 July 1656; also discussed by Savio (1972) 117. On health policies and the plague in Rome, see Sonnino and Traina (1982).

[4] NLS, *Crawford* B.17(24), *Editto per la sospensione dei tribunali*, 21 October 1656.

[5] The following year Gastaldi was appointed general health commissioner for the Papal States. He later wrote a *Tractatus de avertenda et profligandis peste politico-legalis* (Bologna, 1684) when he was bishop of Benevento, published whilst Gorizia was struck by a plague epidemic (Capparoni 1935).

foodstuffs arriving in the city; opening pesthouses for plague victims and safe houses for the quarantine of suspected carriers; forbidding physicians, surgeons and barbers to leave the city; and strictly regulating burials (in common graves outside the city walls) and the closure and disinfection of suspect houses (and possessions).

Gastaldi's strict approach was reassuring. The Venetian Gregorio Barbarigo, appointed health overseer for the *rione* of Trastevere, wrote to his father: 'I am safer here in Rome, even though there has been some sign of the disease, than in any other place in Italy which has business with Naples, since I'm certain that here the plague orders are many and all obeyed.'[6] The measures were harsh. When several plague cases were observed in Trastevere, some 6,000 people of the *rione* found themselves locked in from one day to the next, in the attempt to prevent the plague from spreading to the rest of the city. It was nothing less than a surgical strike. In the words of Cardinal Sforza Pallavicino, the repressive action was aimed at 'cutting off, according to the rules of surgery, all of the foul and base part from the greater and better part of the body; but because great resistance on the part of the Trasteverine population was foreseen, one evening without warning three cardinals were sent there, strong of hand, head and reputation' (Sforza Pallavicino (1837) 6).

Rome became a city of boundaries. A *cordon sanitaire* protected the city, whilst another internal cordon divided it into healthy and infected areas. The city was further fragmented by the boarded-up houses of plague victims, either in quarantine or undergoing disinfection. The urban space was defined exclusively in terms of the struggle against the epidemic (Boiteux (2006)). The Tiber was blocked by chains at either end of the city, at the Ponte Milvio to the north and at the Ripa port to the south, and ships could not pass without authorization. The city's main pesthouse, where plague sufferers were sent, was on Tiber Island. This was a reserved space, already the site of a hospital, convent and church, and conveniently located at the river's halfway point. From here the many deceased would be transported outside the city, to their final resting place in a common grave at St Paul's Outside the Walls.

The measures taken had a certain effect, at least when viewed from the city as a whole. Rome escaped the massacre which struck Naples. The number of victims – plague reduced the Roman population from 122,978 in 1655 to 100,019 two years later – whether in absolute terms or as a percentage of

---

[6] Letter of 8 July 1656 (Bertolaso (1969) 243).

the population, was far less than the figures for Naples.[7] However, from the point of view of the populations quarantined from the rest of the city, the policy was nothing less than calamitous.

In addition to closing off much of Trastevere by means of walls, gates and armed guards, another part of the city was similarly dealt with by the health officials: the ghetto, located just across the Tiber. Special measures enacted in July 1656 sealed off this already segregated space from the rest of the city. It was assigned its own pesthouse and health official and divided into four quarters, each with two further health deputees and a physician and surgeon, all Jewish. Contemporary chroniclers, like Sforza Pallavicino, reported that it was the government of the ghetto which sought this harsh quarantine in order to protect the area, paying for the measures itself (Sforza Pallavicino (1837) 35). Barbarigo noted in a letter that in the ghetto 'every little suspicion [of plague] causes houses to be shut, and the Jews themselves, to escape a worse eventuality, have decided of their own will to shut themselves in'. They did this, Barbarigo concludes, despite the fact that the 'quantity of the people there, the poverty and the small size of the place means sure death for them'.[8]

Despite traditional Jewish concerns for health and sanitation (see Stow, this volume), perhaps we should not take their desire to close themselves in, as stated by these reports, at face value. Did they have much choice? We must consider contemporary Romans' attitudes towards the Jews, as well as to the perceived health risks posed by their poverty and their trade in used clothes – both of which were believed to be causes of plague. According to public health edicts, Jews were believed to threaten the health of the public because 'they live filthily', as did vagabonds, mendicants and Gypsies.[9] The records of the Congregation of Health make no mention of the Jews' own wishes, simply that the area was to be closed off, as Trastevere had been. In any case, the effects of the measures were soon felt. Mortality rates in the ghetto were twice those for the city as a whole, as Eugenio Sonnino's research has shown. Moreover death rates in the ghetto's pesthouse were

---

[7] As Paul Slack reminds us, making retrospective judgements on the health policies undertaken by the authorities in the past is a risky business. We should focus instead on the ways in which the policies were enacted (Slack (1988) 450–1). According to modern estimates around 40 per cent of Naples' population died of plague and related causes, reducing the number of the city's inhabitants from 400,000–450,000 in 1655 to 240,000–270,000 two years later (Galasso (1982) 1.46–7).

[8] Letter of 26 July 1656 (Bertolaso (1969) 247).

[9] 'Delle persone che per habitare sporcamente possono pregiudicare la sanità', in *Ristretto delli ordini più importanti sopra la Sanità*, Biblioteca Apostolica Vaticana, Miscellanea, Arm. IV: 61, fols. 129–30 (Sonnino (2006) 41).

somewhat higher than in the city's other pesthouses. Contemporaries like Gastaldi and Sforza Pallavicino were nevertheless relieved, having expected much worse.

A climate of fear and suspicion reigned throughout the beleaguered city. All inhabitants were required to report cases of deaths at home, even unsuspicious ones, providing all the details to the notary of their *rione* employed for the purpose. Sick people suspected of being plague sufferers were to be visited by physicians appointed to the task and their names likewise notarized. The terminology is revealing. The physicians were known as *brutti*, in the sense of vile, ugly or dirty. A boat on the Tiber was either clean (*pulita*), when rowed by healthy individuals, or dirty (*brutta*), when transporting corpses or the sick. Plague sufferers were 'suspect' people (*sospetti*). When they were identified, their houses were sealed off, their inhabitants were forcibly hospitalized and their habitations – using a metaphor of purification – were thoroughly cleansed or purged (*purgati*), their clothes and mattresses burnt. The measures generated widespread fear: fear of being carted off to the pesthouse, fear of having one's house shut or boarded up (*serrata*), with the notice *Sanità* ominously posted over the door, fear of having one's possessions burnt, 'purged' or stolen, fear of being left alone and poor (Sonnino and Traina (1982) 437).

'Clean' people sought to stay that way. The physicians, most famously of all, went on their visits to the sick – accompanied by a clerk to take note and an ecclesiastic to comfort – with a long stick to keep the sick at a safe distance. The stick was the only contact permitted in order to examine the patient (figure 9.1). The physician's long gown of waxed cotton, broad-brimmed hat, thick eyeglasses, gloves, and a mask with a long 'beak' where they could put miasma-fighting perfumes form an iconic image of the early modern plague. It was meant to separate the physician from the source of infection.

The response of the city's health officials focused on people (and their goods). The emphasis was on separating 'suspect' people from the healthy – both in life (in the pesthouses, where over half died) and after death (when their corpses were sent in barges down the Tiber to be buried in deep common graves outside the walls). Rome's inhabitants were either healthy, 'suspect' (that is, sick or at risk of becoming so) or cured individuals. This was now the only classification, the only hierarchy that mattered. Curiously, in view of contemporary theories about how plague originated and was spread, less was done about 'pollution' in the sense of hygienic and material conditions. Rome already had a programme of street cleaning, administered by the *Presidenza delle strade*. Shopkeepers and tradesmen paid a tax of four *giulii* every three months for the rubbish to be carted away and the streets

L'HABITO con il quale vanno i Medici per Roma à medicare per difesa del mal Contagioso è di tela incerata, il Volto ordinario, con gli Occhiali di Christallo, & il Naso pieno di Profumi contro l'infettione. Portano vna Verga in mano per dare à vedere, e dimostrare le loro operationi. In Roma, & in Perugia, Per Sebastiano Zecchini. 1656.

9.1 'The uniform in defence against the contagious disease with which physicians go about Rome treating [the sick]', to quote the title of this 1656 print, was meant to create a boundary between them and infection. © Trustees of the British Museum.

swept. However, the records of the Congregation of Health indicate that only a few extra payments were made to street-sweepers and carters, like the one *scudo* and 50 *baiocchi* paid to Tommaso di Lorenzo on 29 June 1656, 'for having carted the rubbish near to the river with his horses'.[10]

As Renato Sansa has observed, the health officials were much more concerned with keeping people in place and charting death rates than they

[10] Archivio di Stato, Rome, *Camerale II, Sanità*, 5:2, fol. 13r (Sansa (2002b) 100; see also Sansa (2002a)).

were with the state of the city's streets. The threat was more filthy people than filthy spaces. At its meeting of 22 July 1656 the Congregation noted that 'the streets have not been well cleaned', but did nothing about it.[11] The *Presidenze delle strade* seldom met during this time; but then meetings were discouraged. The fact that this body was not exempted from the prohibition is indicative. One of its officials complained that shopkeepers and tradesmen in the cordoned-off areas of the city, Trastevere and the ghetto, were not paying their tax and so he was short of the money due him.[12] The implication was that these streets were not being cleaned as a result – at the moment when they most needed it.

The response to 'people out of place' was another matter. Gastaldi adopted a hard line, prepared to authorize capital punishments, performed in public at the very scene of the crime, to serve as a lesson. Such was the penalty administered to a clerk who had escaped from the pesthouse on Tiber Island, the gallows clearly visible to the institution's inmates (Pastore (1988) 149–50). Exemplary was the punishment of three men found guilty of charging a fee for the issuing of false declarations of health to 'suspect' people shut up in their houses. One of the three was a physician, Girolamo Rota. By paying him, victims of plague were 'liberated' before their period of quarantine had elapsed. The three men unwisely kept a register of the payments made to them, adding insult to injury by calling it the *Libro delle delitie della peste* ('Book of plague delights'), and when caught by officials, were dividing up the spoils.[13]

Having the right health pass could mean the difference between life and death. One chronicle, by the abbot Ruggero Caetani, reports the tragic case of a man coming from Terracina on horseback. He carried his health pass with him, which certified that he was free from plague, as were the areas he had passed through. He had put it for safe keeping in the lining of his trousers, but when the guards at the gates of Abano stopped him and demanded his pass, he was unable to find it. He searched his clothes and belongings in vain, and began weeping and beseeching the guards to show him mercy. They gave him fifteen minutes to prepare his soul for death and shot him on the spot. When the man was undressed for burial, they found the pass inside one of his boots.[14]

---

[11]   Biblioteca Corsiniana, Rome, MS. 34.C.6, fol. 120r (Sansa (2002b) 100).

[12]   Archivio di Stato, Rome, *Presidenza delle strade*, 10, fols. 211r–v (Sansa (2002b) 101–2).

[13]   One of Rota's collaborators was decapitated; the other, a Dominican friar, was sentenced to life imprisonment. Rota escaped execution and was exiled.

[14]   Archivio Caetani, Rome, *Misc. 161/300*, fols. 245–6 (Rocciolo (2006) 132).

Charlatans and alchemical remedies

At a time when people were dying, when shops were closing and when medical practitioners (and anyone else who could) were fleeing for their lives, those who possessed plague 'cures' occupied a privileged position. On 27 October 1657 the Tuscan grand duke's representative in Rome, Gabriele Riccardi, communicated the following item about the treatment of plague in the city:

> A few days ago a certain Neapolitan alchemist, formerly in the pay of his lordship the duke of Bracciano, entered into the pesthouse to treat the sick with a certain powder of his. On the day he arrived in Rome he caught the plague and, taking his medicament, was cured; and then they presented him with four plague sufferers, to whom he gave his powder, and they are all better, the medicament causing copious sweating. They are now negotiating his contract (*condotta*): he was requesting 500 *scudi* a month, whilst the Congregation was prepared to pay a set amount for each plague sufferer cured.[15]

How can we explain the interest on the part of the authorities in these and other remedies, to the point of being prepared to bargain on the health of Roman inhabitants? What was the role of medical charlatans and their remedies in time of plague? Charlatans were a liminal group at the best of times, and the plague brought with it a pressing need to draw and enforce boundaries, between clean and unclean, infected and healthy.

Charlatans and mountebanks were also prohibited from performing their regular activities. These 'people [who] appear in the squares and sell things with entertainments and buffoonery', according to the brief but remarkably neutral definition by a Roman health official,[16] were generally tolerated by the city authorities, as long as they followed established bureaucratic procedures for their licensing. Once they had been issued with a licence by the Roman Protomedicato tribunal, both the charlatans and their remedies were recognized as legitimate, and selling could take place publicly. The concept of the 'licensed charlatan' might strike us as something of a contradiction, but the tolerant stance of the Roman authorities towards what was perceived as a specific medical category, with precise functions and characteristics, can be found throughout the states of early modern Italy.

---

[15] ASF, *Mediceo* f. 3382. On Riccardi, see Corradi (1973) IV.790–8.
[16] Giacomo Giacobelli, interrogated in 1632, ASR, *Università* 67, fol. 113v. Parts of this section are drawn from Gentilcore (2006).

This was all in sharp contrast to the elite rhetoric of the time, medical and otherwise, which painted charlatans as lying cheats.[17]

For the charlatan who intended to sell his medicine publicly in Rome there was thus a well-established bureaucratic procedure for obtaining a licence, as well as strategies for ensuring a successful recourse to the authorities. This is evident in the many licence applications (*suppliche*) from charlatans that survive in the archives. With all due modesty, one had to stress the originality of one's remedy, but not that it was so new that it went against the accepted rules of pharmacy. One had to remark on its proven efficacy, exemplified by numerous cures. It was helpful if at least a few of those cured were people of high social status, whose cures naturally carried more weight. One had to explain that one was operating out of charity and not for any financial gain, while noting too that one also depended on it for one's living and to support one's (ever-increasing) family. There had to be some kind of appeal to authority, such as previous licensing in the same place or elsewhere. The patronage of the duke of Bracciano, head of the important, ancient noble Orsini family, helps explain the enthusiasm shown by the Roman authorities for our Neapolitan alchemist. At the same time, the authorities were being presented with a *fait accompli* in the form of a medicine, already prepared and ready for sale – a point which the supplicant had somehow to get across without being too assertive. There was always an element of negotiation in the encounter or clash between charlatan and health authority.

Romans knew perfectly well where to find charlatans, whether it was to buy their remedies or to be entertained by their stage routines. When Jacoma Florentia complained that 'my body had hardened up and I couldn't go', her husband brought her an electuary 'which was good to make me go'. The 'electuary was in a small tin container, with the printed handbill, and he told me he had bought it in Piazza Navona' from a charlatan who performed there.[18] The square was one of the city's main centres, with 399 stall-holders registered for its weekly market (held on Wednesdays) in 1614, and the site of spectacles, feasts and games (Gerlini (1943) 27). During his stay in Rome in 1645, John Evelyn spent 'an afternoone in Piazza Navona, as well to see what antiquities I could purchase among the people, who hold mercat there for medaills, pictures and such curiosities, as to heare the mountebanks prate, and debate their medicines' (Evelyn (1955) ii.368).

---

[17] For an example of which, see Mercurio (1645) 264ff. On Mercurio's life and works, see Pancino (2001).

[18] ASR, *Università* 1, 'Processo contro Dionigio Alberti Padovano per il seme di ricino', deposition of Jacoma Florentia, fol. 302r.

The city of Rome itself, universal seat of a continually expanding church, host to regular influxes of pilgrims and visitors, whose numbers reached hundreds of thousands during holy years, occupied a unique position. It may have been due to the city's experience in managing these influxes that the Congregation of Health took the decision to prohibit the entry of charlatans into the city. The reason, according to the historian of medicine Andrea Corsini, was that charlatans were always associated with beggars and vagabonds, and were expelled from plague-ridden cities in the same repressive edicts that 'were repeated often and everywhere' (Corsini (1922) 64). And it is true that charlatans, like other groups of itinerants and outsiders, could become scapegoats, included in clean-up measures enacted during time of plague.

But in fact foreign charlatans were not expelled from Rome because they were like beggars in the eyes of the health authorities, for they were not. Their status and activities were ensured by the licensing regime, which granted them official recognition and differentiated them from the idle mendicants who so preoccupied the authorities.[19] Charlatans, whether Roman natives or from outside, had a sanctioned occupation, a *professione* in the Italian of the time, however suspect it might sometimes be. A detailed study of several hundred edicts, issued by the health offices of the different Italian states to 'govern' the plague, reveals not a single prohibition against charlatans and mountebanks (Jarcho (1986)). Charlatans arriving in Rome were expelled not in the context of anti-vagrancy legislation, passed by the Congregation of Health,[20] but in the context of an edict prohibiting public gatherings of any sort.

Charlatans who were native to Rome or were established inhabitants there continued their activities as medicine-pedlars and healers, avoiding public squares and instead inviting sufferers to come to them or else offering house calls. As always, these activities were complementary to the appearances in the square or their sales campaigns which might take them outside the city. The medicines which charlatans sold were needed more than ever, complementing both those prepared by apothecaries according to the official pharmacopoeias and those made in the home according to the traditions of domestic medicine. The activities of charlatans in Rome did not come to an end. Indeed years later 'l'Idiota italiano' ('the Unlearned Italian') boasted that his 'Angel Oil' against plague had been tested in Milan and Bologna

---

[19] On the figure of the vagabond in Rome during the period, see Rosselli (1996).
[20] NLS, *Crawford* B.17(46), *Editto contro mendicanti*, 19 August 1656.

during the plague of 1630 and 'publicly compounded in Rome in the year 1656 to universal benefit' (Camporesi (1990) 270).

Charlatans did not want to miss the economic opportunity afforded by the plague. In 1657 charlatans who wanted to circulate in Rome petitioned the then-governor of the city 'to be allowed back in the city to move about as before' (Corsini (1922) 65).[21] There was a non-stop hunt for remedies even during times of normal mortality, with people on the lookout for new and improved medicines and therapies. In times of crisis mortality people were even readier to try a multiplicity of remedies, drawn from a wide variety of sources, even putting remedies of limited efficacy to the test. Barbarigo told his father that 'it is crucial now to stay healthy and alive', not to believe 'so easily in doctors'. Like many other of his contemporaries, he had recourse to a variety of medicaments. According to his letters, Barbarigo obtained 'remedies and antidotes, vinegar, quicksilver to wear on his body'; he asked his father to get hold of viper's powder and toad pills, personally recommended by the Jesuit polymath Athanasius Kircher; a Venetian friend from his student days, Giulio Giustinian, gave Barbarigo 'a tin of preservative remedies against plague'; he was constantly 'on the lookout for preservatives, recipes, electuaries' (Bertolaso (1969) 241, 246, 250, 259, 267).[22]

And, then, just when the services of the apothecary were most needed, their shops were forced to close when the apothecaries or their assistants came down with plague or when the necessary simples and compounds, like theriac, became unavailable due to commercial blockades and quarantines. At the end of July 1656 the Congregation of Health warned the city's apothecaries not to take advantage of the situation at the expense of the sick poor.[23] The apothecary of the San Bartolomeo pesthouse was assigned the task of ensuring the supply of medicines, of compounding various medicines and, perhaps most important, of keeping a register of effective cures, about which he had to inform the Congregation, accompanied by an indication of their relative costs.[24]

The response of the authorities to new medicines during time of plague was *sui generis*, given that no one remedy in the traditional pharmacopoeia was considered universally effective. The authorities were

---

[21] I have not been able to verify the source, but the date given by Corsini for it, 1656, is suspicious. The governor it refers to, Baranzoni, was in post only from the following year. Moreover the epidemic continued at full strength throughout 1656, with a brief respite beginning in December.

[22] On Kircher and his 'toad pills', a preservative against plague worn as an amulet, see Baldwin (1993b) 237–8.

[23] NLS, *Crawford* B.16(127), edict of 29 July 1656.

[24] NLS, *Crawford* B.16(137), *Instruttione per lo spetiale Roma*, edict of 1656.

desperately searching for effective medicines to try out, while their fellow citizens were falling like flies around them. The remedies proposed to them were worthy of serious consideration.

I have not been able to find out much about the Roman response, but it cannot have differed greatly from that of Venice in similar circumstances during the plague of 1575–6. To start with, there is the same contradictory attitude towards charlatans during the epidemic. On the one hand, as of November 1575, charlatans were not to 'mount their banks in the square at the Rialto or in other places to sell, nor practise in any way the charlatan's trade [*arte del zaratan*], either from a platform or on the ground'.[25] On the other hand, the government was desperate for remedies for the *mal contagioso*. This included the purported cures of physicians, charlatans, priests, noblemen, as well as anonymous individuals. In all, from the summer of 1576 to the following spring, twenty-two Venetian citizens petitioned that city's Health Office (the Sanità) and the Council of Ten with a variety of secrets, remedies, preservatives, antidotes, precepts and electuaries for plague (Preto (1978) 90).

It is particularly striking how the petitions for the exclusive rights to prepare and sell plague remedies resemble those made for patents and privileges in fields as diverse as mechanics and printing. The petition was always based on the promise that something would be produced which was useful for the community (the city and Republic) and new on the market. The supplicants for Venetian patents made certain offers or guarantees: that the test or trial (*experientia*) of its novelty and usefulness would be made within a certain term, the right of the state to use the invention for public business, and so on. They often stressed their own effort in the development of the particular product (Mandich (1936) and (1958)). But in the case of plague remedies petitioners also sought cash payments, in addition to the more usual exemptions. This was all part of a process of bargaining in which both parties participated.

Trials involving plague remedies were carried out on the sick themselves, of whom there was an all too ready supply. Since the numbers of sufferers was so large and fatality was so high, the authorities must have assumed that there was nothing to lose. And the victims – victims not only of plague but of the non-canonical remedies being tried out on them – could say little. All this helps us to understand the attitude of the Roman authorities before the proposal of our Neapolitan for an alchemical powder which had already showed itself to be 'effective' against the plague. Indeed a sort of trial had

---

[25] Archivio di Stato, Venice, *Secreta*, 95, 4v–5r (Laughran (1998) 223).

already been conducted when he cured both himself and then four other plague victims, who had been provided for him by the personnel of the pesthouse.

As far as the alchemical aspect is concerned, alchemy continued to promise precious metals, medicines, knowledge of the natural and super-natural worlds, stimulating enthusiasm throughout Europe. Seventeenth-century Rome was a city of contrasts, where a pope like Urban VIII could condemn the astrologer Orazio Morandi for sacrilegious activities, whilst celebrating magical rites with Tommaso Campanella (Fiorani (1978); Troncarelli (1985) 11–32). There alchemy, condemned as heretical by Sixtus V in the bull *Coeli et terrae* in 1586, found some degree of support amongst the Jesuits of the Collegio Romano, like Kircher and, later, Francesco Lana Terzi, particularly in its 'spagirical' form, that is the study of minerals, plants and animals for the extractions of their oils, spirits and quintessences, for medical use (Baldwin (1993a)).

In 1655, Rome saw the arrival of Queen Christina of Sweden, learned pro-tector and patron of scholars, artists and natural philosophers, including amongst the latter a few alchemists, such as the Roman nobleman Massi-miliano Savelli.[26] Princely courts routinely made large sums available for the financing of the laboratories and research activities of alchemists. Noble patrons hired alchemists to carry out a range of functions: to compound medicines, make artificial gems or pearls, give advice on mining activities, and produce the tinctures or powders which promised to turn base metals into gold (the so-called philosopher's stone). It is all too easy to assume that these court alchemists were all impostors and cheats. However, linked to the belief that the transmutation of metals was possible (even if difficult) was the belief in the 'virtues' of a series of products related to alchemi-cal operations. As a result, princes negotiated the terms of employment with their hired alchemists in a systematic and practical way, stipulating contracts which clearly outlined the obligations of both parties and for-malized the payments for services performed. Whether the alchemists so employed sought to create the panacea that was the philosopher's stone, or confined themselves to the preparation of chemical medicines along Paracel-sian lines, alchemy offered the possibility of medical marvels. This was a practical, entrepreneurial alchemy, carried out by self-taught alchemists, who were more interested in the production of goods than the production

[26] On Christina of Sweden and her circle, see Bignami Odier and Partini (1983); Åkerman (1991); Di Palma (1990). On the marquis of Palombara, author of a 1656 manuscript guide to the alchemical art, 'censored' by the author himself four years later, see Gabriele (1986).

of knowledge (Nummedal (2007)). Our Neapolitan clearly falls into this category.

Practising alchemy in time of plague brought its own risks, however. In Genoa, three alchemists, hunting for salamanders for their alchemical experiments, were suspected of spreading the plague, and were eventually executed.[27] But the search for the elusive philosopher's stone was worth the risk, as outlined in this poem by a Sicilian natural philosopher, dedicated to Christina of Sweden:

> Sta medicina pri li corpi umani
> Sana ogni morbu ed adequa l'umuri
> Né mai l'effetti soi saranno vani
> Quannu su li materiali veri e puri.[28]

Shame it was not true. The remedies available to seventeenth-century Romans were at best palliatives, and the hunt for a medicament effective against plague continued frenetically. (Queen Christina was wary enough to flee plague-stricken Rome for the relative safety of Paris.) A vast array of different products was tried out, their use and administration sometimes negotiated directly with their originators, like our Neapolitan alchemist. But if the Roman authorities could do little for the victims of plague, they could still take measures to prevent Romans from becoming victims, seeking to limit the spread of the contagion. It was a case of difficult choices, and the authorities opted for the hard line, rooting out filth and controlling the movements of people, within the limits of what was possible. Through the enforcement of boundaries separating the sick from the healthy, drastic measures enacted by Rome's Congregation of Health sought to ensure that people were kept in place and that order – moral and spiritual, physical and material – was maintained.

---

[27] When they fled to Piedmontese territory they were captured and ended up on the gallows by order of the duke of Savoy. Cambiaso (1908); Pastore (1988) 138.

[28] 'This medicine for all human bodies / Cures every disease and balances the humours / Its effects will never be shoddy / When the ingredients are true and pure.'

The poem, known only as 'Opera del siciliano filosofo siracusano', deals with the making of the philosopher's stone. Biblioteca Apostolica Vaticana, MS. Patetta 781, fols. 2r–7v (Bignami Odier and Partini (1983) 276).

10 | Was the ghetto cleaner...?

KENNETH STOW

Let me open with the words of Luigi Maria Benetelli di Vicenza, *docente di ebraico* in Padua and Venice, in 1603. He tells the following story:

> A Jew falls in a well, but it is Shabbat, the Sabbath, and so he says to a Christian who wants to save him:
>
> > *Sabatha sancta colo, de stercore surgere nolo.*
> >
> > 'Observing the Sabbath is my bit; allow me, then, my rest in this vile and stinking pit.'
>
> To which the Christian replies the following day, Sunday:
>
> > *Sabbatha nostra quidem Salomon celebrabis idem.*
> >
> > 'On our Sabbath, too, will you delight in the same foul stew.'[1]

Comment here is superfluous: offal is where the Jew belongs – for is it not his, or her, customary place? – the offal in question being Judaism itself. And what this offal needed most was flushing away, a job fit for the waters of a river, namely, the Tiber, which brings us also to Rome and its Jews. Referring to the Tiber floods, Rose Marie San Juan has said that: 'As a metaphor, water's flow disguises a preoccupation with urban boundaries.' Witness the words of a poem composed right after the flood of 24 December 1598, once again targeting the Jewish dog:

> It [the river] seized at once...the piazza of the Jews, this place is one of a kind...Then it went inside the ghetto to find the false, wicked and fraudulent Jews and if naked, naked made them leave their beds. To them it would do...disrespect, because they deserve it: these Jewish dogs, who are the enemies of all Christians [*perche lo mertan questi Giudei cani che son nimici di tutti i Christiani*].[2]

It is as though the Jews and their filth are flooding the city, which takes its revenge by flooding them back. The conflict between purity and impurity is clearer yet in the words of Filippo Maria Bonini, from 1663:

---

[1] Benetelli (1703) 410.
[2] Cited in San Juan (2001), ch. 4 ('Water's Overflow'), esp. 129 and 282.

This obstinate people got some benefit from the flood. They refuse to wash off the filth of the soul in baptismal waters; but at least those of the flooding Tiber wash off the dirt and stench of their bodies. [So obstinate are they, that they would rather] trust in their God [than have a priest come to their aid].[3]

It sounds as though Bonini had read the story just cited above. Jews are like the dirty waters they sit in: filthy dogs, who prefer (to eat) (even their own) filth, rather than, by implication, to consume the Eucharist of purification. Jews are also depicted as oblivious to the regulations about public sanitation that had been in force in Rome from no later than the mid-sixteenth century, testimony to which is visible as one walks through the streets of old Rome and sees these regulations inscribed clearly on building walls.[4]

To round out the picture are the words of David Pike, expanding (as it were) on San Juan's image, that sewers are perceived as boundaries: the threshold between the 'above' and the 'below'.[5] The description fits perfectly common *perceptions* of the ghetto as the 'sewer' *par excellence*. As depicted by drawings in San Juan's book, during plagues Piazza Giudea was entirely sealed off. The ghetto with its filth was also likened to a den of crime. In 1749 – the images had staying power – Anna del Monte, held captive in the Roman Casa dei catecumeni and steadfastly refusing to convert, was told to go back to the *canaglia* of a ghetto and its filth. Indeed, a century later, in 1871, with the ghetto walls down, there was fear that 'the filth' might spread throughout the city unhindered. Did not Pius IX lament that *i cani*, as Jews were also historically named, no longer fenced in, were going about *per le vie latrare*. As Gregory the Great had once observed, it would be better if these Jews 'went back to their vomit'. No wonder that truly to call a person low, one called him (or her, too) *il rifiuto dei Giudei*.[6]

---

[3]   Cited in San Juan (2001) 280: *Ritrasse tuttavia quella gente ostinata qualche beneficio dall'inondatione del fiume, mentre dove ricusa di lavar nell'acque del battesimo le sozzure dell'anima, videsi da quelle del Tevere mondar le spurchezze del corpo, e delle stanze, che per la puzza, e immonditia loro si rendono in qualche parte poco meno, che impratticabili...* [One miserable Jew refuses the help of a priest, telling him] *che non haveva bisogno del soccorso de Christiani, bastando a gli Hebrei invocar il nome di Dio... Fortuna fu di questi infelici che non fusse l'inondatione in giorno di Sabbato, perche si sarebbero d'ostinatione lasciati perire.* (emphasis added).

[4]   San Juan (2001) 129–37.

[5]   Pike (2005b). One also can discuss filth as a kind of conduit or reflection of passages in the city, much as the prostitute, in nineteenth-century London and Paris, was actually called a 'sewer' or a 'seminal sewer'. Yet the issue here is also one of ambivalence. Christians entered the ghetto, for business, and sometimes for the thrill of it. There were also the *neofiti*, converts, who constantly re-entered, despite laws that prohibited this. The ghetto threatens and thrills, therefore, simultaneously (as in nineteenth-century London and especially Paris, Pike adds, one took part in sewer tours). How much the more, accordingly, was there a need for borders.

[6]   See Stow (2006) xvii and 137, and the literature cited there. For Anna del Monte, see Del Monte (1989).

Nonetheless, as Pike adds, sewers fascinated, much like prostitutes, who, too, were often likened to sewers – just as Jews were often assimilated to prostitutes. It was here that the line between pure and impure was drawn, impurity being the traditional image associated with Jews, especially in Rome, where Jews had first lived centuries *before* there were any Christians or Christianity.

Jewish 'perversity' as encapsulated in the sewer image nonetheless reminded people that separation of the clean from the 'dirty' was often more ideal than real. Going far afield, we note Puebla Mexico in 1700, where just 14 per cent of the houses, in a city of 50,000, had running water; 'sewage flowed in the streets, mixed with excrement and loaded down with the refuse of butcheries, including animal blood, bones, and sometimes animal organs'.[7] Even at its worst, conditions in the ghetto were never this bad. Indeed, Rome's 3,500–4,000 Jews struggled to avoid anything like it, certainly from the time of the ghetto's inception, and probably from even before. They may have been forced to live tightly packed, and often among fouls smells, as Ferdinand Gregorovius reports, although much later, in 1853:[8]

> When I first visited [the ghetto], the Tiber had overflowed its banks and its yellow flood streamed through the Fiumara, the lowest of the ghetto streets... The flood reached as far up as the Porticus of Octavia, and water covered the lower rooms of the houses at the bottom. What a melancholy spectacle to see the wretched Jews' quarter sunk in the dreary inundation of the Tiber!

[7] Loreto and Cervantes (1994). Discussions of urban cleanliness and waste disposal in this period proved, to me, at least, hard to find.

[8] Gregorovius (1948) 89–90. It is worth bringing in more of this description, as follows: 'the ghetto houses in a row, tower-like masses of bizarre design... The rows ascend from the river's edge, and its dismal billows wash against the walls. It is only a few steps from the bridge to the ghetto, whose level is extremely low... the foundations of whose houses serve as a quay to hold the river in its course... Each year Israel in Rome [he is playing on the idea of Israel, in bondage, in Egypt] has to undergo a new Deluge, and like Noah's Ark, the ghetto is tossed on the waves with man and beast.

The physiognomy of the ghetto environment struck me as penetrating the atmosphere with gloomy imaginings... [the darkened aura of] the Porticus of Octavia... its ruined and blackened arches gaping upon the stinking Pescara, that crowded and dark fish market where the Jews' fasting fare is laid out on stone slabs... [or] The Place of Tears where stands an old palace [that of the Cenci]... If we now enter the streets of the ghetto itself we find Israel before its booths, [another play, this time on the Fall harvest festival of Sukkot or Booths] buried in restless toil and distress. The [Jews] sit in their doorways or outdoors, on the street which affords scarcely more light than their damp and dismal rooms, and tend their ragged merchandise or industriously patch and sew. The chaos is... indescribable.'

But this did not keep the Jews from 'fighting back'. The 'sanitary reality' they themselves cultivated, or more precisely strove to cultivate, was the reverse of what prejudice said.

That prejudice, however, could not, nor would it, be overcome. Even had the ghetto become sparkling pure in its physical aspects, in the eyes of most Christians, it would have remained as spiritually contaminated as the poems and other texts cited above pictured it. The 'Jewish stain' was sensed even when lawyers and legists strove to incorporate Jews into the civic body. For while these people regularly said that Jews were *de corpore civitatis ubi degunt*, they also affirmed that this membership was in the civic body alone. Membership, in its spiritual complement, hinged on receiving those baptismal waters which had the power 'to regenerate' – read: purify – and confer complete citizenship and civic rights; and even then the result was more often than not illusory. Converts were never trusted, and, in Rome especially, they were forced to belong to a distinct confraternity, where all their activities came under close scrutiny. The stain of being Jewish did not wipe easily, if ever, clean. However much, therefore, Jews – in real life – laboured to sanitize their precincts, their successes had no effect, indeed, were irrelevant, in moderating Christian views. Regardless of the physical reality, Jews and Judaism remained both polluted and polluting. In Poland, wrote Pope Benedict XIV in 1751, 'places [where Christians once lived and where they have been replaced by Jews] are now overcome with a stench . . . leaving Christians destitute'.[9]

Jews themselves, to be sure, saw matters from a contrary perspective. It was they themselves who were pure, the outside world a source of impurity. In their mind's eye, the ghetto walls, or so I have surmised, created a (kind of) walled Jersualem, realized on foreign soil, but still a pure place, a holy place, whose inhabitants these walls actually sanctified, set off, as God's people. That Jews and Christians competed on the subject of who of the two was pure, who impure, who was the contaminator and who needed to keep defences up is a theme I have examined elsewhere at length.[10] The urge to keep the ghetto physically clean, much as Katherine Rinne (this volume) speaks of the sparkling, if not holy, water flows nurtured in Rome by the popes, would have been an obvious extension of such thoughts. Jews, that

---

[9]   *A quo primum*, in Lora and Simionati (1994) 580: *Loca . . . diruta modo, situ et squallore foeda . . . Christianorum . . . destituta reperiantur.*

[10]   See Stow (2006), esp. ch. 5, where I discuss these matters in terms of Mary Douglas, whose purity discussions I perceive as very much modelled on Pauline perceptions of a spiritual world that needs defence against a contaminating physical one.

is, were impelled 'to keep the ghetto clean' by more than the municipal regulations which, as we will see below, they scrupulously observed.

It is easy to be misled. Gregorovius was writing when the ghetto was at its 'lowest' ebb, and from every possible aspect. At the start, in the mid-sixteenth century and for some decades thereafter, things were considerably less dismal. The attempt at order, especially with respect to waste, water and disposal, was sincere, conforming, I believe, to papal standards set for the entire city. Rome's Jews, ghettoized from July 1555, were as aware as anybody of the need to dispose of waste. The ghetto seems even to have possessed downward flowing toilets (as did Venice), which most probably were also the Roman norm. The consciousness Pius IV and Pius V (the latter's normally drastic policies towards Rome's Jews notwithstanding) had sought to instil throughout the city of the need for running water and adequate sewers had penetrated the ghetto, too.[11] Jewish concepts of urban cleanliness were up to date, and Jews took advantage of the infrastructures the above popes had built. This, Rome's Jews tell us themselves, in their own words, as they deposed (in Italian) before their own Jewish notaries,[12] speaking of toilets, baths, butcheries and smoke – as well as of the work entailed in constructing sanitary facilities, and sewers. It is the tone and affect reproduced below and the heavy documentation in the notes that tell their story best.[13]

There was, to start with, an awareness of 'clean' air, however Rome's sixteenth-century Jews understood the term. In 1537, two decades prior to the ghetto, one husband brought his sick wife to her brother's home, so that the she would have a 'change of air'. For the same reason, about a year later, a father and brother asked a son-in-law to bring a sick daughter to live with them.[14] By the same token, smoke was *scomodo e disgusto*, truly annoying. In 1580, Jewish communal officials obliged the owner of an *osteria* to clean the locale's chimney once a month, and similarly, they forbade opening a window next to a fireplace.[15] Fire itself was a potential hazard: one man

---

[11] Delumeau (1957–9).

[12] On these notaries, see Stow (2001), esp. ch. 3; see also (1995 and 1997): these volumes contain the summaries and often text of 2,000 of the Roman Jewish notarial acts, housed in the *Archivio Capitolino Storico*, Rome. The archival references in subsequent notes are constructed using the format: F., fascicle, l., book, f. folio. Almost all are outside the years covered by the two volumes above, 1536–57.

[13] The body of the essay will thus bring summaries, the notes replete with the original texts.

[14] F.11, l.2, f.20v, 1 May 1537, and F.11, l.2, f.89r, n.d. September 1538.

[15] F.8, l.3, f. 107v, 1 December 1636: *Essendo che Donato Menasci heb. a fatto una finestra al suo camino per pigliar luce p. d.o camino, la qual finestra e vicino alla loggia de Giacobbe q.m Vitale heb. di Lattes, la qual finestra d.o Donato non la posseva fare senza il consento del sud.o Iacobbe di Lattes, perchè dalla d.a finestra non se pol sperare se non scomodo e disgusto, per il fume che da quella uscirebbe e sarria disgusto e scomodo al detto Iacobbe, hora sono le sud.e parte convenute*

made un *tramezzo di pietra* to avoid damage such as fires had caused in the past.[16]

Foul smells were repugnant: a certain Ricca threatened to denounce to the governor's tribunal and to the *maestri delle strade* a woman neighbour who refused to fix a *necessario maleodorante*, a smelly toilet.[17] Smell, dirt and detritus were to be avoided, particularly in commercial areas, and especially in the production of meat, quite the opposite of what was noted above regarding eighteenth-century Puebla. The tenant of one butchery promised to keep it free of all *sporchitia*.[18] The owner of another promised to make a roof, to keep rain from making a mess.[19]

Cleanliness extended to bath-houses, used for bathing and barbering,[20] and, even more, to the *miqve*, the ritual bath used for *ritual* purification

---

d'acordo in questo modo, che d.o Donato promette e si obbliga non fare ne far fare in d.o camino di giorno foco di legna, ne meno metterce qualsivoglia sorte di cosa che possa far fume, overo si obbliga fare una vetriata a d.a finestra acciò non abbia a venire fume fora e che dasse fastidio a esso Iacobbe o suoi piggionanti, e non seguendo d.e patte come si è detto tra di loro convenuti, tra tre giorni prossimi d. Donato promette e si obbliga riserrar d.a finestra e mai più aprirla.

[16] F.10, l.2, ff.90r, v, 18 November 1607: *essendo che detto Pellegrino [q.m Angelo di Rignano] fa istanza de voler fare un tramezzo di pietra nella sua bottega, ove oggi c'è di tavoli, ad effetto che non li succedi e venghi alcun danno in casa sua, come altre volte le havenuto assucedere per conto di foco, per il che detti heredi [di q.m Salomone di Rinozza] e detto m. Davit [q.m Efraim Corcos] pretendono de non havere a contribuire in tal spese che vol fare detto m. Pellegrino, atteso che loro si contentano de starsi nel modo che stanno oggi et che non selli debbi restregnere ne rimovere niente del lor sito, conforme al modo che si trova oggi, e m. Pellegrino pretende che loro debbino participare alla spesa che lui farra.* See also F.3, l.2, f.1r, 5 December 1580: Angelo Capovano, Fattore and Sabato q.m Moise di Livola agree that once a month Sabato will clean the *camini* of the *ostaria* in order to avoid any possible damage.

[17] See Feci (1998) 580. The *maestri delle strade* began functioning in 1452, with statutes from 1480. For various papal letters regarding streets and sanitation, see Greenhalgh (n.d.).

[18] F.6, l.1, n. 344, 7 January 1587: Conzi q.m Meluccio, together with his son Meluccio and Menacuccio *pizzicarolo*, rent Meluccio di Haronetto a *loco de macello* for two years. Beside the rent, *Meluccio promette dare a Conzi la mita della carne de far le locaniche d'ogni bufala che ammazzara durante detta locatione, et Conzi pagarcela secondo il solito, et cosi' Meluccio si obbliga de tenere pulito detto loco e non ce tener sporchitia in nessun modo.*

[19] F.14, l.1, ff.113v–114r, 24 April 1589: The butchers are Christians, who, for reasons of Jewish dietary laws, can slaughter no meat, and the owner is a Jewish baker, a perplexing arrangement: Simone Veneziano (*di Veneziano*), Jewish baker, rents to Gironimo Romano, butcher, and to his partner Giamaria norcino *una parte del suo cortile, cioe 20 palmi de muratore, et si obliga detto Simone de coprircelo et farceli un tetto ben coperto, che non ci piova, et parato dinanti di tavoli, con la sua porta da servir con sarratura et cose necessarie, nela quale detti m. gironimo et compagni ci possino tenere la carne morta che loro farranno per uso del loro macello, ma non ci possino amazzare niuna sorte de bestia, anzi ce debbino mantenerci polito di continuo.*

[20] There were boxes or drawers to keep clothing, and money – under lock and key, for that matter – while people were bathing. One entered by ringing a bell, suggesting a certain concern over privacy, although nothing is said about times for men or women, who no doubt were never served at the same time. See F.10, l.1, ff.186r–v, 1 July 1597: Leone Trevis and Moise di

(most frequently) following menstruation.[21] The free-running water in the bath for bodily cleansing was kept hot with cauldrons, with water drawn from a special *pozzo*. The *miqve*, in turn, where women bathed nude and fully immersed, had to be flushed clean, including walls and rafters, and also

---

Cascian Ascarelli agree that *Leone da la sua stufa al detto Moise che la debbi essercitare detto Moise da oggi sin tanto che sia finita la lite de la detta stufa et che sia dechiarato a chi debbi restare, al detto m. Leone, overo alli tre stufaroli che la essercitavano p.a, che erano Iseppe di Pellestrina, Salamon Abiglia, et Moise di Cascian che hora la piglia*, for the price of $1\frac{1}{2}$ scudi a week; Cascian receives also the following items for the use of the *stufa*: *4 bacili di ottone, 10 concutelli di rame,14 succatori, tre cimarri, una di cotone, un altra compra dalli stufaroli, valutata sc.* $1\frac{1}{2}$, *un altra di panno vedaccio, un lavamano di rame con il suo ferro che sta giu nella stufa, un altro lavamano di ottone, un lavatore di testa grande di rame con il piede, nove rasoli, tre para di forbici napolitani, una sede granna, doi banchetti, un ferro de appiccare il lavamano, un specchio, 4 lenzolicchij da assuccare le persone, nove cassetti da porci li vestiti in tre pezzi, un focone de legno, un vigliatore da tenere la candela de legno, un campanello.* Also F.2, l.6, ff.251v–252r, 10 October 1603: Inventory of utensils for running the *stufa*: '*Robbe per bisogno della stufa: Caldari murati n. 4, vetriati con rete n. 2, un lavamano di rame, una traglia di ferro, doi secuti di legni con cerchi di ferro et manichi di ferro, un coperchio di ferro della fornacella, sette concoline di Rame, quattro banconi di legno da tenere li vestiti con sei chiavi e sei serraturi, un focone granne con soi spurtelli et cuperchio, otto ammattonati, quattro inpannati rotti e vecchij, una cassetta murata da mettere dinari senza chiave, con una serratura, doi cappoli neri di legno; una chiave d'ottone grande da dar l'acqua alla vasca della stufa.* And F.10, l.2, f.27r, 22 April 1607: a woman rents a barber her *stufa* and rooms; the *Descreta m.a Giuditta, moglie di Moise q.m Livuccio di Piperno, detto Rossetto*, in the presence of her father Leone Trevis, rents her *stufa* to Samuele q.m m.o Giuseppe di Palestrina, barber, plus two rooms and a 'half hall', *e ditto Samuelle possi godere le quattro caldari murati et il ferro de la stufa et doi ritricati et reti che stanno in le finestre.* F.10, l.2, ff.40r–v, 3 June 1607: Leone q.m Rabbi Vito Trevis receives back from Samuele q.m m.o Giuseppe di Palestrina, barber, the following things, *le quale Robbe sono tutte le Collatelle ch'io dovevo havere, che servivano per la stufa, li cassetti da tenere robbe, li coppolinci, il focone grande, le secche di legno con ferri, la taglia di ferro, la corda che serve per il pozzo* [leased by him for the use of the stufa] *e mi contento che detto Samuelle possi godere il ferro grande che se tiene avanti la bocca che scalda la stufa durante il tempo della locatione fattali da mia figliola.* Finally, F.10, l.2, ff.144v–145r, 27 April 1608: Moise q.m Leuccio di Rosetto (Levuccio or Leone di Piperno, called Rossetto, or Roscietto) and his wife Giuditta, not present, rent to Leone q.m Rabbi Vito Trevis, father of Giuditta, *la stantia al paro di cortile onne oggi ci sta dentro Samuel q.m m.o Giuseppe di Pelestrina, ove ci tosa et essercita la barbaria et anco li alloca la stufa et la Cantina al paro alla stufa*, for 25 scudi a year.

21 The *miqve* serves as a place of ritual purification, frequented at the end of the menstrual period by a wife who has seen no blood for a week, not even a stain, meaning also that the husband is obligated by religious precept to satisfy her sexual needs on the evening she bathes. Publicly flaunting one's going and coming from the *miqve* was a way of flaunting ongoing (marital) sexuality, for sexuality's (not procreation's) sake, as indeed Judaism prescribes. That the ritual purification of the *miqve* should be combined with physical cleanliness and aesthetic enhancement tells us much about how Jews viewed their own persons, at least in Rome at this time. One might mention, in passing, though, that more than one wife seeking a divorce, for reasons including spousal violence, simply refused to declare herself 'clean', that is, clean of blood for a week. Only the menstruate can decide this, no one else. In practice, this meant no sexual relations, and the ruse worked, more than once; see here Stow (2001) 79.

properly lit.[22] Each synagogue had a *miqve*, and the women who operated that of the Sicilian Scola were to provide clean linen, apparently for both the *miqve* itself and the synagogue.[23]

But what of waste and disposal? It seems to have required a special permit to throw water in the street.[24] Toilets and basins connected to sewers were the norm. Toilets were assumed in every rented dwelling,[25] or they were shared, as rental contracts specify. This included the right of the one without the toilet to pass through the dwelling of the other; the concept of privacy so cherished today, let us recall, is modern. Refusal to share could end in arbitration.

Toilets also seem to have had an emptying mechanism; for instance, in the phrase '[they will construct] a *cataratta nel Corritore*'. Lacking this, we read that one Raffaele also had the right to empty the *Cantaro* in the toilet that stood in the room occupied by Yehudah once a day, although no more than that.[26] Where there were no toilets, owners had to install

---

[22] F.10, l.1, f.14v, 23 April 1595: *il lavatore et bagno delle donne di detta scola . . . et si obligano dette donne et Gabrielle de evacuar allor spese il lavatore quando fara bisogno et mantenerlo detto loco netto e pulito*; F.1, l.5, ff.47r–v, 31 January 1627: The depositories of the Scola Siciliana rent out the cellar and bath of the Scola for one year. Angelo di Diodato da Sezze, *depositario della cassetta di Scola Ceciliana*, and Giuseppe (Iosef) q.m Sabato Saadun (Sagadun), also depository, rent to Sabato di Leone Sacerdote *tutto il sotterraneo di scola Ciciliana, cioè cantina, bagno e tutto il contenuto in essa per un anno*, [and Sabato] *si obliga de mantener detta cantina netta, pulita, e se reserbano ancora li suddetti a favor di d.a Scola Ceciliana di poterce metter e conservar legnami, travicelli o altri simili in detta cantina*; and F.10, l.1, f.42v, 25 August 1595: the Camerlenghi of Scola Catalana and Aragonese lease a house and *miqve* to a banker's wife and his sister. Moise Soschin Tedeschi and Emanuele q.m Anselmo Sacerdote, *camerlenghi* of the Scola Catalana–Aragonese, lease to Laura, wife of m.o Giuseppe (Iseppe) di Palestrina, banker, and to madonna Marchigiana, his sister, *il bagno con la casa dove sta detto bagno della detta scola . . . per il che dette donne si obligano de mantenere detto loco netto e pulito allor spese et simelmente lavare le lampe et candele di vetro della detta scola quante volte farra bisogno lavarli.*

[23] F.10, l.1, ff. 187r–v, 12 January 1597: The Camerlenghi of Scuola Siciliana, Abramo q.m Rubino Avdon and Angelo del Presto, rent again to Gentilesca q.m Isach Negri and Gentil Donna, wife of m.o Giuseppe di Palestrina, *il lavatore de lavar li panni et il lavatore delle Donne detto da noi hebrei miqve* [for one year].

[24] F.14, l.1, ff.8r–v, 24 January 1588: Palomba q.m Ventura di Segni and sons Isach and Gabriele rent to Giuseppe q.m Rabbi Benedetto di Camerino, for two years, *la lor bottega con il solaretto e la cantina, cioè che m.a Palomba possi tenere un banchetto dalla banda verso casa Agnelo de Scazzocchia e buttar laqua nella strada et andare in la Cantina quando li fara bisogno.*

[25] With regard to rentals, Jews at most held long-term and inheritable leaseholds, known as *cazagà*, from the Hebrew, to hold, or a presumption, but codified in Roman civil law as *ius cazagà*, guaranteeing rent stabilization. See Stow (2001) 11–12.

[26] F.10, l.2, ff. 23r–v, 26 March 1607: *se per caso se farra in detta casa una cataratta nel Corritore* [but in its absence] *che li sia lecito de andar a buttare il suo Cantaro nella stantiola dove cè il necessario e buttarlo pero ogni giorno et la mattina avanti pranzo*; the *Cantaro* being some kind of recipient, bucket, etc. F.13, l.1, f.45v, 25 April 1613: *per essere che il s.r Giulio di Maggistri nobile Romano vol fare nelli soi stantij, posti nella torre odel passatore, ovve oggi ci sta il detto*

them;[27] and denying someone else permission to use one's toilet during the construction of their own was a cause for legal complaint. Similarly, a barter agreement – for instance, a loan in exchange for free rent – stipulates toilet privileges. A sale, too, is made contingent on installing a drainpipe, the *sciaquatore*.[28] The presence of sanitary devices was *de rigueur*. A proper apartment had a *necessario*, a *sciaquatore* and a *focolar*, a fireplace.[29] These, I note, are clauses in contracts of 'execution', which describe not what was theoretically expected, but what actually was. These toilets and drains were truly being built and used; whether the standards of cleanliness were identical to ours, of course I cannot say.

But what was to be done with waste materials from *necessarii* or the dirty water carried by a *sciaquatore*? Apart from odours (and it has been said that in Britain even in the nineteenth century, odours rose to living rooms),[30] waste created havoc in cellars, the *cantine*. One Sabato thus agreed 'to make a drain, either a trench or a barrel [vault, as in the ancient Cloaca Maxima]... for the *necessarie* [toilets] were damaging Sabato's *cantina*'[31] (in Venice, downward tubes in the houses led to the canals). The waste from

---

Pineca, alcuni necessarie, li quali li conviene darli la strada che vadino giù nelle grotte o sia cantina che possede il detto Sabato, parimenti beni del detto s.r Giulio. Alternatively, a rental specifies: F.3, l.2, f.77r, 17 February 1583: Mastro Giovan Maria di Lorenzo Marini falegname, who lives near the river, rents to Samuele di Sabato Cappellaro and his wife Vittoria an apartment under the roof, for 8 scudi; the owner will pay for improvements, including a toilet (text in Hebrew).

27 F.9, l.1, ff.97v–98r, 30 April 1579: 'The room [exchanged] is the one Angelo lives in. If Angelo wants to switch rooms Yoav must agree, on condition that there be free use of the toilet in the other room' (text in Hebrew); and F.5, l.3, foglio incollato tra f.184v (a) and 185r (b), 5 February 1617: witnesses on a *scomunica* concerning the use of a water-closet: (a) *Benedetto de Veletre disse che lui avisto tutti quelli de casa de Moretto annare a quel destro in casa de detta Graziosa e non cera ne tramezo e ne scala e quella de moretto ce teneva li panni e li dava sc. 1 lando a detta Graziosa*, and (b) *Io Dunato de Miele hebreo fo fede come moretto e li figlioli se servivano de un neciesario di che sta alla ientrata den casa Gratiosa quano vi era la alavorare con Liucio de Latore e non ciera trameso ne la deta casa.*

28 The *sciaquatore* was a drainpipe of some kind, although later it might be used for a wash basin, perhaps connected to a drain. Another rental, F.9, l.1, f.125v, 18 February 1580, specifies that Raffaele and his family have 'the right to throw only water in the *sciacquatore*' and cannot use it for any other reason without Yehudah's permission (text in Hebrew).

29 F.13, l.1, f.109v, 6 December 1613: *et potendoce Ioseph farci in detta cam.a Il necessario, saquatore e focolar*; and F.10, l.3, f.46v, 15 May 1611: the costs for improvements in a house are shared by tenant and leaseholder. Simone di Aron di Segni states that Isach Zarfati, procurator for his sister Ester, has paid 6 scudi for improvements made by Simone in the house owned by Isach, *che sono un camino, un necessario, un sciaquatore*.

30 See on this Pike (2005b).

31 F.13, l.1, f.45v, 25 April 1613: *tanto che ci facci chiavica quanto fossa o botte...[since]... facendo danno alcuno li detti necessarie nella cantina di detto Sabato... si obliga Pineca di farli accomodare.*

toilets was carried to the river by sewers.[32] This continued the tradition of the Cloaca Maxima (see Hopkins, this volume) and made Rome special in this regard, even in the sixteenth century, or so I believe. Jewish documents give us a good idea of how at least one or two such drains or sewers (connected to private houses and often shared) were built and of the procedures necessary before construction could begin. With an eye to sharing expenses, the owner of a bath-house (not a *miqve*) agreed to two neighbours linking their *necessario* to his *chiavica*.[33] Angelo di Segni and Robino Samen did the same.[34]

Once built, drains, sewers and their ambient had to be kept clean and in good repair.[35] No doubt the same materials used in construction would be kept on hand, including *calcia, pietra e pozzolana*.[36] The whole process was supervised by the *maestri delle strade*.[37] One prospective constructor emphasized that the *maestri* must first issue a licence before works began (sounding modern indeed).[38] More prosaically, the builders' pay had to

[32] On drainage and sewage in Rome in general at this time, see Rinne, this volume, as well as her website, www3.iath.virginia.edu/waters (accessed July 2011); and Pecchiai (1948) 455–59, where he notes that Clement VIII levies a tax *per raccomodare la volta sfondata et muri della chiavica di Borgo vecchia . . . e per rinettare la detta chiavica sino al fiume.* Hence, the drainage projects in the ghetto correspond to drainage in Rome as a whole, which is my essential point in this study.

[33] F.10, l.3, ff.159r and v, 23 May 1612: the owner of a *stufa* gives permission to two neighbours to connect their water-closets to his sewer. Leone q.m Rabbi Vito Trevis agrees that Daniele di Tivoli, represented by his son Moise, and Giuseppe (Ioseph) q.m Strugo Bises *possano inboccare li lor necessarij nella bocca della mia chiavica della stufa.*

[34] F.5, l.4, ff.24r–v, 21 February 1622: Since Angelo q.m Isaia di Segni has made *la chiavica nella sua cantina, del quale ce reiescie il necessario de Robino Samen q.m Gusef Samen*, they agree to share the building expenses.

[35] F.13, l.1, ff.207r and v, 14 December 1614: David q.m Abramo Sacerdote and Isach q.m Salomone di Efraim Corcos, *homini deputati dell'Università dell'hebrei di scola Castigliana* to see to the rental of houses whose leases are held by the *scola*, rent a house for three additional years to Rabbi Giuseppe (Ioseph) q.m David di Bezalel di Cori, and *bisognanno acconciar il tetto de detta habitatione Rabbi Ioseph debbi pagarela spesa della fattura e calcia che ce bisognara e la scola pagare tevoli e canali che ce andranno; . . . Item convengono e vogliono che bisognanno nettare il necessario o la chiavica di detta casa, ne debbi pagare un terzo Rabbi Ioseph . . . dovendosi fare la salciata avanti detta habitatione ne debbi pagare Rabbi Ioseph tutto quel tanto che aterra a rabbi Ioseph alla sua habitatione, Conforme alla tassa delli s.ri m.i di Strada.*

[36] F.1, l.5, f.111v, 17 September 1628: Azriel and Leone q.m Simone Veneziano *se son convenuti fra loro daccordo che nel nettar della cantina del d.o Azriel e far la chiavica, conforme al hobligo della devisione fatta fra lor fratelli, il d.o Azriel a pigliato sopra di se,* [which included] *de farce tutte le spesi tanto de Architetti, muratori, calcia, pietra e pozzolana* [a hydraulic cement discovered by the Romans and still used in some countries, made by grinding pozzolana (a type of slag; volcanic ash)] *eche il d.o Leone non sia hobligato di metterce più che scudi dieci.*

[37] On the *maestri*, see n. 17.

[38] Again F.10, l.3, ff.159r and v, 23 May 1612: *possano inboccare li lor necessarij nella bocca della mia chiavica della stufa. Impero ciò sia con la licentia li Ill.mi M.i di strada impero ciò sia con la licentia li Ill.mi M.i di strada.*

guaranteed,[39] tenants pacified, and the owners, under whose homes a new sewer might pass, insured against damage; the tenants are Jews and the owners Christians.[40] Then, both *architetti* and *muratori* (architects and builders) had to be hired; the work had to be professional.

One text provides great detail, and, more to our point, it reveals the deep roots of the Jewish concern for sanitation. I refer to the drain and other sanitary devices installed at the home of the Compagnia del Imparar li putti hebrei (di leggera) (the Talmud Torah Society, the equivalent of an elementary school), which all Jewish children whose parents could not afford proper tutors attended. This means probably the majority of Rome's Jewish children. The task was to hook up a trench (or line, or pipe) leading from various toilets and wash basins to a sewer running down to the river; it was undertaken to free the school from 'the constant expense the current [open] trench caused, not to mention damages'. However, it was first necessary to get the agreement of the tenants, the people living in the homes and stores, under which the *chiavica* (the drain) was to pass. The communal leadership, as well as the heads of the Talmud Torah Society, thus were empowered to deal with these people, and the deal was made.

The society's heads agreed that 'excavations, columns and conduits were to be dug and placed in the manner the architects and builders judged best'. Otherwise, the pipe and the sewer would be improperly aligned. The conduits were to run 12 palms (lengths) under the earth, no more and no less (less being too close to the surface, more no doubt being too close to underlying water). They were to begin in the garden (surely a very small one) and come out again in the street – no doubt meaning to pass under the street or there would be a need for a sump pump, which, of course, did not exist. Then the conduits were to pass under a second building in the same way. The tenants promised 'to pose no objections or block the works', including when pillars had to be built to link the conduits to what is called the *chiavica maestra*, a large sewage line running through the ghetto to the Tiber, whose existence, it appears, has recently been discovered.

The next clause guarantees tenants and owners, as well as their heirs, against loss, specifying that any and all expenses were the Talmud Torah Society's obligation. And, finally, water from a fountain belonging to the

---

[39] F.10, 1.3, ff.159r and v, 23 May 1612, upon payment of $4\frac{1}{2}$ scudi made to m. Leone di m.o Iacomo *muratore suave*.

[40] F.10, 1.3, ff.159r and v, 23 May 1612, *se venisse danno alcuno al detto m. Leone* [Trevis – a second Leone – the owner of the *stufa*] *per tal inboccatione, che siano obligati contribuire alla spesa*.

Talmud Torah would be diverted 'to ensure that sewage be carried away and the drain and eventual sewer be kept clean'.

The agreement sounds as though it was made yesterday, its precision being the fruit of well-tried practice, not improvisation. Basins, toilets, latrines and waste removal were a regular part of ghetto life, and it is clear that the entire procedure accorded with Roman norms, including the licensing.[41] Yet as I have stressed above, reality and image when it came to Rome's Jews were on a path of constant collision. No matter how the Jews

---

[41] The entire text follows; quotation marks in the body of the essay concerning this episode refer back to here. F.1, l.5, ff.114r and v, 4 October 1628: *Per esser che la compagnia del Imparar li putti hebrei di leggera de bisogno de inboccar una Cantra con diversi necessari e sciaquatori in chiavica che corra al fiume, ad effetto libberarse dalle molte spesi e pericoli che giornalmente la d.a cantera a dato e da, de qui è che avendo li hoffiziali de detta compagnia cercato de voler accomodar detto loco, e per questo anno linfrascritti hoffiziali data autorità alli cammorlenghi che se toccano in hoffizio al giorno di hogge, che sono m. Sabato q.m David di Graziadio e m. Samuel del q.m Salvator di Sorano hebrei, de concordarse col li eredi del q.m Robino de Capranica heb.o, ad effetto loro diano il passo da casa loro ho dalla lor bottega alla Immondizia che dalli necessari e sciaquatori e cantera della casa dove hogge reside la d.a Comp.a, e essendo che il passo che averranno da la casa ho bottega delli suddetti sia la più comida e con maggior comodità, e minor spesa che loro possano avere, con aver li suddetti hoffiziali fattane scrittura, e da cinque di loro sottoscritta, e qui da verbo in verbo copiata nel modo e forma seguente. Copia: Al nome de Dio, adi' 4 di 8bre 1628, in Roma, Noi infrascritti humeni deputati dalla congrega e dalli sig.ri fattori della nostra comonità heb.a a dar strada e via al nettar la Cantra e inboccarla e far hogne spesa necessaria e opportuna nella casa che tiene la compagnia de Inparar li putti heb.i, avendo visto che non vien bene la d.a inboccatura di cantera se non passa per sotto la casa del sig. Camillo Mignanelli, hogge abitata dalli eredi e figlioli del q.m Robino di Capranica heb., da qui è che vedendo noi il d.o bisogno e necessità, danno autorita e facola a m. Sabato del q.m David di Graziadio e a m. Samuel del q.m Salvator Sacerdoto da Sorano heb.i, al presente Camorlenghi e ministri della detta compagnia del imparar li putti, di concordare con li suddetti figlioli et eredi del q.m Robino di Capranica, e de poter con loro istrumentare, e tutti hoblighi e scritturi che in tal caso farranno aprobbamo e vogliamo abiano loco e fermezza . . . In prima li suddetti Robino e Sabato di Capranica concedono licenzia et danno facoltà alla detta compagnia et per quella alli suddetti camerlenghi che possano inboccar tutti necessarij, sciaquatori e canteri fatti e da farse per benefizio e comodità della detta casa, e che possano passar sotto la lor bottega, dalla banda che confina con m. Moise Achim, con darli facoltà che a hogne lor beneplacito e per qual se voglia tempo possano cavar e far pilastri et tirarce condotti nel miglior modo che alloro ho allor architetti e muratori parera; con questo che detti condotti debbano passar dodici palmi sotto terra e non meno e più, allor beneplacito e che possano prencipiar dal giardino e venir fora in strada, con passar detti condotti per la d.a casa di Capranica nel modo sud.o e promettano li suddetti Sabato e Robino non dar alcuno inpedimento ne molestia, tanto per far pilastri bisognosi quanto condotti e inboccar in chiavica maestra e hognaltra cosa che bisognara per la detta Chiavica per qualsivoglia tempo davenire, e in ricompenza del d.o benefizio li suddetti camorlenghi in nome della detta compa.a promisero di far tutta la detta fabrica a spesi della d.a Compagnia, in modo tale che li suddetti eredi no ne abiano spese, danno ne nocumento alcuno . . . e concedono da hogge li detti camorlenghi, nome come sopra, che li suddetti di Capranica e lor successori possano inboccar in detta chiavica da farse da loro come di sopra tutti necessari, sciaquatori et hognaltra cosa che per lavenire ce inboccaranno e promiserò ancora li suddetti camerlenghi, nome come sopra, che laqua del ritorno della fontanina che se trova al presente in detta casa della d.a Conpagnia, passara per la detta chiavica acciò porti via la immundizia e che mantenghi netta la d.a chiavica.*

laboured to cleanse the ghetto, the image of it as a repository of filth refused to go away or even to be lessened. But since when has any Jewish behaviour or any other Jewish reality had the capacity to dismantle the Jews' negative image? The ever-polluting 'Jewish dogs' had to be cordoned off.

Yet there was also a lighter side, as the following story, which has nothing to do with sanitation *per se*, makes wonderfully clear, and from which – to make an awful pun – we learn just how *necesarii* latrines might be. Signora Ricca was apparently having an affair with one Signor Yequtiel, who had come to visit. Ricca felt safe, since her husband was at work. However, his unexpected return at midday left her and her lover with their pants mutually down. To pull herself and her costume back together, Ricca escaped into the toilet.[42] In another version, Yequtiel himself escaped through the toilet window. *Necessario*, indeed.

---

[42] F.1, l.1, f.76v, 12 August 1560 (in Hebrew): Yehudah di Rabbi Sabato declares that he took a statement from Mrs Bensivenuta, the wife of Moise, concerning all she knew about what he had asked her, namely whether she saw Mrs Ricca, wife of Yehudah Zaccai, daughter of Raffaele Zarfati, do something ugly with a young man by the name of Yequtiel, son of Mrs Nora. She answered that one day, on one occasion, after 12 noon, 'I saw Mazliach standing close to my bed; and at that moment Yehudah came in and said "I have already found you here", and Yequtiel did not answer anything'; F.1, l.1, f.77v, 18 August 1560: in front of the Congregation and the Memunim of the Scola Siciliana the parties try to solve the problems of the alleged case of adultery involving Ricca, wife of Yehudah Zaccai; F.1, l.1, f.78v, 23 August 1560: Ricca, wife of Yehudah Zaccai, states that her husband had called Yequtiel to help at the market; they lived in the house of a widow, Sara di Segni, and when the fact happened there was a question of an open door to the bathroom; the husband apparently came home at midday and called her, but she delayed, saying she was in the toilet; Yequtiel was at that time in the house, to borrow a pair of scissors; yes, her husband went fishing at night, but the children were in the house; she does say that Yehudah is the suspect adulterer; she states that since the hour that she was *be qjnian* (*ba-shuq*) (shopping, *la spesa*) with her husband a silly spirit came unto him; that is, she does not know if her husband did that out of love or because he wanted to go away and leave her as the 'living widows' (a technical term for desertion); besides, weren't her children at home; and if there is a question of their saying anything against her, one child was only eight years old; F.1, l.1, f.80v, 29 August 1560: in the Scola Catalana–Aragonese the arbiters Rabbi Yehudah di Yoav, Yehudah di Rabbi Sabato and Abramo di Aron give their decision concerning the case of the supposed adultery committed by Ricca, wife of Yehudah Zaccai; they state that the husband has not proved his case, so he must live with her and provide for her and the children; he has to sleep with her and may not hit her or leave Rome by more than 40 miles without her permission; she must obey him, and she is forbidden to be alone with the suspect (Yequtiel). In a further text, F.1, l.1, f.81v, 1 September 1560, the story ends 'well': Ricca, wife of Yehudah Zaccai, appears to state that Samuele Zadich spoke to her father saying that her husband intends to travel to Turkey. She says that thirty days after she gives birth (she is now pregnant), she will go with him 'everywhere he wants, and this I will do, to go with him to all places, and abide by his commands, and do his will as modest women do who serve their husbands faithfully and happily'.

11 | Urban ablutions: cleansing Counter-Reformation Rome

KATHERINE RINNE

Wracked by sin and corpulent from greed, the Catholic Church was a diseased body that had grown polluted from 'the long putrefaction of vices' that it had encouraged over several centuries, leading to its disgrace and ultimately to the Reformation.[1] Martin Luther and other Protestant reformers claimed that the Church's spiritual and moral decay was the outcome of a religious structure in which popes operated like kings; cardinals lived like princes; monastic communities were unregulated; clergy were corrupt; pastoral care was capricious; indulgences were sold like lottery tickets; and immorality was rampant in the general population. Pope Paul III convened the Council of Trent (1545–63) to respond to these allegations. When Pius IV ratified the Council's decrees in January 1564, he instigated strict disciplinary rules that were intended to abolish heresy, expand the powers of the Inquisition, subjugate and convert the Jews, and encourage church attendance and sacramental practices in order to reform the spiritual lives of Christians, while reforming the administrative practices of the Roman Curia and clarifying the doctrines and rituals of the Church. As though attending to an invalid recovering from a serious and long-lasting disease, Pius IV and other Counter-Reformation popes, especially Pius V, were determined to prevent a relapse in their patient and to ensure that it atoned for its sins.

Rome, as the centre of the morally depraved Church, was also polluted. As the primary locus of decay it too was clearly rotten to its core – 'an asylum of the wicked' and the centre of abominations – a foul environment that nurtured degeneracy (Rossetti (1894) 2, 10). No longer could one marvel at 'Rome and all her noble works' as Dante had done (31.34). Rather, one saw that in general the streets were impassable, houses had collapsed, drains were clogged, aqueducts stood in ruins, fountains were dry, the ports were in disarray, melancholic ruins had crumbled, the banks of the river Tiber were foul and the threat of outside attack was constant. Meanwhile prostitutes, beggars and bandits roamed through the city more or less at large (see figure 11.1). This disorder encouraged and also reflected the corruption of

---

[1] Cyprian, *On the glory of martyrdom*, in Wallace (1869) 233, writing about the putrefaction of the individual soul.

V effigii del Castello dell'acqua Martia terretigie, the cui è doto è lugta dove si vedinauano l'acque delli aquedutti, qual servitua per distribuire detta acqua in diverse parti della città, sopra queste edifitio doua ciguota A, si si aggua logge doi trophei di marmoro, quali intauda alcuni autori sono quelli dirigati in honore di Mario, p la victoria che hebbe contro li Cimbri popoli, stella parte signata B. e la strada Tiburtina di en altia porta do s.º Lorenzo, si nella parte signata C. si la strada Prenestina che va a porta Maggiore

11.1 Standing ruins of the so-called 'Trophies of Marius', the once monumental terminal fountain for one of Rome's eleven ancient aqueducts. The line of some standing arches can be seen on the far left. Etienne Du Pérac, 'Vestigii del Castello dell'acqua Martia', in Du Pérac, *Vestigi dell'antichità di Roma* (Rome, 1575). © The Vincent Buonanno Collection.

the Church. As the Counter-Reformation Church literally fought for its life by instigating strict rules among clergy and parishioners alike, it also waged war on pollution in Rome in an organized and systematic fashion not seen since antiquity. These efforts to cleanse the morally polluted Church, as embodied in the decrees of the Council of Trent, provided the impetus for cleansing the polluted city systematically.

It was in this hostile and corrupt milieu that Counter-Reformation popes sought to protect fragile souls (including thousands of pilgrims who flooded into Rome each year) by controlling corruption and vice in the public realm – in particular by sheltering these souls from marginal populations like Jews, vagabonds and prostitutes; shielding them from heretical ideas; and removing temptations that could lead to sinful thoughts and acts. It became an urgent concern to control contact with these abject groups through segregation. Paul IV had already sequestered the Jewish population, which was seen at that time as morally tainted (see Stow, this volume). In 1555 he had literally herded them into a *serraglio* – a walled and gated enclosure that in this case was later known as the Jewish ghetto – in order to isolate them from Christians. There was also a deep-seated fear that

Jews, who were considered by the Church to be physically unclean, carried and passed diseases such as the plague (Byrne 2004). Vagabonds, unlike religious mendicants who begged alms, were seen as thieves (San Juan (2001) ch. 1), and were routinely excluded from strategic or ceremonial public spaces, as were prostitutes, who offered temptation at nearly every corner. The free-ranging presence of prostitutes, like that of vagabonds and Jews, caused extreme anxiety to the Church, so Pius V expelled the women from Rome in 1566, and later placed them in their own *serraglio*, known as the Ortaccio, along the banks of the Tiber near the Mausoleum of Augustus (Delumeau (1957–9) i.424–5).[2] Many aspects of daily life that the Church saw as immoral or dissolute were rigorously policed and punished. Games of dice and cards were banned, fortune telling was prohibited, pilgrims were discouraged from sleeping under porticoes or begging, members of religious orders were forbidden to ride in carriages, and sumptuary laws were enacted to limit luxurious display. It was through these and other measures that the Church could forcibly demonstrate its control over a hostile urban environment as it sought to rebuild its spiritual integrity.

During the first thirty years directly following the Council of Trent the 'Counter-Reformation ideal of social discipline was pursued with greatest determination' (Storey (2008) 240). And it was exactly at this time that the most fervent efforts were undertaken by these same popes to cleanse Rome physically so that it would mirror a newly reformed Church and the redeemed souls of its inhabitants. It was incumbent upon Rome, as *caput mundi*, first to 'sanctify itself . . . in order to offer its visitors neither reason for scandal nor material or spiritual annoyances' (Gregory XIII, 1575),[3] because in Rome even the 'appearance of the ground . . . [and] everything offered to the eyes' had to impress 'the soul with something sacred' (St Carlo Borromeo, 1574).[4] But how could tender souls be nourished, transformed and redeemed in a city that had fallen from grace and stood in ruin?

Rome's physical decline over nine hundred years had begun to slow once Martin V returned the papacy to Rome in 1420. Beautiful new palaces like the Cancelleria (1489–1513), and churches like Sant'Agostino, begun in 1483, were constructed on an enormous scale in the new Renaissance style that favoured hierarchy and symmetry. Treatises by Francesco Filarete, Francesco di Giorgio, Leon Battista Alberti and other architectural theorists offered

---

[2] For studies of prostitution in Rome, see Cohen (1998).

[3] Pope Gregory XIII, 'Bull of Indiction, 1575', paraphrased in Levillain and Malley (2002) 727 ('Holy Year'). Gregory XIII is misidentified as Gregory XII.

[4] Carlo Borromeo, 'Pastoral letter, 10 September 1574', quoted in Levillain and Malley (2002) 727 ('Holy Year').

detailed plans on which to organize cities and to create wholly new ones – including how to integrate systematic sanitary measures, like sewers and freshwater supplies. New, straight streets were cut through the contortions of Rome's medieval neighbourhoods, and Martin, Nicholas V, Sixtus IV and Clement VII all put legislation in place that was meant to maintain a sense of decorum in the public realm, particularly through regulations regarding the disposal of human waste, debris and rubbish. Yet, as noble as these efforts were, they were not linked as an integrated system at the scale of the city.

Much of the progress that had been made since 1420 was rendered null by a series of setbacks that occurred in the early sixteenth century. The 1527 sack of the city and a ruinous Tiber flood in 1530 – the worst in its history – ravaged and depopulated Rome. These events not only caused physical destruction and loss of life from drowning, but also left the city vulnerable to disease from standing water and lack of food. Another sack was threatened between 1556 and 1557, but then only one day after peace was concluded in September, the city was invaded by another exceptional flood. That flood severely damaged most of Rome's bridges including Ponte Sant'Angelo and Ponte Santa Maria, caused widespread devastation in the Campo Marzio and the Borgo and killed hundreds of people. The northern flank of the Leonine Wall was threatened with collapse, while the inner embankment of Castel Sant'Angelo was completely destroyed, as was its recently constructed defensive bulwark begun by Paul IV in 1556 (Frosini (1977) 166–70). Nonetheless, the city grew rapidly after 1527, creating a water and sanitation crisis that was exacerbated by the deplorable physical conditions of the buildings, bridges, streets and drains.

There was also a water crisis in Rome. Most inhabitants relied on the squalid, dangerous and fetid river Tiber, which flowed through the city, for their domestic supply. One ancient aqueduct, the Aqua Virgo (built by Marcus Agrippa in 19 BC to serve the Baths of Agrippa and by this time called the Acqua Vergine), still functioned and sporadically served the Campo Marzio, while the Acqua Damasiana, a small, early Christian aqueduct, served the Vatican and St Peter's. This meant that little fresh water flowed into the city for drinking and cooking, and even less water was available to flush the *chiaviche* (drains) that were filled with debris thrown into the streets. From the fourteenth century on, the Statutes of the Roman Comune required private individuals to maintain the street directly in front of their own property and to remove all their rubbish and debris and take it to Piazza Navona or to Porta Settimiana in Trastevere, depending on where the individual lived. From these points, official rubbish collectors in turn disposed of everything along the banks of the Tiber (Re (1880) 172). A 1452 edict from the *maestri delle strade* (civil servants with responsibility for public works) specified that fishmongers and butchers dump their waste in the Tiber, as opposed to public streets and piazzas (Pastor (1938–61) XVII.108). Although these regulations had been in place for centuries, they were seldom enforced – a fact made clear by the number of official edicts that were published on a regular basis condemning the repeated disregard for the *salus publica*, 'public health'. This was still the case in the mid-sixteenth century. Because of lax enforcement, the streets and piazzas were typically mired in rubbish, excrement and debris, and awash with standing water. Over time, many of the drains were transformed into sewers (*cloache*).

Ancient Rome's once substantial system of drains, originally built to drain surface streams, barely functioned in the sixteenth century (Rinne (2010) 193–203). It is never easy to keep drains clean, and this was especially true during the medieval period in Rome when many of them became completely blocked with rubbish, debris, subsiding soils and collapsing structures. It is clear that some ancient drains, which by default became sewers, were restored and functioned during the high medieval period, but most of them must have been rendered inoperable in 1084 when the invading army of Robert Guiscard sacked Rome. His departing mercenaries burned and pillaged huge areas of the city, generating tons of architectural rubble that fell into and over the drain mouths, effectively burying some of them for hundreds of years (Lanciani (1988) 337).

Not only the drains were buried, but also natural springs (Corazza and Lombardi (1995) 198–9). Constrained by gravity to flow downhill, and trapped under earth and debris, the water saturated the surrounding land, making it an ideal incubator for mosquitoes and malaria. This was the case with the ancient Spring of Juturna, which once bubbled in its own fountain in the Roman Forum near what was later the Church of Santa Maria Antiqua. But with rising ground levels it was lost under several metres of earth. The ancient Cloaca Maxima had been built to drain this area and carry the water from the Forum Brook – into which Juturna's water flowed – to the Tiber (see Hopkins, this volume). But the ground had become so boggy by the sixteenth century that a *ponticulus*, a 'small bridge', was built to facilitate crossing over *dell'inferno* (as this marshy land was known) to reach Santa Maria Antigua. Other areas were also notoriously damp because of buried springs and streams or damaged drains. The Campo Marzio between Porta del Popolo, the foot of the Pincian Hill and Via del Corso was often mired in mud, as was much of the area known as the Pantano that extended from

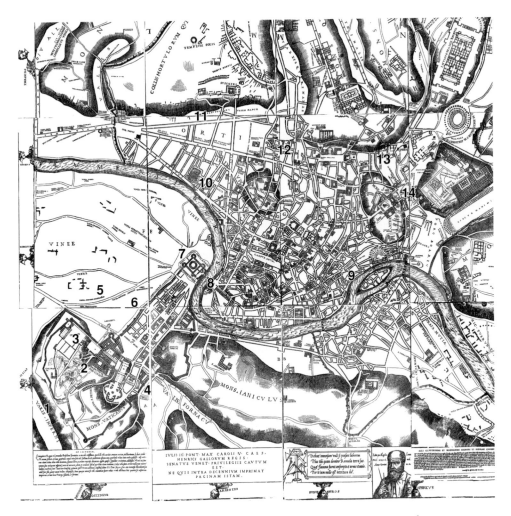

11.2 Public works projects carried out by Popes Pius IV and Pius V between 1560 and 1572 in many areas of Rome: 1. Vatican gardens; 2. Vatican palace; 3. Belvedere; 4. Porta Cavalleggeri; 5. Valle d'Inferno; 6. Borgo Pio; 7. Castel Sant'Angelo; 8. Ponte Sant'Angelo; 9. Jewish ghetto; 10. Ortaccio; 11. Foot of the Pincian Hill; 12. Trevi Fountain; 13. Pantano; 14. Roman Forum. Reproduced from Leonardo Bufalini, *Roma*, 1551 (1560 reprint).

the Forum of Trajan to the Basilica Julia and the foot of the Velian Hill (Lanciani (1990) II.227–8; see figure 11.2).

Drains were clogged from both the street-level opening and their river mouths. The motive force of the river during flood-state could push debris into the drains (which had never been provided with traps), and then at times cause water and waste to back-flow and spill out into streets and

11.3 Although the Column of Trajan was treated with respect as a venerable ancient monument throughout the medieval period, the area around it did not fare as well. Several metres of earth and debris were removed from its base in the 1560s as part of the *risanimento* (drainage) effort in and around the area known as the Pantano. Then new drains were built and linked to the ancient Cloaca Maxima. From Etienne Du Pérac, 'Disegno della Colonna Traiana', in Du Pérac, *Vestigi dell'antichità di Roma* (Rome, 1575). © The Vincent Buonanno Collection.

piazzas located quite far away, just as it did in 1548/9 when Tiber water pushed into the area around the Fountain of Juturna. Penelope Davies, in this volume, makes clear that this was frequently the case too in antiquity. Because of this, the drains were essentially inoperable and it is likely that many of them remained permanently clogged for decades (if not centuries) at a time. As drains clogged, street levels also rose dramatically from natural causes like flooding and man-made causes like fires. By the sixteenth century, levels in the Roman Forum, for example, had risen as much as ten metres since the Republican period, while at the same time the ground continued to become saturated by the trapped water (Lanciani (1988) 115; see figure 11.3).

Because of the rising street levels during the medieval period, it was easier to construct a parallel drainage system several metres above the old ones than to excavate through the accumulated soils to reach ancient drains. Without a reliable water supply to flush debris, it was in any case dangerous to connect to ancient drains. The easiest and most common tactic to deal with waste involved creating a surface-level channel down the centre of the typically unpaved street to carry away rainwater and any debris that was illegally thrown into the street. But there were exceptions. High-traffic areas like the Borgo and the Portico d'Ottavia, where Rome's fish-market had been located since the medieval period, were among the few areas that had been served by underground drains and brick-paved streets that had been sporadically maintained at least since the thirteenth century (Lanciani (1989) I.32, 74). The constant flow of pilgrimage traffic, whether on foot, on horseback or in carriages, generated filth, dirt and rubbish in the Borgo, while the fishmongers who worked at the Portico d'Ottavia produced enormous quantities of waste that had to be cleaned out on a daily basis. The Portico and its immediate surroundings had been home to many Jewish residents since at least the fourteenth century (Calabi (2006) 27) and the nearby Piazza Giudea was the centre of Rome's 'rag trade' (San Juan 2001). In 1555 this area became the walled Jewish *serraglio*, which symbolically sealed its identity as an unclean environment, in spite of efforts to clean and maintain its streets and drains (see Stow, this volume).

Clearly this piecemeal method of building and maintaining drains would not support an agenda of cleansing Rome of pollution. It was essential to develop a systematic approach at the scale of the city that would address the movement of fresh water into the city and the manner in which waste water flowed to the Tiber. Even before Pius IV ratified the decrees intended to cleanse the Church, he had already set this rational strategy in motion with three initiatives to cleanse Rome. First on his agenda was a project to restore the Acqua Vergine so that the area of Rome known as the Campo Marzio could receive a reliable freshwater supply and not have to rely on polluted Tiber water for drinking. Second, he devised a project that would fully integrate freshwater distribution and wastewater drainage infrastructures at the Vatican and the Borgo while at the same time protecting the area from outside attack. The Vatican project was linked to his third initiative: to control Tiber flooding, which caused regular and serious damage.

Pius acknowledged that there was still a real threat of attack, both from outside armies and from the Tiber. It was his goal to make the Vatican safe from flooding and self-sufficient during siege, and to accomplish this he had the defensive wall around Castel Sant'Angelo rebuilt and resumed construction on the bastions and moat envisioned by Antonio da Sangallo the Younger in 1537 (Millon and Smyth (1996) 113; see also figure 11.4). But that was only one part of the plan. Integral to it was a complete reworking of the freshwater distribution and wastewater drainage systems at the Vatican,

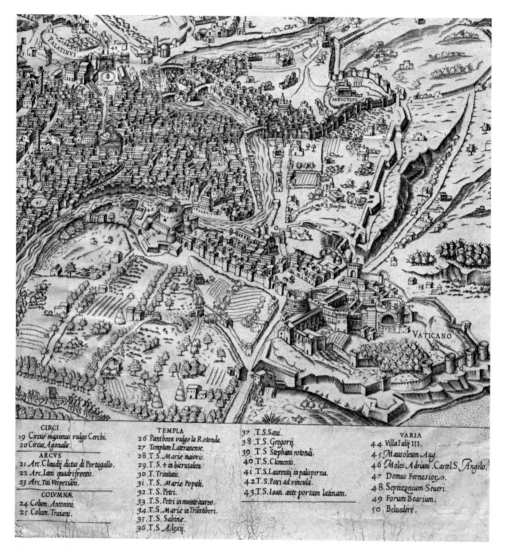

CIRCI
19 Circus maximus vulgo Cerchi.
20 Circus Agonalis.

ARCVS
21 Arc. Claudij dictus di Portugallo.
22 Arc. Iani quadrifrontis.
23 Arc. Tit Vespesiani.

COLVMNÆ
24 Colum. Antonini.
25 Colum. Traiani.

TEMPLA
26 Pantheon vulgo la Rotonda.
27 Templum Lateranense.
28 T.S. Marie maioris.
29 T.S. + in hierusalem.
30 T. Trinitatis.
31 T.S. Marie Populi.
32 T.S. Petri.
33 T.S. Petri in monte aureo.
34 T.S. Marie in Trastiberi.
35 T.S. Sabine.
36 T.S. Alexij.

37 T.S. Saui.
38 T.S. Gregorij.
39 T.S. Stephani rotundi.
40 T.S. Clementis.
41 T.S. Laurentij in palisperna.
42 T.S. Petri ad vincula.
43 T.S. Ioan. ante portam latinam.

VARIA
44 Villa Iulij III.
45 Mausoleum Aug.
46 Moles Adriani. Castel. S. Angelo.
47 Domus Fernesior.
48 Septizonium Seueri.
49 Forum Boarium.
50 Beluedere.

11.4 The Vatican, St Peter's, the Borgo and Castel Sant'Angelo in 1561, shortly after Pius IV embarked upon a systematic infrastructure project. The Belvedere (50) can be seen to the lower right while the Fountain of Santa Caterina (R) is seen in the piazza in front of St Peter's. The New Borgo Pio will be built on farmland just outside the Leonine Wall, while the moat around Castel Sant'Angelo is under construction. From Giovanni Antonio Dosio, *Roma* (Rome, 1561), detail. © Biblioteca Hertziana, Rome.

St Peter's, Borgo Pio and Castel Sant'Angelo that was conceived of as a unified project. Work began at the level of the individual latrine and extended to ornamental fountains and the defensive moat. Cleaning Vatican latrines, including those of the Horse Guards who lived next to Porta Cavalleggeri, was the first order of business between 1560 and 1562. Shortly thereafter, Pius sponsored the construction of a new public fountain at the same Vatican gate, which provided water to the guards and their horses and to the public who entered through Porta Cavalleggeri. The run-off water was used to flush the drain of the new latrine. The new fountain took advantage of a spring known as the Acqua Pia, which flowed near the Villa Cesi at the foot of the Janiculum. The spring had been discovered in 1562 during the excavation of the clay beds located in the Valle dei Fornaci – clay that was used, in part, to make the new drains at the Vatican (Corazza and Lombardi (1995) 191).[5]

The various freshwater cisterns and the conduits leading to existing fountains in the Vatican palace, Vatican gardens and Belvedere were cleaned and repaired at approximately the same time. Beginning in June 1561, the first payments are recorded for a major initiative to clean and repair existing drains that carried run-off water from the Belvedere, the garden fountains and the palace to the *chiavica grande*, the 'large drain', which passed through the Borgo to the Tiber. The Borgo drain was thoroughly restored between 1562 and 1564.[6]

In early 1563 work began to construct a new *conserva dell'acqua* (water tank) for the Giardino Segreto, the pope's private garden in which he was completing the magnificent Casino Pio that Paul IV had begun. Next, the Acqua Damasiana was refurbished. Originally built in the fourth century under Pope Damasus I (366–83), the aqueduct had been restored by Bramante in 1504 when he also restored the Santa Caterina fountain in Piazza San Pietro (see figure 11.5).[7] Pius IV also provided water to this fountain (which in turned served an animal trough with its run-off water) after restoring the Acqua Damasiana. Another aspect of his larger plan dealt with overcrowding in the Borgo. Shortly before his death, Pius sponsored the construction of a new residential area known as Borgo Pio just to the north

---

5 See for example Archivio di Stato di Roma, *Camerale 1, Busta* 1521 *Fabbriche* (1560–8), 9r, 30 September 1560; 14v, 31 December 1560; 16v, 31 January 1561; 19v, 31 April 1561, for cleaning latrines. For the spring, see Ceselli (1873) 105; Corazza and Lombardi (1995) 191.

6 Payments continued into June 1564. Archivio di Stato di Roma, *Camerale 1, Busta* 1521 *Fabbriche* (1560–8), 9v 30 September 1560; 61v, 7 August 1562, for restoring cisterns; 21r and 22r, 31 May 1561; 25r, 30 June 1561; 26r, 31 July 1561, 33r, 30 November 1561, for cleaning Vatican drains; 21r, 31 May 1561, for restoring the existing Belvedere fountains; 70v, 9 March 1563, for remodelling the lower court of the Belvedere and removing the Fontana Belvedere; 57r, 15 June 1562, for Giardino Segreto; 51r, 30 April 1562; 59r, 3 July 1562; 61v, 7 August 1562; 78v, 10 June 1563; and 111v, 2 June 1564, for restoration of the Acqua Damasiana; 44v, 28 February 1562; 50v, 18 April 1562; 52r, 10 May 1562; and 55r, 25 May 1562, for Chiavica de Belvedere; 61r, 7 August 1562, for Piazza San Pietro.

7 Infessura (1890) v.254; Lanciani (1989) i.87 and 239–55.

11.5 Bramante restored the Santa Caterina Fountain (served by the Acqua Damasiana) in front of St Peter's in 1504 for Pope Julius II. Pius IV also restored the same aqueduct and fountain. An animal trough that received the run-off water was located just out of the picture to the lower right of the fountain. Etching by Martin van Heemskerck, c. 1535. From C. Hülsen and H. Egger, *Die römischen Skizzenbücher von Marten van Heemskerck im Königlichen Kupferstichkabinett zu Berlin, mit Unterstützung der Generalverwaltung der Königlichen Museen zu Berlin* (Vienna, 1911).

of the *passetto* (the wall connecting the Vatican to Castel Sant'Angelo). He had three new streets laid out there, with drains to link to the *chiavica grande*, and also had a defensive wall built to protect the new quarter.[8]

By 1565 when Pius IV died the defensive moat around Castel Sant'Angelo was nearly complete. This undertaking involved more than simply building a defensive ditch. Rather, this was an earthwork project intended to reshape the landscape outside the Vatican walls to the north. In addition to digging the actual moat, it also required building impressive banked earth defensive walls, diverting the Acqua Valle d'Inferno (a stream that flowed to the Tiber), and reconfiguring all the irrigation ditches in the orchards and vineyards that flanked the entire length of the stream. In addition to increasing the defensive potential of the Castel Sant'Angelo, Pius also sought to use the moat as a means to protect the Borgo from flooding. He envisioned that

[8] Archivio di Stato di Roma, *Camerale 1, Fabbriche* (1560–8), 1521: 211r, 31 May 1561.

the flowing water would 'ornament' the Borgo while eliminating *mal aria,* bad air, thus rendering the entire area more salubrious and habitable.[9] Although the moat remained unfinished at the southwest corner until the pontificate of Urban VIII, Pius V finished the restoration of Ponte Sant'Angelo in 1566, thus linking the Campo Marzio more firmly to the Vatican (Ceccarelli (1938) 9).

Henry A. Millon and Craig Hugh Smyth ((1996) 112) point out that it was the 'coincidence of fortress and bridge' in this location that 'urged measures for fortification and flood control to be considered and undertaken together'. But this project is clearly much larger than that 'coincidence' as it entailed a large portion of the Vatican area both inside and outside the walls; and we are impressed not only by the scope of the work, but also by its careful sequencing. All existing water infrastructures – from drain spouts on the roof of St Peter's to the latrines of the Horse Guards and the drainage ditches of orchards outside the Vatican walls – were integrated and linked to connect the farthest reaches of the Vatican gardens and the Belvedere to the banks of the Tiber, Castel Sant'Angelo, Borgo Pio and Ponte Sant'Angelo.[10] Water management, the development of amenities for pilgrims, urban beautification and defensive projects were seamlessly integrated into a plan to cleanse the Vatican and protect it from attack.

Pope Pius V has rarely received credit for his urban interventions. When compared with the grand rhetorical spaces and straight streets that other popes inserted into Rome's urban fabric, his projects are relatively modest. Whereas Julius II sponsored the Via Giulia, and Paul III initiated Michelangelo's Campidoglio, Pius sponsored hidden conduits and drains – urban amenities that provided a framework for public works projects related to the *salus publica.* Rather than carve out a piazza, he provided water for new fountains that served the populace and animated existing public places; and rather than insert new straight streets, he provided the conduits and drains beneath them that kept them clean and afforded a more healthful environment. Rhetorical spaces are clearly seen to be political expressions of papal absolutism (Nussdorfer (1997) 161). But what could conduits and drains, hidden beneath the streets, express about papal intentions? As we shall see, they provided Pius V with the water that was necessary for Rome's ablutions and represented means to implement a Counter-Reformation agenda

[9] Biblioteca Casanetense, Editti 18, 2: 525, 1565, 'Bulla S. D. N. D. Pii Papae IIII Confirmationis iurisdictionis...', in particular, 'Ordinationes Nove'; and Archivio di Stato di Roma, *Bandi* III, 22 August 1565.

[10] Archivio di Stato di Roma, *Camerale 1, Mandati* 921: 195r (payments from May to September 1566; and 922: 3r, 5 April 1567).

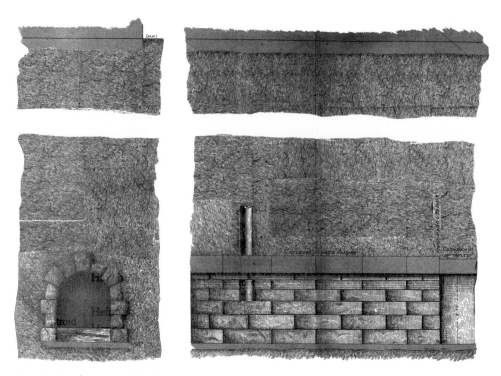

11.6 Section drawings of the Cloaca Maxima make explicit the amount of alluvium that had accumulated in the Roman Forum since antiquity, when the ground level in this area was about 11 metres above sea level. In this nineteenth-century survey drawing the ground level is shown at 20.61 metres above sea level, while the base of the drain sits at 6.65 metres. From P. Narducci, *Sulla fognatura della città di Roma* (Rome, 1889), plate 6.13 (top).

in which a clean city would mirror and support a newly purified Church and the larger Christian community.

The Vatican project offered Pius V a model that he could apply at a larger scale; along the Tiber, in the Campo Marzio and in the district called 'Pantano'. Conditions along the Tiber were particularly vexing – and they were directly related to the health of the drains. The ancient historian Livy (1.22.27–8) had once praised the Cloaca Maxima as a marvel of ancient Rome because it carried away all the city's filth (*receptaculum omnium purgamentorum urbis*) in an underground channel (figure 11.6; see also Hopkins, this volume). In antiquity as in the sixteenth century, that filth all flowed to the Tiber, which was a foul and unregulated environment; so foul that Pius V invoked Livy's words to describe the Tiber with contempt, not praise (Lanciani (1990) II.32). For him, the Tiber was nothing but a sewer that symbolized Rome's decline from a city of marvels and faith into a city filled with immorality and corruption.

It was clear to Pius V that conditions in the Tiber could not be remedied without first dealing with the drains that led to it, and also that it was also necessary to increase the amount of water that was available to flush them so that debris would not remain on the banks of the river, but would be pushed into its current and washed out to sea. To increase the water supply meant developing an integrated regional system that linked all the way from freshwater aqueduct springs located outside the city to conduits and fountains inside the city, and then to drains that led to the Tiber. Thus his primary task was to complete Pius IV's restoration of the Acqua Vergine, which was Rome's major source for pure water at the time although it probably served only three public fountains: at Piazza di Trevi; at the Church of Santa Maria Maddelena near the Pantheon; and perhaps at the Church of Santa Maria della Canella in Via dell'Umiltà (Fea (1832) 27). There were also a few public cisterns and small springs that provided water to public fountains like the one that Pius IV sponsored at Porta Cavalleggeri. But Tiber water, although highly polluted, was always abundant and was used for cooking, drinking, kitchen gardens, laundry and industrial purposes by the majority of the population. *Acquaeroli*, 'water-carriers', collected water from the Tiber (decanted after a week, which allowed the sediment to fall to the bottom), which they then sold throughout the *abitato*, the densely inhabited areas of the city located in the low alluvial plain on both sides of the river.

The healthfulness of the Tiber was contested although strong arguments were offered well into the sixteenth century in support of its salubriousness. In a 1552 book dedicated to Pope Julius III, its author, Alessandro Petronio, argued that Tiber water was preferable to that of all others including the Acqua Vergine, and that it should be used universally for drinking and cooking, because it was without equal for smell, taste, smoothness and clarity. In 1556 (the year following Julius' death) Giovan Battista Modio, a Florentine doctor living in Rome, published a treatise on the Tiber that opposed that of Petronio. In this treatise, which he presented to Cardinal Ranuccio Farnese, the nephew of Pope Paul III, Modio repudiated Petronio's argument, saying that the 'Tiber was useful for some things, but that for drinking . . . it was the ruin of the city'. Andrea Bacci countered Modio's treatise in 1558 by arguing that the Tiber's water was exceptionally healthful.

The Aqua Virgo/Acqua Vergine had been damaged in 537 during the Goth attack on Rome. It was restored several times until the tenth century but not again until perhaps the fourteenth century. Nonetheless, it had sputtered along, but sometimes failed altogether. Piecemeal repairs had been ongoing since 1453 when Pope Nicholas V sponsored a partial restoration of the channel – but not of the springs that lay sixteen kilometres outside the city. That work quickly proved to be insufficient and further repairs were undertaken over the next one hundred years, by which time water shortages were critical. Pius IV resumed its restoration in 1560, yet, for reasons too complex to address here, little work had been completed when he died in 1565.[11] Pius V took up the mantle from his predecessor and finally brought the project, which included a restoration of the catchment basin at the springs, to completion in August 1570. It would not be immoderate to claim that this restoration initiated Rome's most comprehensive water distribution plan since antiquity – a plan that was shaped under Pius V's direction and which provided the basic framework for future public works projects.

It is important to note that the effects of the restoration of the aqueduct had ramifications beyond the obvious ones of increasing the water supply and making it available for general use through new ornamental and utility fountains that were located in the most important public piazzas. Just as the restoration of the Acqua Damasiana was essential to Pius IV's systematic programme at the Vatican, so it was the new, reliable Acqua Vergine supply that made possible the restoration of drains in the Campo Marzio. Without having completed the Vergine restoration there would not have been the necessary motive power to remove waste from city streets and to allow it to flow freely into the river current from the underground channels. The run-off water from fountains of all types was essential for cleansing Rome and creating a healthier environment for its inhabitants.

Shortly after Pius V became pope, Rome was ravaged by a particularly brutal epidemic (perhaps typhoid) that was, of course, felt most severely in the Campo Marzio. The situation was aggravated by standing water that collected at the foot of the Pincian Hill, where springs seeped from the hillside and where some water also issued from the broken aqueduct channel. As Pius V's agenda for a healthy city included draining marshy areas, this increased his fervour to complete the aqueduct restoration initiated by his predecessor.[12] Restoring the aqueduct would doubly serve the city by

[11] Long (2008); Karmon (2005).

[12] Biblioteca Apostolica Vaticana, *Urb. Lat.* 1040: 286v, 31 August 1566; AC, *Decreti*, 23: 6v, 6 July 1566; and 22v, 9 July 1566; and Archivio Storico Capitolino, *Decreti*, 24: 26r, 9 March 1569. *Et ex cod. S. C. sanitù extitit unia voce S. summenda esse scuta quinquanginta et . . . aqua' Salonis pro expurganda et removerela aqua putrefacta subtus montem S.ta Trinitatis, ad manudendù Aerem salubrem pro bono publico.* Pius V established the *Congregatione cardinalizia super viis*

improving the water supply and minimizing standing water because the inner surfaces of the aqueduct were sealed with waterproof cement.

Closing open drains that ran along the edges of streets, restoring ancient underground drains and constructing new ones were all part of Pius' programme to rid Rome of 'putrid water and to maintain salubrious air for the good of the public'.[13] The Chiavica di San Silvestro, an open drain that ran through the Campo Marzio, had been a notorious health hazard since at least the fifteenth century.[14] It ran for almost two kilometres from the Trevi Fountain to the Tiber through the area that was among those most affected by the 1566 epidemic. For Pius, it became the focus of a major public health initiative that he linked to the restoration of the Acqua Vergine and the Trevi Fountain that terminated it. Once the aqueduct restoration was completed in 1570, it was possible to tackle the open drain. In 1571, in order to implement this plan, Pius V had some of the run-off water from the Trevi Fountain diverted before it reached the open drain. He had it rerouted it to a new fuller's facility for the manufacture of woollen cloth and then replaced the open drain with a new closed drain.[15] Like the original, this one followed the shortest route to the Tiber, which respected the topography of the Campo Marzio and the existing street pattern. The closed drain ultimately linked to an ancient drain-cum-sewer at the Mausoleum of Augustus (now known as Chiavicone di Schiavonia) that already flowed in the same general direction, but deeper underground (Narducci (1889) 14–15).

For Pius V, creating access to pure water, improving drainage and removing stinking wastes were of paramount importance, perhaps equal to that of controlling prostitution, reforming clerics and cleansing parishioners' souls. But, unlike moralizing sermons that could be delivered anywhere, the delivery of fresh water was constrained by topography. This meant that although Pius ordered that water be piped to 'all the squares in the City' it could in fact only flow to certain piazzas.[16] Thus, the environment of

---

pontibus et fontibus (Cardinalate Committee in charge of roads, bridges and fountains) on 31 July 1567 to oversee and facilitate the restoration of the aqueduct. See Archivio di Stato di Roma, *Libri Congregatione*, 1: 3, 31 July 1567.

[13] Lanciani (1990) iii.31; Pastor (1938–61) xvii.106.

[14] Re (1920) 69, 78. See also Archivio di Stato di Roma, *S. Silvestro in Capite*, 4996:3 11–12r, 9 September 1463. The drain was cleaned and restored in 1538, but in 1569, it was still described as giving off a foul odour and causing death.

[15] Archivio di Stato di Roma, *Congregatione*, 1: 20, and Cassio (1756–7: 1.284). Narducci (1889) 14–17, pl. 28: 3 (who surveyed tens of kilometres of ancient and early modern drains at the end of the nineteenth century) cites several examples of newly constructed drains linking to restored, ancient drains.

[16] Archivio di Stato di Firenze, Medici Granducal Archive, MdP 3080: 22008, folio 825, 30 September 1570.

only some of Rome's inhabitants would improve, while that of others would not. Why? Because water flows downhill, so obviously only those persons who lived near to or at a lower elevation than the water source could derive immediate benefit from it, while others might have to walk great distances or simply go without. But Pius was concerned about the whole city, not only the Campo Marzio. He used the same strategy in other parts of Rome that were as yet unserved by aqueduct water but nonetheless susceptible to standing water. The first was a waterlogged zone quite near to the Roman Forum that was made up of two connected marshy areas: the Pantano Spolia Christi, around the Column of Trajan, and the Pantano di San Basilio, stretching from the Forum of Augustus to the Basilica Julia. In 1562, Pius IV had already ordered the *maestri* to clean and repair completely a *chiavica* that had once drained the Pantano Spolia Christi, from the Forum of Trajan to a point where it intersected with the Cloaca Maxima as it ran under the area of the Forum Transitorium from the north. Five years later, under Pius V, work began to build a drain for the Pantano di San Basilio, which linked near the same juncture.[17] Once the *risanimento* (drainage work) was complete, the ground level was raised and two new streets were built: Via Alessandrina and Via Bonella (Lanciani (1889) 30–1; see also figure 11.7).

Between 1569 and 1570, Pius V initiated another drainage project at the foot of the western slope of the Quirinal Hill in the area of the Piazza San Marco and Piazza Altieri. Here, the *maestri delle strade* had the Cloaca Minerbe et Camiliani (the Cloaca Minerva) cleaned and restored and then linked to another ancient drain that flowed south to the Tiber. The Chiavica Massima per il Rione di Pescaria, the drain for the fish-market that stood outside the Jewish ghetto, was also restored at this time. This drain connected to another barely functioning drain that originated inside the ghetto.[18] Neither of these areas yet received Acqua Vergine water (because of their distance from its terminus at the Trevi Fountain), although fountains were already planned for Piazza di San Marco and Piazza Giudea.[19]

---

[17] For Spolia Christi, see Archivio di Stato di Roma, *Presidente delle Strade*, 445: *Taxae Viarum*, 425r–6v and 428–32, 1562. Cardinal Bonelli (Pius V's nephew) oversaw the work at Pantano San Basilio. See Lanciani (1992) iv.29–30.

[18] For the Cloaca Minerva, see Archivio di Stato di Roma, *Presidente delle Strade*, 445: 499, and 502r–14v, 15 January 1569; and for the ghetto, see 445: 106r–8v. Pius V also issued a *motu proprio* 'Dei nostri almae urbis' for draining the Campagna. For this, see Lanciani (1992) iv.14.

[19] See Rinne (2010) 45, 72. The San Marco fountain was completed in 1587 and the Giudea fountain was completed in 1593. Rome's hydraulic situation at this time was similar to what it had been in the late first century BC, when the city's water infrastructure had suffered a long period of neglect, and Agrippa turned to the task of its restoration. See Blake (1947) 159; Favro (1992) 61–84.

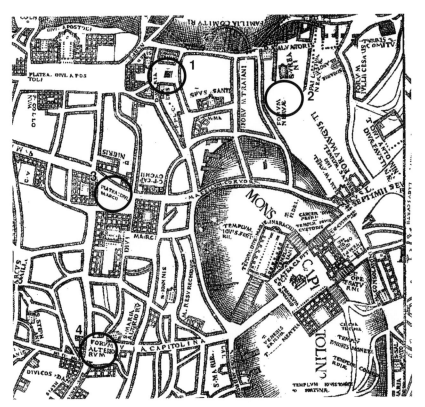

11.7 New drains were constructed and ancient ones restored in order to help dry out the saturated land in areas that were not yet served by the new Acqua Vergine. The Column of Trajan (1) and all the area between it and the Pantano (2) linked to the ancient Cloaca Maxima. The Cloaca Minerva, between Palazzo San Marco (3) and Palazzo Altieri (4), was cleaned, restored and linked to an ancient drain that led to the Tiber. From Leonardo Bufalini, *Roma*, 1551 (1560 reprint), detail.

With this valuable infrastructure now in place below the surface of Rome's streets, it was imperative to protect it. To accomplish this, streets were paved with stone, especially in the areas surrounding the new fountains and directly over their conduits. As water was piped throughout the Campo Marzio, stone-paved streets protected the conduits from carriage traffic. The first newly stone-paved street was Via Paolina (now Via Babuino), under which Acqua Vergine conduits carried water to the new ornamental fountain, animal trough and public laundry basin in Piazza del Popolo and to private persons who lived nearby.[20] Other stone-paved streets followed

[20] See Rinne (2010) ch. 3 for a discussion of water distribution, and ch. 9 for a discussion of stone-paved streets. For a history of stone paving in Rome, see Cibin (2003).

11.8 Small stone pavers known as *sampietrini* were used to pave over and protect Rome's newly constructed water conduits and drains as part of Pius V's integrated public works project. The stones are still used in Rome's historic centre. © Katherine Rinne (2009).

until Pope Sixtus V outlawed them in 1588, insisting on brick (Cerasoli 1900). But after Sixtus' death in 1590, stone once again became the Roman paving material of choice, as it had been in antiquity, since it was clearly a superior material for protecting the conduits and drains (figure 11.8).

Unfortunately, Pius V died in 1572 and was not able to see all his plans through to fruition. Most lamentable was his inability to correct the abuses along the Tiber. But at least the necessary infrastructure that would provide water to clean the drains that flowed into it was beginning to be put in place. The framework that he developed would be amplified many times over by his successors, in particular Sixtus V, Paul V and Urban VIII. Scores of new fountains were built over the next thirty years and beyond. These beautified the city and also provided fresh drinking-water fountains and dedicated basins for animals, laundry and industries, all of which were necessary components of a healthful city.[21]

The early years of the Counter-Reformation in Rome were characterized by efforts to reform the Catholic Church as an institution, to restore personal piety through moral cleansing, and to renovate Rome's urban fabric. Cleansing started with the individual soul and the body of the Church and, as we have seen, extended to the streets and piazzas of the city and to the river Tiber. But was this attempt to clean Rome different from earlier efforts carried out by Martin V, Nicholas V and Sixtus IV, all of whom had actively promoted urban projects intended to improve sanitary conditions? It can be argued that the principal difference is that earlier efforts were not typically effective beyond a limited area – a piazza or a street – and could not systematically address the entire body of the city as did the projects instigated in the late sixteenth century. It seems clear that the Counter-Reformation provided a moral imperative, lacking before the ignominy of the Church had been fully acknowledged, to cleanse the city systematically just as the decrees of the Council of Trent were intended to cleanse the Church systematically. Cleaning Rome was no longer simply about aesthetics, but was about absolution and the resuscitation of a dying city. For this, Rome needed more than a poultice: it needed a transfusion, which was provided by the new freshwater supply that coursed through the city in conduits and fountains, and then systematically flushed the city with drains and sewers. While it is unclear whether polluted Christian souls were transformed through the decrees of the Council of Trent, there can be no doubt that Rome's integrated water infrastructure system helped to cleanse the polluted city physically, and to initiate its transformation from a squalid nest of filth and mud into a city where once again we can marvel along with Dante at 'all her noble works'.

---

[21] For the relationship between cleansing souls and cleansing linens, see Rinne (2001–2).

The clash of picturesque decay and modern cleanliness in late nineteenth-century Rome

TAINA SYRJÄMAA

> After leaving the Piazza [St Peter's], we get a glimpse of Hadrian's Mole, and of the rusty Tiber, as it hurries . . . under the statued bridge of St. Angelo, – and then we plunge into long, damp, narrow, dirty streets. Yet – shall I confess it? – they had a charm for me. Twilight was deepening into dark as we passed through them . . .
>
> It was dirty, but it was Rome.
>
> (Story (1871) 4–5)

> To pass through the poorest quarters [of Rome] is sufficient to have an idea of the dreadful conditions in which the under-privileged classes are [living].
>
> The families crowd in humid and insalubrious rooms, where not only public morale is seriously endangered, but also hygiene and public health.[1]

Both these quotations describe Rome during the second half of the nineteenth century.[2] The first was written by an American sculptor, William W. Story, who was a long-term resident of the city. In his book *Roba di Roma*, he recalls his first arrival in the city in 1856. The city he describes – and was enthralled by – was seemingly composed of discordant elements, grandiose monuments and ramshackle alleys. He writes of the dirt, but not in the negative sense of filth. Indeed, he hastens to state that 'the cleanliness of Amsterdam would be the destruction of Rome in artists' eyes'.[3] In this context, dirt is a natural and indispensable part of the essence of Rome, not an external disturbance or nuisance to be removed.

The second quotation dates from 1873, soon after Rome had become the capital of the recently unified Italian kingdom, and is drawn from a report delivered by the police superintendent to the prefect of Rome. The

---

[1] A quarterly political report from the police superintendent to the prefect of Rome. Rome 14 January 1873, Archivio di Stato di Roma, Prefettura Gabinetto, busta 9, fascicolo 322.

[2] On the history of the city of Rome, see, for example, Insolera (2001); Bartoccini (1985); Vidotto (2002).

[3] Story (1871) 5.

living conditions and the appearance of the city had not experienced any dramatic change, but the description stands in stark contrast to Story's words of praise. The Italian police official saw a squalid neighbourhood, in which poverty, disorder and dirtiness threatened the lives of its inhabitants and social stability, whereas Story saw a fascinating spectacle.

A huge gulf exists between the way Rome was perceived by the American sculptor and the Roman police official. Mary Douglas' definition of dirt as 'matter out of place', to which this volume is heavily indebted, also offers a pertinent and stimulating starting point for studying this evident conflict.[4] To William W. Story there was nothing 'out of place' in Rome and he did not see anything that was filthy or disgusting, whilst the police superintendent saw plenty that was 'out of place' and in need of reparation. It is evident that we here encounter co-existing but divergent systems regarding how to judge what a city embodies in terms of its desirability, acceptability and repulsiveness. Thus, at one and the same time Rome was viewed as being both valuable and precious *and* extremely menacing and perilous.

The definition offered by Douglas is well suited to the paradigm of lived space stressed by humanistic geographers. The geographer Edward W. Soja, for example, building upon the Lefebvrian tradition of the social production of space, notes that human beings live in a 'real-and-imagined' world, in which physical conditions and human perceptions are inextricably intertwined. In other words there is a continuous and complex interrelation between physical existence and human agency, with the latter observing, interpreting and describing the former.[5] Thus, in this chapter we do not seek to discover or describe the physical condition of the city of Rome as such, with its streets, ruins and homes; instead we address a more intricate question centred on how the city was conceived to be both dirty, backward and morally corrupt and authentic, innocent, picturesque and unpolluted by modernity. These views were linked to the physical city, but cannot be explained by this factor alone. Instead, what matters are the outlooks of contemporaries. This does not constitute a superficial difference of mere preference, as we can obtain highly different sets of ideals on society and human living through an examination of the debate on Rome's perceived dirtiness or cleanliness.

In this chapter, the focus is on the second half of the nineteenth century, but the discussion is also connected to the preceding period as well as the

---

[4] See M. Bradley's discussion of Mary Douglas in this volume (ch. 1, pp. 11–13). See also Douglas (1966), esp. 36.

[5] Soja (1996). For a historical study on lived space see Syrjämaa (2006).

wider European context. The chapter is divided into two major sections, the first of which examines contemporary views – expressed in visual arts, fictional literature or travelogues – that emphasized the picturesque qualities of the Roman scene. This section shows how a number of observers accepted the prevailing conditions of the city and harboured nostalgic aesthetic considerations regarding the presumed authenticity of Rome and Roman society. The second section concentrates on those individuals who criticized Rome's perceived filth and anti-modern character. Here the focus is on the severe criticisms of contemporary Rome by anti-Catholic travellers and by the representatives of the new Italian administration. The physical conditions of the city were condemned as appalling symptoms of backwardness and moral corruption that were caused by papal rule. Those who espoused such views argued that major renovation was needed to modernize and thus purify the city.

Wondrous spectacles of decay

Since the eighteenth century numerous artists and travellers have described Italy in a romantic fashion. In the European cultural imagination it was *the* South, Europe's 'other': a backward, decadent, but also picturesque country.[6] Many observers delighted in depicting the juxtaposition between the grandiose remnants of antiquity and the presence and daily toils of the poor townspeople (see figures 12.1a–b).[7] They were impressed by the decay and the mixture of death and disappearance, which were surrounded by the daily hubbub of the supposed simple life of the Roman *popolino*. Many were also deeply worried about the future of this idyll and laced their comments with bittersweet concern.

As John Towner has pointed out, a thorough reassessment of the practices of the Grand Tour took place when the romantic gaze gradually became paramount. Emotive involvement became crucial, instead of the encyclopaedic ideal of education.[8] This had a remarkable influence on preferences and judgements regarding Rome, concerning the contemplation of sites, such as ancient (moonlit) ruins, the so-called medieval (and previously despised) urban structures or admiration for the pastoral aspects of the cityscape (including the much admired flora at the Colosseum). The decaying ruins, the forlorn houses and the wild disarray of crooked alleys

---

[6] See, for example, Moe (2002) 13–31, 153–5, 211–13. See also McLaughlin (2000); Dickie (1999) 83–119. It is worth noting that the border between the North and the South proved to be relative: to many northern Italians, Rome was suspiciously 'southern'.

[7] See also Visino (1994).   [8] Towner (1996) 98–128.

12.1a and 12.1b The picturesque gaze cherished the contradiction of the
decayed signs of past glory and humble daily living in Rome. These
engravings were reproduced in Vittorio Bersezio's volume *Roma: la capitale
d'Italia* (Milan, *c.* 1872), 388 and 392.

fascinated romantically inclined observers and were eulogized as being picturesque in literature and visual arts. As the English author Anna Jameson wrote in her partly fictitious travelogue in the 1820s: '[c]ivilization,
cleanliness, and comfort are excellent things, but they are sworn enemies to
the picturesque'.[9]

The Roman painter Ettore Roesler Franz is a well-known example of
an observer who viewed his home city in picturesque terms. He made a

---

[9] A. Jameson, *The diary of an ennuyée*, cited in Moe (2002) 18–19.

12.1 (*cont.*)

series of watercolours of *Roma sparita*, 'vanished Rome', which aimed at immortalizing the 'authentic' character of the city. This approach actually involved incorporating images of studio models into an accurate depiction of the physical urban environment. In most of his watercolours, the detailed description of the built environment forms a background to various assorted Romans – adults, children, chestnut sellers, gossips, knife grinders, traders and ecclesiastics, as well as dogs, mules and sheep – depicted carrying on with their everyday lives. The city was presented as a harmonious and safe haven, where down-to-earth townspeople were able peacefully to live their ordinary lives that were far from luxurious, but were also devoid of vice and the threats of modern, industrialized metropoleis. Ragged clothes, ruined walls and puddles in cobblestone streets are recurrent items, which, however, do not signify sadness or pity. Instead, Roesler Franz even depicts the fish-market in the remnants of Octavia's Portico as an inviting, picturesque attraction with no hint that it could also be filthy and foul smelling (see figures 12.2 and 12.3).[10]

The fish-market was also presented as a worthy attraction in guidebooks: 'Through the brick arch of the Portico we enter upon the ancient *Pescheria*,

---

[10]   A large collection of Ettore Roesler Franz's watercolours is housed in the Museo di Roma in Trastevere. See also Brizzi (1978); Bonsegale and Biagi (1993).

12.2 Watercolour: Ettore Roesler Franz, *Vecchie case in via della Lungaretta* (*c.* 1882) – an attempt to eternalize 'authentic' Rome before it was destroyed by modernity. © Comune di Roma – Sovrintendenza beni culturali – Museo di Roma.

with the marble fish-slabs of imperial times still remaining in use. It is a striking scene – the dark, many-storied houses almost meeting overhead and framing a narrow strip of deep blue sky, – below, the bright groups of figures and rich colouring of hanging cloths and drapery.'[11] It continued to be described in guidebooks even after it no longer existed.[12] When a new, more hygienic, market was opened in a different location, travellers

---

[11] Baedeker (1877). [12] Baedeker (1881).

12.3 Watercolour: Ettore Roesler Franz, *I venditori del pesce al portico d'Ottavia* (*c.* 1880). © Comune di Roma – Sovrintendenza beni culturali – Museo di Roma.

could only imagine the scenes that would have unfolded at the once busy market.

Francesco Adriano de Bonis was one of those who eulogized the beauty of the old centre and was worried that its character and charm would disappear. He felt that Rome was once replete with views of 'pictorial beauty', which he defined as 'an agreeable and pleasant sensation produced in us by the sight of an harmonious ensemble of highly varied, and at times irregular, parts which themselves are not beautiful'.[13] However, according to de Bonis 'the Vandals', that is, the authorities of the capital city of the young, secular nation state, were destroying these views by undertaking a programme of renovation. He criticized the monotonous quality of newly erected buildings, which he viewed as eyesores. He also regarded the construction of new river banks as an act that was destroying the 'wild beauty' of the city's waterway. Furthermore, he regretted how the Colosseum, 'this living poem', had been transformed into a cadaver by archaeological excavations. He acknowledged that the new quarters could be perceived as clean, but he considered this to be an insignificant merit in view of their 'petty and cheap' appearance.[14]

[13] de Bonis (1879) 33.   [14] de Bonis (1879) 23–4, 31–3, 39, 44–5.

These romantic souls depicted the poor living environments of central Rome as an idyll, while the new quarters that had started stretching beyond the old centre in the 1870s were considered to be devoid of spirit and meaning. Indeed, in many ways Rome in the 1870s still gave the appearance of being an *ancien régime* city, despite an increase in the number of lamp posts and newspaper stalls. This perception was especially accentuated when Rome was compared to London or Paris. Georges Haussmann's tenure as the prefect of the Seine, from the early 1850s to 1870, had witnessed an era of major upheaval in central Paris. The city's old centre was practically torn down and replaced with wide boulevards.[15] In comparison, Rome appeared to be an oasis of the past in the rapidly changing landscape of nineteenth-century Europe.[16]

Whilst the unquestioning adoration of Rome was shared by the likes of Story, Roesler Franz and de Bonis, another influential narrative scheme emphasized surprise and the bizarre: how initial shock and disdain turned into enchantment as an observer learned to appreciate the fascinating qualities of the city. Francis Wey, a French traveller, emphasized the contrasting elements of the city by describing how he proceeded along 'narrow and rural' streets in the centre, whilst being forced to suffer 'the pungent smell of turnips' and the odour of soup made from rotten cabbage. He told his readers that anyone staying in the city would have to get accustomed to 'sulphurous emanations . . . since the paving and the black mud of the streets are saturated with this smell'. Significantly, he then continues his story by stating how he was suddenly 'struck by a special vague sound like that of waves'. Rome took him by surprise as the unpretentious streets led him to the Trevi Fountain, which dazzled him with its volume of water and sculptural phalanx. Wey later continued his wanderings on the other side of the river in Trastevere – one of the poorest areas of Rome. The scene described by Wey was full of unpleasant details: the habitations were in bad repair, the people in poor health and their clothes mere rags. Yet Wey did not despise either Rome or Trastevere; indeed he was fascinated by the bizarre panorama.[17]

One of the basic elements of the picturesque is the seemingly opposite tensions that merge together in a panorama. As Luc Boltanski has noted, the concepts of the 'sublime' and the 'picturesque' allowed certain views to be perceived as being above and beyond the merely beautiful. Boltanski describes how the 'sublime' directed attention towards 'the terrifying,

---

[15] Schlör (1998).
[16] See, for example, Henry James' comparison between Paris and Rome: James (1987) 197–8.
[17] Wey (1879) 4, 51.

dreadful, horrible and painful', whilst the 'picturesque' focused on 'the triv-
ial, exotic, popular, caricatural and carnivalesque'. In this context, 'aston-
ishment at the ugly, disgust and rejection is transformed' in an instant into
something enchanting. He emphasizes how notions of the picturesque are
linked to the abilities of the gazer.[18] Those who praised old Rome willingly
identified themselves as civilized, artistic men of taste, and in many cases as
the defenders of the morals and values of the 'original' society.

Many travellers and artists viewed Rome from this angle. The decayed
and timeworn appearance of the city, with its narrow, dark streets and
wildly unco-ordinated habitations – its so-called medieval characteristics –
evoked the idea of a backward city in which time had stopped. Those with a
picturesque gaze created the illusion of a 'true' Rome, which was thought to
have remained unchanged for centuries. To them, the decay and disorder of
Rome were signs of a traditional society marked by innocence, simplicity and
happiness. These viewers were afraid of losing what they considered to be
the beauty of Rome when houses were whitewashed and sewage networks
were constructed. More importantly, they also feared for the traditions
and values of Roman society itself. Understandably, many became fervent
conservationists.[19] From their point of view, the urban renovation and new
practices were leading to the contamination and ultimately destruction of
an authentic Roman society.

The picturesque gaze also perceived urban poverty as an aesthetic spectacle. Affluent observers delighted in describing the nobleness of the simple
life. I have argued elsewhere that the presence of beggars, for example,
aroused many different responses, which included aesthetical assessments
devoid of pity or compassion. A mother and children in a Roman street,
for example, who were probably posing for an artist, could be described
as beggars forming a plastic group.[20] Such comments emphasized calm
contentment – not unjustifiable misery.

A late nineteenth-century short story by Enrico Torrioli describes how
bourgeois discourse of the poor in the city presented them as satisfied with
their lot as long as they could maintain their simple lives in 'old' Rome. The
narrator tells the story of a fictional cobbler, Augusto Palestrini, who had
to give up his humble workshop and home in the old centre after his wife's
death. The couple had lived in rudimentary conditions as 'they scraped
along somehow as they did not have to pay a penny for lodging: they had
formed the bedroom and kitchen by pulling a piece of curtain at the back
of the hovel, where he mended the shoes of servants and workmen of the

---

[18] Boltanski (2004) 120–3.   [19] Pasquali (2002) 332.   [20] Syrjämaa (2007b) 148–9.

quarter'.[21] Their life had been harmonious, however, with small pleasures such as an abundant plate of his favourite dish – *maccheroni* – on Sundays. Unwillingly Augusto had to move to the new quarters, which were rising on the outskirts of the old centre, to seek employment. He became the concierge of a brand new block of flats, but did not feel any emotional connection with his new abode. His old home had been dishevelled but also homely – a quality lacking in his new flat, which was situated along a wide and well-paved street.

As Richard Wrigley has argued, the question of dirt and the picturesque had aesthetic, but also medical and political dimensions.[22] The aesthetic evaluations were based on a set of fundamental values relating to good living and a sound society. Although those with a picturesque gaze did not call themselves nostalgic, their attitude was precisely typical of nostalgia: they longed for an idealized past and romanticized traditional society.[23] They also shared the critical attitude of nineteenth-century culturalists, who attacked the widespread belief in material and technical progress, industrialization and commercialization. In the most industrialized countries, in particular, such opposition railed against a future which they foresaw as being an inhumane, engineers' world, with standardized and rationalized solutions. Some looked to an alternative vision formed from an idealized conception of the organically developed nature of medieval society.[24] Furthermore, the question of the picturesque quality of the old centre of Rome was complicated by religious and political matters. The supporters of theocracy, who heartily regretted that Rome had become the capital of a secular state, associated any transformations with destruction and degeneration.

Yet it was a complex matter. Even for a single person it could be difficult to determine whether to admire or despise the physical conditions of the city. An interesting example is provided by Vittorio Bersezio, a Piedmontese author and politician, who was torn between two conflicting ideals. In his volume *Roma: la capitale d'Italia* (*c.* 1872), Bersezio initially describes Roman backwardness in negative terms: 'puddles of stinking black water'; 'solid and liquid missiles raining down from the balconies and windows'; 'the most intimate acts of everyday existence which are calmly fulfilled on the stones of public streets'. He blames irresponsible papal policies for this corruption, but states that he firmly believes that 'the inevitable march of progress' would soon reach Rome. Yet, at the same time, he kept alive the romantic tradition by stating that '[t]o tell the truth, some things which,

---

[21] Torrioli (1992) 194.   [22] Wrigley (2007) 157–80.
[23] Lowenthal (1986) 4–13; Boym (2001) esp. 6, 16.   [24] Choay (1965) 12–22.

when estimated with the standard of other cities, would be smutty, create an artistic impression in the special [Roman] milieu that surrounds them'. He also referred to William W. Story, quoted at the beginning of this chapter, and to the declaration that if Rome resembled Amsterdam in cleanliness it would no longer hold any artistic fascination.[25]

Vittorio Bersezio had mixed feelings as he gazed at Rome. He admired the picturesque quality of the old centre, but running counter to this admiration was the desire to see Rome become a modern, national capital and an international metropolis. This in turn implied different criteria for judging dirt and cleanliness, disgust and pleasure.

Visions of purity in streets and society

During the last years of papal authority in Rome in the 1860s a British traveller – with Italian ancestry – George Augustus Sala, wrote that he was well aware of 'the purlieus of Westminster Abbey and Notre Dame de Paris' but yet he considered the worst to be in Rome: 'up to this writing I have seen nothing so forlorn and so revolting, so miserable and so degraded, as the "humbler classes" of Rome'. Sala described how the townspeople were 'deplorable creatures, fluttering in rags, wallowing in dirt... these mothers, who from sheer lethargic carelessness suffer their babes to become humpbacked and bow-legged'. He argued that the poor conditions and filth of the homes were 'unfit for human beings, and scarcely fit for hogs' and were a cause of 'Roman fever', or in other words malaria.[26] Thus, Sala saw the rudimentary living conditions of the city as, at best, miserable and, at worst, fatal.

Sala's criticism was directed pointedly towards the Catholic Church. He was disgusted by the huge contrast between the luxury of the Church and the everyday misery he saw in the city. He compared Rome to other Catholic cities and provocatively asked: '[w]ould you tell me, if you please, why it is that the most orthodox Catholic cities always stink so intolerably?'[27] His strong views were openly anti-Catholic as he blamed the papacy for encouraging physical decadence and moral and social corruption.

This attitude was not only restricted to non-Catholic foreigners. As a matter of fact, many Italian, and thus Catholic, arrivals in Rome who championed the Italian cause also blamed the governance of the papacy for the

---

[25] Bersezio (c. 1872) 11–15.   [26] Sala (1869) 450.
[27] Sala (1869) 447. On anti-Catholic concerns in evaluations of Rome see also Janes, this volume.

lamentable state of the city. Contemporary discussion abounded with critical descriptions of disorder, decay, filth, poor living conditions and outdated infrastructure. Many of those who identified themselves with the national cause expressed disgust and disappointment at what they saw in the city. For example, Giuseppe Manfroni, a police officer from Genoa, who arrived in Rome only six days after the takeover of the city in September 1870, was shocked at the conditions he saw. He recalled his feelings decades later in his memoirs: 'the impression the railway station and its environment made on me was disastrous . . . I was confused to see the entrances of the station full of filth, the streets leading to the centre almost dark and blocked by ruins and vegetable garden hedges [and] people lying on church steps'.[28]

Rome was not chosen as the capital city for what it was at the time, but for what it had been.[29] Thus, the conditions faced by many new arrivals in the city shattered their fervent belief in mythical Rome. The new authorities were convinced that the poor conditions of streets, habitations and market places posed a serious problem for both public and moral health – not to mention anything about the national and international credibility of the new capital of the young nation state (see figures 12.4a–b). The low credibility of the city as a modern capital can be seen, for example, in an ironical essay that dared to joke about the city's dirtiness:

> With impatience to see Rome, the capital city of Italy, I rush down the steps four by four and here I am at the main door.
>
> Before crossing the threshold a sweeping excitement deeply affects me; it is the joy of finally seeing one of my sweetest dreams come true . . .
>
> Thus, I wandered about the streets without even deigning to glance, if you will excuse my saying so, at some urinals not worthy of the capital, from which streamed rivulets of fair water the colour of the Tiber. But I, superior to that kind of trifle, said to myself: don't worry about them, just hold your nose and pass by.[30]

The satirical journal *La Raspa* ('The Rasp'), which published this essay, was not an anti-Italian publication, but it did criticize the administration of the new capital city for its handling of the current problems. Comparing Rome to other contemporary capitals was a painful experience for the new patrons of the city, even though it was far from being the only city with problems of overcrowded quarters and of waste disposal.[31] Yet it was also the time when

---

[28] Manfroni (1920) 3.
[29] Rome was more of an ideology than a city in the political debate: see Chabod (1951) 179–323.
[30] 'Roma capitale' (*La Raspa*, 2 July 1871).
[31] For example, public urination was forbidden in Paris in 1850: Barnes (2006) 79.

12.4a and 12.4b The new rulers of Rome wished to clean the city in both a physical and a moral sense. These photographs, taken in the 1880s, depict the area of Vicolo del Pavone and that of Via delle Azimelle. The former area was demolished because of the construction of the wide Corso Vittorio Emanuele, whilst the latter belonged to the planned project that aimed to tear down the entire ghetto. © Comune di Roma – Sovrintendenza beni culturali – Museo di Roma.

huge public attention focused on the large sewage network of Paris, which was even advertised as a tourist attraction.[32]

---

[32] Gandy (1999) 23–44.

12.4 (*cont.*)

According to many of the new patrons, papal policies had stifled progress, that is, human development towards increasingly higher standards. This was an extremely serious accusation when one takes into account the fact that progress was the favourite axiom of the century and that continuous and swift change was expected. The concept was repeatedly used, but its contents were vague and were rarely defined in explicit terms. The belief in progress was most evident in a scientific and technological context, but was by no means restricted to this 'hard Enlightenment', to use the concept developed by Sven-Eric Liedman. Instead, it also permeated other fields of human life – the 'soft Enlightenment' – touching upon morals, art and religion.[33] According to those who detested the current condition of Rome, papal governance had led to an overwhelming sense of apathy and corruption among the city's inhabitants. In this context, decay meant filth, poverty, ignorance and suffering. It symbolized stagnation and retardation instead of the picturesque. The new authorities wished to transform Rome into a modern, clean, safe and dignified capital. Luigi Pianciani, who twice served as the mayor of Rome, declared the lofty objectives of this transformational process: 'free administrative institutions, serving science and art, schools and

[33] Liedman (1997). On the history of the nineteenth-century popular belief in progress, also see Syrjämaa (2007a).

workshops', along with 'comfortable and salubrious homes', 'wide, neat, safe and well-illuminated streets' and 'adequate sewage systems at market places and manifold amounts of water'. He declared these objectives to be new Italian monuments, which would 'distinguish the Italian Rome' from that of the emperors and the popes.[34]

Urban planning was strictly linked with public health policy. David S. Barnes has shown how the notion of a public health policy developed in France in the 1830s and 1840s. Although the nineteenth century witnessed a series of new biological findings and transformations in medical thinking, Barnes pointed out that the authorities notably increased their efforts to establish a clean and safe local environment, irrespective of whether the causes of diseases were considered to be found in miasmas or in germs, or something in between. Thus, the changes were not directly or solely linked to medicine but also to changing social ideals. The attitude was, in fact, closely related to an optimistic belief in 'progress' and technocratic regulation.[35] In the light of Barnes' observations, I would also like to emphasize that 'medical materialism' is not of interest here.[36] In our case, it does not matter whether medical explanations of the time were 'correct'. What does matter is the way the debate of the time could provide concepts and values that were used to think about the city, and how it could legitimize some arguments over others.

On this basis, the condition of Rome at the time was evaluated in highly negative terms of filth and degeneration, whereas the vision of the city in the future was depicted in glowing terms of cleanliness and modernity: with newly whitewashed facades, smooth-running traffic, efficient illuminated streets, well-equipped, open market places, neatly weeded and scientifically excavated ruins and, last but not least, huge river embankments to prevent flooding from the Tiber. The city council noted as early as the spring of 1871 that '[t]he new constructions can prove to be of great benefit, not only for the decoration and dignity of the city, but also for the convenience of the private townsmen and among them the less well-off'.[37] On the grounds of these comments it would be easy to imagine that Rome had never been cleaned. However, the city undoubtedly experienced many cleanliness campaigns over the course of centuries and these practices had been firmly encoded in

---

[34] Pianciani (1874) 6.

[35] Barnes (2006) esp. 13, 65–104, 194–228. On the concept of miasmas, see also Thorsheim (2006) 10–18.

[36] Cf. Douglas (1966) ch. 2.

[37] *Atti del Consiglio Comunale di Roma*, 6 March 1871, 157–8. See also in the same context attachment Lett. B. 162.

12.5 The new quarters along Via Nazionale, *c.* 1875, constructed according to the principles of modern international urban planning, prioritizing wide, straight and paved streets furnished with regular street lighting and a subterranean sewer system. Photograph: Carlo Baldassarre Simelli. Reproduced by permission of Fondazione Marco Besso.

regulations. Moreover, as K. Rinne has shown in the previous chapter of this volume, the papal renovation policies tightly linked together both physical and moral aspects in their urban projects.

Nineteenth-century urban planners and many politicians of the time believed in a well-structured, systematic society and in a city in which rational planning would create a well-ordered environment. They saw conformity as a basic quality of a smooth-functioning and healthy society (see figure 12.5). They believed not only in the power and capacity of planners to create a city responding to this ideal, but also in their right and duty to mould the city and to define a 'correct' way to live in it.

In 1873 the city engineer Alessandro Viviani presented the first version of a new urban plan for Rome, which included proposals to improve the old centre, whilst also constructing an entirely new, elegant and salubrious quarter on land which had been practically uninhabited since antiquity. In technical terms, it was not too difficult to construct a new quarter on land that had recently been used predominantly as vineyards, vegetable patches and pasture.[38] Improvements to the old centre, however, posed different challenges. Its seemingly disordered structure limited the supply of fresh air and sunlight and was a hindrance to traffic. At the same time, its crooked streets were lined with ancient monuments, grand palaces and ostentatious churches.

Viviani and Pianciani, among others, were following the ideals of international planning discourse, in that they placed their trust in the rational, the systematic and the scientific. The key concepts of the time were 'to better', 'to enlarge' and 'to regularize'. The first two concepts were repeatedly advocated in the Roman debate, whereas the need 'to regularize' was only implicitly included. In Rome, no urban planner dared to suggest seriously such radical renovation of the urban structure as had taken place in Paris. Yet almost any sort of planning based on the old centre of Rome anticipated some degree of demolition.[39] The most notable project, which was only partially realized, was to demolish the entire mazy ghetto. The planned demolitions partly aimed at creating a healthier urban environment, but were also dictated by questions of prestige and power. The authorities particularly wanted to control the immediate surroundings of the most famous antique monuments and symbolic places.[40] These attempts were not a complete novelty as such, because many similar projects had been enacted in the preceding centuries that were designed to eliminate 'unfit' elements from around the major monuments. Thus in effect the authorities were merely repeating past papal attempts to push the signs of everyday life beyond the cultivated gaze of those wishing to focus solely on the major historical sites of Rome. Thus, the promoters of the 'Italian' Rome actually continued the policy of their often discredited predecessors by attempting to purify the sacrosanct ancient monuments. This was carried out by isolating these monuments from the 'ordinary' urban built environment and its daily bustle.

Alessandro Viviani's statements clearly show how aesthetic qualities were also linked to the new constructions. When presenting the city plan, Viviani enthusiastically described how the construction of the river banks would not only save the centre from disastrous floods, but also improve its appearance. The project would enable 'the row of protrusions' and the 'indecorous and grimy facades of houses' that blotted the cityscape to be removed. Moreover, the project would entail eradicating the 'sludgy steep banks, which produce

---

[38] I have argued elsewhere that contemporary decision-makers, who professed to adore antiquity, were not able to appreciate the historical character of this area as it did not correspond to the prevailing idea of monumentality: see Syrjämaa (2008).

[39] See for example Del Prete (2002).

[40] See for example *Atti del Consiglio Comunale di Roma*, 14 September 1871, 1006.

a most sad look on the shores of the Tiber'. In the place of this decadent view, Viviani envisaged two wide streets running along the river embankments that would be 'beautified with brand new edifices of regular appearance'. Furthermore, he suggested that the streets should be adorned with 'spacious arcades', which would protect passers-by from inclement weather. Viviani assured his audience that whilst straight arcades could be 'tiresome to the eye', there was no such danger in this case as the river bends would guarantee alternating points of view.[41]

The new administration copied foreign models when designing areas of public space fit for the new Roman bourgeoisie. The plans included large-scale constructions, but also considered minor spatial details, such as the creation of flowerbeds and the setting of exotic plants. The administration tried especially hard to smarten up the area around the Stazione Termini, the principal railway station, which was now the main gateway into the city. Thus, despite concentrating on technical matters and the control of urban space, the authorities did not avoid aesthetic questions. Yet, looking back to late nineteenth-century visual arts, it is clear that the gaze of artists did not linger too long on these new achievements. Urban cleanliness was a definite trend in urban planning and architecture, but it does not seem to have led to many pictorial descriptions. Picturesque views, on the other hand, continued to be produced regularly.

In one important respect, the dirtiness of Rome paled in comparison with that of its foreign rivals: no dense conglomeration of industrial smoke entrapped the city. Smoke was a daily problem in London and Paris, even though it was considered as only a nuisance in the early years of industrialization. It was only later in the century that people started to become conscious of its dangers to public health. The air in the non-industrialized parts of Mediterranean Europe especially pleased the British bourgeoisie, who desired to escape foggy London.[42] However, Rome's relatively pure air merely strengthened the image of it as a backward – and even rural – place, rather than a modern metropolis.

The idea of cleanliness was not restricted to physical or aesthetic conditions, but came to be considered as a basic requirement for a healthy social body. Pamela K. Gilbert has described the great boost in 'medical mapping' in mid-century London by stating that 'the mechanisms of sanitary inspection' were a means to institutionalize surveillance. In the background lingered the idea of uniform, rational citizens and the systematized usage

---

[41] Viviani (1873) 22.
[42] On the gradual evolution of the idea of industrial pollution, see Thorsheim (2006) 10–19.

of space.[43] Similarly in France '[t]he urgent need to civilize the savage elements in French society for the sake of the nation's physical health'[44] led
to an intensified health policy. Broadly equivalent trends can also be seen
in Italy: the authorities struggled to re-fashion the Roman *popolino* into
Italian citizens and to produce a transparent, easily controllable, systematic
society. Many attempts were undertaken – more or less consciously – to
create an Italian national feeling and identity, or *fare gli italiani*, during
the nineteenth century.[45] Consequently, urban planning and health policy
formed intrinsic elements within this overall agenda.

Conclusion

How we define the filthy, the disagreeable and the disgusting is a part of
a historical phenomenon, intimately related to the cultural history of the
senses.[46] Sensing is not a simple physiological act, but always embraces a
more abstract, cultural layer of values and assessments. In the case of Rome,
those who saw, smelled, or even sometimes had unfortunate tactile contact
with, matter that they considered to be unacceptably dirty demanded rationalization, hygiene and the control and standardization of urban space.
They saw themselves as the protagonists of purity, progress and modernity. Thus, they were not only purifiers in the physical sense, but also saw
themselves as cleaners of the moral corruption for which they blamed papal
governance. Opposed to this group were the defenders of the picturesque,
who despised this new, international and modern way of living and constructing. To them, it seemed to constitute a menace to the authentic nature
of Roman society. Paradoxically, the ideas arriving from the North, which
were intended to be innovative and to clean the city, were perceived from
this perspective as a pollutant that would destroy the authenticity of Roman
society.

Eileen Cleere has argued that the popularity of picturesque ideals in
England rapidly collapsed at a clearly definable point of time, namely in
the 1840s, in the wake of the publication of an influential sanitary report.
She explains that 'reform-minded Victorians . . . saw dirt as a social embar-
rassment and a material danger, a sign of oppression, disease, and moral

---

[43] Gilbert (2004) 78–9; cf. 85–7. See also Cohen (2004).   [44] Barnes (2006) 196.
[45] See for example Tobia (1991); Syrjämaa (2006) 177–211.
[46] On the history of the senses, see for example Corbin (1986); Smith (2007); cf. Bradley, this
     volume, ch. 1, pp. 38–9.

degradation'.[47] I have not attempted to make a clear-cut chronological distinction here. One can note the waning popularity of the so-called picturesque gaze and the rise of the discourse of cleanliness, but they did co-exist in temporal terms. The phenomenon was linked to a wider European transformation of ideals and practices, but there may have been some differences in the speed and timing of various stages. In the case of Rome, the picturesque arguably received stronger support later than in some other European metropoleis, as Italy was the common playground of European romantics.

Both systems of gazing – as described above – and the discourses surrounding them were aspects of cultivated society in Italy and Europe. We know very little about the views and feelings of those Roman townspeople who actually lived in the much-debated conditions of the old centre. There are, however, some traces of both discontent and affection. When Giggi Zanazzo collected proverbs in late nineteenth-century Rome, for example, he heard people saying that 'every hour the doctor goes to a house without sunshine'.[48] Thus, people not only connected overcrowded areas with rudimentary sanitary conditions and discomfort, but also linked them with danger. Yet a shanty was a home for many Romans. Better housing alternatives were few and far between, and naturally the home became the focal point in people's lives. It was the pivotal point from which they formed social networks and undertook their daily spatial itineraries. In practice, there was very little movement to the new quarters. In the bourgeois imagination, as seen above, the reason for this was that the poor disliked alien environments and held the old quarters in great affection. Another, more prosaic, explanation for the limited movement of Romans is that only a small number of poor inhabitants could find the financial and social means to acquire a flat in the new quarters. In any case, many contemporaries described how there appeared to be two Romes, as the new and old quarters were so markedly different in architectural terms, in their overall infrastructures and in their conditions as well as in the manners, practices and dialects of their inhabitants.

Here we have focused on the debate about dirt and decay in Rome during a relatively short period of time, whilst in her pioneering study Mary Douglas outlines sweeping and overarching patterns of purity and dirt, moving in a subtle and pervasive manner from one century to another – or even from one millennium to another – and from one continent to another. To a

---

[47] Cleere (2004) 133.

[48] Zanazzo (1908) 83. Cf. also Barnes (2006) 74–8; Thorsheim (2006) 22–8.

historian, such a method of analysis is unfamiliar but inspiring. It may be dizzying, but it also creates ideas and helps to develop new approaches. In its turn, history also has its contribution to make. Mary Douglas warned that '[t]he anthropologist falls into the . . . trap if he thinks of a culture he is studying as a long established pattern of values'.[49] In this respect, a historical case study, which offers a close and detailed reading of a specific historical situation, can bring to light a number of significant aspects: the manifoldness of perceptions, the co-existence of varying, even conflicting, systems or subsystems and the intertwinement of change and continuity.

Douglas' views have also been criticized quite recently by the philosopher Olli Lagerspetz, who has argued that her key concepts are vague. He notes that Douglas, for example, does not make a distinction between dirt and mess, jumble and raggedness.[50] In our case, this does not seem to be such a critical distinction. On the contrary, there is no reason to try to isolate 'dirt' from other related forms and conditions. The interesting thing is how contemporaries described and interpreted their environment and how they named and evaluated their own Rome. The key to reading the city can be found in the well-known saying about beauty, which Mary Douglas revised by stating: '[t]here is no absolute dirt: it exists in the eye of the beholder'.[51] The nineteenth-century Roman case of the clash between the picturesque and cleanliness provides a fruitful example of this, as well as of the co-existence of highly divergent cultural systems. The city is a lived, real-and-imagined space, which does not have an independent existence separate from those who inhabit it. The very same 'real' physical conditions could be classified as inappropriate, as dirt, as 'matter out of place', and as the true essence of Rome, which had to be safeguarded against modern transformations that would pollute and ultimately destroy it.

---

[49] Douglas (1966) 5.   [50] Lagerspetz (2008) esp. 104–9.   [51] Douglas (1966) 2.

13 | Vile bodies: Victorian Protestants in the Roman catacombs

DOMINIC JANES

"'Hilda, have you flung your angelic purity into that mass of unspeakable corruption, the Roman Church?'"[1] This cry from Nathaniel Hawthorne's *The marble faun* (1860) is echoed, with varying degrees of histrionic exaggeration, through a vast corpus of Victorian literature in English. This is not simply a question of Protestant anti-Catholicism, although of course it represents just that, but also reflects a deeply seated set of anxieties about pollution, dirt and disgrace. Hilda, a pure New England Protestant woman, is depicted before her religious indiscretions as inhabiting a tower standing high over the city of Rome. Descending from the fresh, pure air of the physical and moral heights she encounters the dank, interior seductions of the Catholic confessional. Religious hygiene, therefore, for Hawthorne, was expressible in connection with the exterior and interior forms of the city via a vertical axis of moral and physical interior descent. It is important to emphasize the key role of Mary Douglas' work in relation to the task of understanding such an overlap between perceptions of moral and physical pollution. Douglas enabled us to understand pollution as a cultural construction and, hence, to focus on the forms by which it was specified and classified. The perceived moral importance of the division between Roman Catholicism and Protestantism led many visitors from Northern Europe to view Italy in general, and Rome in particular, as a *locus* of peril because they feared that their desires might lead them into doctrinal transgression.

Such perceptions were heightened in relation to areas of the city which were considered polluted for medical reasons, such as burial grounds. Valerio Valeri, as explained in the introduction to this volume (p. 13), has developed Douglas' work with particular reference to the corporeal. Such insights, when combined with those of Kristeva, encourage us to think of the corpse as fascinating and uncanny refuse that is disturbing and polluting not simply because of the dangers of disease, but also because it problematizes the boundary between the human and the inhuman. This chapter builds on other work in the current volume on the sanitary zoning of the city of

---

[1] Hawthorne (2002) 284.

Rome by focusing on the murky depths of the city, notably the catacombs, as visited and imagined by Victorian Protestants.

Britons had been marching about in at least some of the catacombs and crypts and reporting back on their experiences since the fifteenth century.[2] The impact of the Reformation was to render the city a place of moral and physical peril for ardent Protestants, a fact that did not deter those who came to the city primarily for its classical heritage. The catacombs did, occasionally, feature on the eighteenth-century Grand Tour; however, most were still unexcavated and those that were known were lacking in monuments that appealed to the tastes of the time. A further reason for this neglect derived from the fact that 'despite its educational focus the Tour was often motivated by the desire to possess and display antiquities', and the boney treasures of the catacombs were not widely regarded as desirable elements of domestic decoration.[3] It was once common to think of the tradition of the Tour as having been ended by the Napoleonic wars, but recent work has flagged up connections with nineteenth-century patterns of visiting and tourism.[4] Mariana Starke (1761–1838) was one of a series of figures whose writings provided the basis for Victorian travel guides. Such writers offered 'perceptive comments on the everyday life and social fashions around them'.[5] Thus, Victorian travellers continued to visit the ruins of classical antiquity, but their attention was increasingly drawn both to contemporary society and to religious institutions. The Christian revivals of the time, notably that of the Oxford Movement in the 1840s, stimulated a wave of religious tourism which encompassed both the sincere devotions of Catholics and Catholic converts and the anti-pilgrimages of Protestants seeking the self-validating delights of outrage and denunciation.

One of the prominent early English-language guides to the catacombs was the Roman Catholic James Spencer Northcote's (1821–1907) *The Roman catacombs; or, some account of the burial-places of the early Christians in Rome* (1857). In the next decade he published further on the subject and incorporated the important new archaeological researches of the Italian expert Giovanni Battista de Rossi (1822–94), the result of whose discoveries was to establish the modern understanding of the catacombs which ensured that, as one modern historian has put it, 'the taste for fantasy [in interpreting these remains] had run its course'.[6] However, my chapter will subsequently consider traditions of interpretation in which religious and moral

---

[2] Gaston (1983) 145.   [3] Wilton-Ely (2004) 152.   [4] Wilton-Ely (2004) 161.
[5] Wilton-Ely (2004) 158.   [6] Gaston (1983) 160.

speculation, which we may understand as culturally constructed, predominated over archaeological analysis.

According to Northcote, visitors had to obtain their ticket from the cardinal vicar and might then proceed to meet the guides. However, 'these *custodi*, being mere labourers, are not able to explain things': hence the need for the book.[7] He attested that the mode of burial, holes bored into soft tufa rock, was sacred, since this was how Jesus himself was buried.[8] Nevertheless, he noted that air holes were necessary otherwise the atmosphere would have become nauseating.[9] Those who were farsighted enough to peruse the range of Catholic advertisements at the back of his volume might have stocked up with a bottle of Eau de Cologne 'brewed up by a religieuse' against just such an eventuality. Yet outweighing any physical disgust there was, he assured the reader, moral and intellectual enlightenment to be garnered.

Northcote had published a series of explanatory articles in the *Rambler* during 1848–9, but he was keen to emphasize that it was not his own writings, but those of Nicholas Wiseman (1802–65), first Roman Catholic archbishop of Westminster, that 'effectually destroyed that indifference to the subject on the part of English visitors to Rome of which I had previously complained'.[10] Cardinal Wiseman's *Fabiola* (1855) was intended by its author as being simply the first in a series of volumes illustrating the history of the Church: three hundred years each were to be allotted to the Church of the catacombs, the basilica, the cloister and the schools.[11] The fact that the other volumes were never written left an unwarranted impression that the cardinal, whilst consolidating the resumption of the Catholic hierarchy of bishops in Britain, held the formative post-apostolic years of Christianity in especial regard. In fact his presentation of the catacombs, influential though it was, is remarkably matter-of-fact. He regarded these caves as places of rest before the resurrection and he eschewed sensationalism.[12] His attitude was one of pious and measured reverence.

Interest was, however, rising in Britain even before Wiseman's novelistic intervention, as one can see, for example, from the appearance of William Henry Anderdon's (1816–90) *Two lectures on the catacombs of Rome* (1852). Anderdon was an ordained Anglican who converted to Rome in 1850 and was to become a Jesuit in 1872. In his lectures he evoked the catacombs as a stimulating place for the contemporary visitor. He assured his listeners that 'we will not linger above ground'.[13] We must go from crumbly sand

---

[7] Northcote (1857) 66.   [8] Northcote (1857) 20.   [9] Northcote (1857) 40.
[10] Northcote (1857) x.   [11] Wiseman (1855) vii.   [12] Wiseman (1855) 142.
[13] Anderdon (1852) 1.

to a 'deeper story', the harder tufa below.[14] He tells us that we must follow those ancient excavators whose need for 'utmost secrecy' drew them to delve deeper, three or four storeys down.[15] He argued that what we would find down there must either be monuments to Catholic faith, or moral 'man-traps set on the premises', into which Protestants might fall unawares. But surely, he asks, these caves are not the result of the fakery of Jesuits?[16] He compared these humble excavations to those priest holes that did such vital work under the English Protestant persecutions.[17] To argue that the cata-combs said anything of the reformed faith was, for Anderdon, as paradoxical as claiming that inscriptional evidence proved that the Midlands had been colonized by Hindus.[18] For him, as for Wiseman, the catacombs exhibited the ultimate sacrifices of the martyr saints and were thus thoroughly and gloriously Roman Catholic.

The efforts of Northcote, Anderdon, Wiseman and others ensured that a steady stream of Roman Catholics arrived in Rome from Britain seeking to enter the catacombs. However, other visitors were simply there for entertain-ment. One such specimen of insouciance was the Royal Navy commander Henry T. Ellis, who made a hobby of going around tombs and catacombs, from Ripon bone-house and the catacombs of Paris to the mummies of Palermo. He delights in telling us that 'a lady is said to have died of fright a few years ago among the standing up "Dried Monks of Malta", through the inconsiderate thoughtlessness of a practical joker who had pinned her dress to one of them, and consequently when she moved away, brought him down upon her out of his niche'.[19] Or again, he tells us with relish of a set of exhumed corpses which had for some reason not decayed after burial in Bordeaux, one of which was that of a 'wretched little boy, who had been buried alive; the limbs and hands drawn up'.[20]

Gothic tropes provided a consistent mode in which the sceptical observer approached the catacombs. Gothic, as a literary genre, had developed through the imagination of foreign (often Italian) locales as settings for moral and sexual transgression. In Charles MacFarlane's *The catacombs of Rome* (1852) we read that the catacombs became truly terrible dur-ing the Middle Ages when, 'after each of these spoliations, devastations, and desecrations of these barbarians, the catacombs were neglected for a season . . . The void chasms were left to bats and obscene birds and beasts, to runaway debtors, thieves and banditti.' It was in 'dread' that peasants

---

[14] Anderdon (1852) 8.   [15] Anderdon (1852) 24.   [16] Anderdon (1852) 48–9.
[17] Anderdon (1852) 65.   [18] Anderdon (1852) 78.   [19] Ellis (1871) 18.
[20] Ellis (1871) 21.

passed the mouths of the caverns.[21] In the catacombs today, 'nothing can be more solemn than this subterranean gloom, and the effect produced by the objects brought to life as you advance. The yawning tombs on either side of you, and before and behind you, – skulls, skeletons, crosses! Nothing is here but speaks of persecution or of death' (note the mention of 'Catholic' crosses as objects of horror).[22] MacFarlane says that 'it is a relief, never to be forgotten by those who have once felt it', to have light and fresh air in those few places where holes have been cut up to the surface.[23] Monsieur Perret, an architect working under Napoleon, spent six years excavating there and was, we are informed, during that time 'in a manner, buried alive in these dark crypts'.[24]

A similar mood of delicious Gothic despair is evoked in Emma Raymond Pitman's *Vestina's martyrdom: a story of the catacombs* (1869): 'the sight of the place that was to be for an indefinite period their only home, struck a chill into the hearts of our two young friends. Nursed in the lap of luxury as they had been, it was but natural that they should feel somewhat despondent in this gloomy underground dungeon.'[25] Such sentiments leaked out into a number of the guidebooks, such as W. H. D. Adams' *The catacombs of Rome: historical and descriptive, with a chapter on the symbolism of early Christian art* (1877), which emphasizes the mysterious extent of the catacombs as a labyrinth (figure 13.1).[26] Shut your ears to the guide, he tells us, who would have you bow before gore: 'every bottle, you are told, was once filled with human blood: do you not see the stains of ineffaceable red?'[27] Thus it can be seen that the reception of the catacombs as sensational spaces was very much an attribute of non-Catholic writers. Roman Catholics, by contrast, adopted a matter-of-fact viewpoint and regarded these chambers as straightforward evidences of faith, despite their physically insanitary aspects.

The importance of doctrinal viewpoint for the perception of pollution can be seen from the change in John Henry Newman's (1801–90) attitude to the catacombs and their contents after his conversion to Roman Catholicism from Anglicanism in 1845. Wendel Meyer has explained how before this time 'the spirit of the primitive Church was tangibly present for Newman in the Christian monuments of the city, but that spirit was encased in the corpse of contemporary devotions of the Church of Rome which were racked with doctrinal decay'.[28] Assonitis (this volume, pp. 139–52) has

---

[21] MacFarlane (1852) 37.   [22] MacFarlane (1852) 61; Janes (2009b) 111–17.
[23] MacFarlane (1852) 65.   [24] MacFarlane (1852) 181.   [25] Pitman (1869) 327.
[26] Adams (1877) 51.   [27] Adams (1877) 55.   [28] Meyer (2010) 748.

A WALK THROUGH THE CATACOMBS.

13.1 W. H. D. Adams, *The catacombs of Rome: historical and descriptive, with a chapter on the symbolism of early Christian art.* London, Nelson, 1877, p. 54.

shown, through a discussion of the Church of Rome as a diseased body in the thought and rhetoric of Savonarola, how such thinking has a very long pedigree. Elsewhere, I have explored how Charles Dickens, in *Pictures from Italy* (1846), was very much a Protestant Briton of his time in viewing Roman Catholicism as an animated corpse; as a zombie religion, terrifying and profoundly polluted and polluting in its confusion of purity and sin, splendour and filth, ancient death and contemporary life.[29] As we have seen above, one response to this was to encounter Catholic Rome and its monuments as a species of Gothic horror show; however, many other Protestant visitors aimed not simply to sensationalize what they saw but also to moralize on the experience.

---

[29] Janes (2009a) 184.

The Protestant rhetoric of pollution

Whilst sheer spectacle and sensationalism undoubtedly played a role in the rising popularity of the catacombs amongst British visitors to Rome, I want to proceed by discussing a further group of early Victorian visitors, namely Protestants whose beliefs had played an important role in leading them below the streets of Babylon in search of moral instruction. Vincent Lankewish has raised the interesting notion (which connects with George Haggerty's theorization of the Gothic as an essentially queer genre concerned with the elaboration of sexual transgression) that the reason for much of this excitement was that it was suspected that there was something queer going on down in the dirt.[30] Lankewish argues that Wiseman's *Fabiola*, amongst other works, raised the spectacle of the glorious virgin who was married to Christ (and so challenging the Protestant belief in the moral supremacy of the married state).[31] Moreover, Lankewish's article also raises the image of the catacomb and the closet as places of (dangerous) secrets. Such speculations can be related to the long-running concern with cities as places of sexual deviance, as explored by Salvante (this volume), since one of the attractions for a certain kind of Northern European tourist was the reputation of Italy for sodomy. But whether or not we wish to follow these interesting speculations, I do believe that it is unquestionable that certain Protestants accepted the Catholic view of the underside of the city as being exemplary of Roman belief and practice and, therefore, applied to those caves the full range of sexual and moral innuendo that they habitually applied to Catholics in general.[32]

The firebrand Irish Protestant Michael Hobart Seymour (1800–74) visited Rome with his wife on a sort of extended Protestant honeymoon after which he wrote a series of widely read publications which detailed the abuses of the city and its religious life. He tells us in his *Mornings among the Jesuits at Rome* (a title meant to send a shiver down the spine, equating, as it did, for its then readership to latterday 'afternoons with the IRA', or 'evenings with Al-Qaeda'), that he was shown down into the Roman catacombs by an animated Catholic corpse, a monk who 'looked like a moving plague – a personification of the malaria – a walking pestilence. There he was, an attenuated thing, a living skeleton.'[33] The nightmare 'other', in this account, is that horrific thing, a 'stagnant spirit' animated to life and power. What drove these monsters of corruption was greed. It was an abomination with

---

[30] Haggerty (2006).   [31] Lankewish (2000) 247.   [32] Janes (2009b) 117–24.
[33] Seymour (1850) 300.

a terrible appetite. The sale of relics, Seymour tells us, 'a traffic of the most disgraceful and degraded nature', led to frenzied figures in cowls and wimples pillaging the underground caverns, ripping bones out of graves and pounding them into splinters. And when these seams of perverse gold were worked out and the supply of 'lawful relics of ecclesiastical merchandise' ran low, 'the monks who were the merchants in this matter' manufactured fake relics, and thus 'the demand of the market called forth an adequate supply'.[34]

Hobart Seymour is not an isolated figure amongst Protestants in his use of the language of bodily pollution in relation to Catholicism at this date. Edward Bellasis (1800–73) was a leading parliamentary barrister who converted to Roman Catholicism on 28 September 1850. In February 1851 he published anonymously on the terminology used by Church of England bishops in relation to the so-called Papal Aggression of 1850 (the restoration of the hierarchy of Catholic bishops in England). He noted that the most frequently used adjective was 'corrupt'.[35] It was widely alleged by these Anglican divines that the pope was out to foster an outbreak of Catholic corruption within the geographical boundaries of England, and that the 'Puseyites' of the Oxford Movement were active agents capable of taking that corruption within the boundaries of the Church of England and, so, into the moral heart of the nation. Perhaps the most frequent image that was used in relation to such Anglo-Catholicism was that of pollution. It was feared that those who favoured the introduction of Catholic ritual into the Anglican Church were sullying the purity of Protestant worship by the importation of ideas and material culture that were tainted. Perceptions and descriptions of the Reformation were a key battleground. The relationship between England's past and present was very much at stake in a situation in which the Reformation was presented by many as having been a crucial step towards the recovery of pure religion. The key notion was that, by cleaning off the dross of centuries, a state of grace had been reached to match the one that had been lost after the apostolic age. This vision comes through strongly in Hobart Seymour's 'The English Communion contrasted with the Roman Mass', preached (on 5 November, no less) at St George's Church, Southwark, in 1843:

> like those who would restore or renovate some ancient temple, removing the dust that defiled it, the mould that tainted it, the moss that covered it, and yet retaining all that was beautiful and useful; so the Reformers of the

[34] Seymour (1850) 330.   [35] Bellasis (1851), discussed by Wheeler (2006) 29–30.

Church of England, longing to restore and renovate her in all her original purity, removed all the corruptions that had crept into her doctrines, and all the abuses that had crept into her practice, while they retained all that was scriptural in doctrine, and all that was holy in practice; all that was conducive to the beauty, and all that was essential to the order of her services.[36]

The Reformation, thus, purified the body of the Church and those of its people and 'restored the Church of England to her primitive objects and her primitive beauty, treading in the steps of our Lord'.[37] Speaking a few weeks later, Dr Lee, professor of Hebrew in Cambridge, attacked the Oxford Movement using similar invective in which terror of the tyranny of tainted flesh was intensely expressed. Shall we, he asks, again 'succumb to the rule of an hierarchy, at once the most usurping, tyrannical, oppressive and cruel; again to incrust itself as it were, in the unintelligible and useless jargon of the schools, the traditionary trash of useless religious fraternities; the dust and darkness of the monastery, the cell or the hermitage!' In this miasmic world of physical and spiritual dirt, pollution and danger, we look round with Lee and see a horrid vision of 'the flesh mortified under the vain hope of thereby purifying the spirit; [with] the rotten and rotting bones of saints, pictures and images of the Virgin' lying strewn about.[38]

The movement to emphasize the Catholic tradition in Anglicanism was demonized for sullying the purity of Protestant faith. The very persons of the Anglo-Catholic clergy were also held to be tainted. The Anglo-Catholic priest Arthur Douglas Wagner (1825–1902) of Brighton was described by one Protestant polemicist as 'as oily a piece of fluid tallow as ever mistook its way into a surplice, or a *sewer*! . . . Puseyites will pole their smooth chins into all kinds of filth, so carefully avoided by honest men!'[39] It is important to emphasize, however, that Ritualists and Tractarians felt that they too were on a mission to clean up Church and society. This even comes across in the anti-Tractarian skit, *A paper lantern for Puseyites*, in which the High Churchman on arriving in a new parish opines that

> the surplice is covered with stains old and recent,
> and I am sure it looks neither 'comely' nor 'decent'
> . . . every corner and cranny is smothered with dust,
> which shows how these 'questmen' [churchwarden's assistants]
> have heeded their trust.[40]

---

[36] Seymour (1843) 376.   [37] Seymour (1843) 382.   [38] Lee (1843) 526.
[39] Anon., 'Adam Bede' (1860) 5.   [40] Anon., 'Will o' the Wisp' (1843) 8–9.

British and Irish clergy of the early Victorian period were, therefore, highly attuned to notions of purity and impurity that spanned the physical and the spiritual and, thereby, contributed to the construction of the contemporary moral culture. Much of the energy was generated by disputes within Protestantism in general and Anglicanism in particular, which was becoming split between High and Low Church opinion. In the ensuing battles, both sides were fascinated by purity, but had very different views of the role of (supposedly) sacred objects in the achievement of that aim. What for one side were expressions of symbolic and sacramental reverence were transformed by the other into an obsession with the seductive bodies of priests, worshippers and their idols. The diversity of Protestant gazes was deployed upon Roman Catholic Italy not simply with the aim of revealing its faults, but also as a means of distinguishing between true and false Protestants and so revealing obsessions with purity back at home.

Urban danger and disease

Syrjämaa, in the preceding chapter, has provided a fascinating contrast between the picturesque viewpoint of William Story and the censorious line taken by George Sala in relation to the state of the poor districts of Rome. The very period when the catacombs were being excavated was precisely when sanitary reports were giving rise to increased fears of urban dirt and leading to a wider application of Gothic, rather than picturesque, modes of visual reception (pp. 209–10). For British visitors to Italy the ultimate example of urbanism was not Rome, but a London in which moral and physical pollution and degradation were now understood to be combined. For instance, on 14 December 1850, Charles Dickens published a 'December vision' in his periodical *Household Words*. 'I saw a poisoned air', he wrote,

> in which Life droops. I saw Disease, arrayed in all its store of hideous aspects and appalling shapes, triumphant in every alley . . . I saw, wherever I so looked, cunning preparations made for defacing the Creator's Image . . . I saw from those reeking and pernicious stews, the avenging consequences of such Sin issuing forth, and penetrating to the highest places. I saw the rich struck down in their strength, their darling children weakened and withered, their marriageable sons and daughters perish in their prime.[41]

---

[41] Dickens (1850) 266.

This Gothic and moralizing vision of the London slums can also be seen in the denunciation issued by the newly appointed cardinal of Westminster, Wiseman, in the same year, who, referring to the 'poor, the really poor', said that, 'in our proud metropolis we seek to drive them from the thoroughfares and streets through which fashion passes, and pack them still closer together in hidden corners, where the corruption will fester more sorely'.[42] As the Metropolitan Sanitary Association put it:

> surrounded by every noxious influence; they were exposed to every deadly pestilence . . . The water they drank, the air they breathed, the surface they walked upon, and the ground below the surface – all were tainted and rife with the seeds of disease and death.[43]

British Protestants and Catholics, therefore, were alike attuned to connecting pollution both with the urban landscape and with moral and medical dangers and transgressions. It is, therefore, notable that Catholics played down the physical dangers of ancient Roman graveyards.

This is remarkable because a particular anxiety was affixed at this time in Britain to bodies in urban churchyards. The constant churning up of the soil as more corpses were squeezed in led to remains surfacing, and when a new grave was dug 'frequently the circumstances are so revolting as to outrage common decency, distress the feelings of relatives, and to exercise a demoralizing influence on all who witness the internment'. People did know that bodies rotted and 'during decomposition give out deleterious gases and liquids'.[44] It was asked, 'what must be the condition of the atmosphere affected by exhalations from that [disordered] surface?'[45] Bearing in mind this context, it is hardly surprising that Catholic enthusiasm for the burial labyrinths of the ancient Roman catacombs was a matter for considerable controversy.

The physical dangers of the catacombs were, in this analysis, very real even for those who were not buying into what they regarded as Roman Catholic delusion. The medieval system of Catholicism that seemed somehow to have persisted in undead animation after the therapeutic attentions of the Reformation brought together dirt, disease, hunger and horror in terrifying juxtaposition and confrontation. It was in this light that Ernest Hart (1835–98), doctor, editor of the *British Medical Journal* and chairman of the National Health Society, drawing a 'sanitary contrast' between London old and new, was able to comment that medieval clerics:

---

[42] Wiseman (1850) 15–16.   [43] Metropolitan Sanitary Association (1850) 14.
[44] Metropolitan Sanitary Association (1850) 56.
[45] Metropolitan Sanitary Association (1850) 15.

gloried in their filth, and rejoiced over and admired their own disgusting wretchedness, so that the monks and fathers of the Church, believing pollution of the body to represent purity of the soul, not only lived in the odour of sanctity, but in others of a most noisome and objectionable character.[46]

The cult of moral and physical modernity had its ironies. For instance, one reason why the catacombs could be confidently projected as a place of medical as well as moral danger and pollution was that the causation of many illnesses was disputed and little understood. A crucial issue was the association of decaying matter with pestilence. The notion of disease inherent in an unhealthy location is known as the miasma theory. This idea fought a long rearguard action against the scientific discovery of microbial causation. For example, Henry Maccormac (1800–86), professor of the theory and practice of medicine at the Royal Belfast Institution, said in 1851 that in circumstances in which one could not pin it down to infection, as for instance was then the case with yellow fever, diseases should be regarded as springing from poisonous decayed organic matter. And even in instances where one could identify a specific infective agent, bad air made an infection much worse.[47] It was claimed that 'if the inhabitants of Great Britain and Ireland would but consent, day and night, to live in a pure untainted atmosphere, it would put a total close to the ravages of consumption'.[48] The problem in the houses of the rich, as well as the poor, was, supposedly, lack of ventilation. It was such ideas as these that drove windows to be opened in sickrooms, since it was held that brain fevers emanated from tropical heat or its re-creation in 'stifling' rooms.[49]

Notions of miasma, sometimes referred to in the modern literature as 'anti-contagionism', were very strong, perhaps because they did not get in the way of free trade and cut earnings which might otherwise accrue but for the quarantining of trading vessels.[50] Quarantine had been normal in early modern England, with its duration depending on the origin of the ships in question. Opposition to quarantine arose partly because it seemed incapable of preventing the spread of disease, or indeed, according to the miasma theory, because it caused disease through enforced confinement.[51] In reality, cholera, for example, was stirred up out of the Indus by just such mechanisms of long-distance exchange. It might, therefore, be thought that contagionism was the progressive doctrine and, scientifically, it was. On the other hand, it led to that great later Victorian obsession, degeneration due to social mixing:

[46] Hart (1884) 8.   [47] Maccormac (1852) 4.   [48] Maccormac (1852) 6.
[49] O'Connor (2001) 29.   [50] Hilton (1988) 159.   [51] Maglen (2002) 413 and 417.

central to the idea of contagion was the peculiarly Victorian paranoia about boundary order... the poetics of contagion justified a politics of exclusion and gave social sanction to the middle class fixation with boundary sanitation, in particular the sanitation of sexual boundaries.[52]

In other words, few people felt secure from physical contagion from disease, or from contagion from moral and sexual pollution. One could move away from bad air, but contagion follows you. The catacombs presented to the imagination the ultimate horror of the lair of miasmic decay which would then follow you in contagious fashion, thus producing such Gothic nightmares as Seymour's 'moving plague'. To enter the catacombs was, therefore, to place the boundaries of your body in peril even as you penetrated the body of the city.

Psychoanalytic readings of fears of boundary transgression can help us to understand why such concerns should focus so obsessively on the body and its health, since it is the location of the self. As discussed in the introduction to this volume (p. 17), Julia Kristeva has provided a prominent discussion of the phenomenon of abjection in her book *Powers of horror* (1982). She delved fearlessly into rejected materials and abased contexts, arguing that 'refuse and corpses *show me* what I permanently thrust aside in order to live'. There is a process of misrecognition here. It is 'not lack of cleanliness or health which causes abjection but what disturbs identity, system, order. What does not respect borders, positions, rules.'[53] It is for this reason that 'the criminal/crime is abject, as is the corpse, because it threatens, exposes the fragility, of what people want to be immutable'.[54] We can think of the catacombs as being, for Protestants, doubly polluting in that they were containers of decaying bodies the supposed propagandistic purpose of which was to penetrate the Protestant body with polluting Catholic beliefs, but which could also kill those very bodies via the agency of disease. For British Catholics, the Catholicity of the cemeteries appears to have overcome concerns about physical contamination, whilst for many British Protestants, the catacombs were clearly positioned as places replete with the dangers and Gothic excitements inherent in viewing transgression.

The defeat of the vile

Visitors regularly regarded a visit to the catacombs as being an intense experience in which the boundaries of their civilized self-assurance were, at

---

[52] McClintock (1995) 47.   [53] Kristeva (1982) 3–4.   [54] Kristeva (1982) 53.

least temporarily, shaken. Why did they go, then? Thrill-seekers, as we have seen, could enjoy this terror as a form of the Burkean sublime. Yet there was another corpus of Protestant opinion that took a different view of the catacombs. This rethought the conceptual body of the city of Rome in terms of Protestant notions of the history of Christianity and, looking beyond the surface layer of physical pollution, opined that the (superior) Protestant gaze could penetrate to the purity of the soul of the city of God hidden deep within. Peril to the physical body was trumped by the opportunity for the soul's transcendent understanding. In this, they took up a position strangely similar to that of their Catholic opponents in that they regarded the dead of the catacombs as being, ideologically, of their own flesh rather than a potentially contaminating 'other'.

The ambiguities of the account of William Ingraham Kip (1811–92), Episcopal missionary bishop of California, are revealing. Like Hobart Seymour, he was introduced to the catacombs of St Sebastian. A monk (not, this time, like a plague) took him down into the ground and, dwelling on romantic thoughts of the past, Kip thought that the very air seemed holy, yet he was glad to ascend and told his readers the story of a group of thirty youths and their teacher who vanished into the catacombs: 'the passage through which they entered, and which has since been walled up, was pointed out to us'.[55] Yet, despite this flash of Gothic horror, he favourably contrasted the evidences of what he found in the darkness with what he had left on the surface: 'it is therefore with a feeling of relief that we turn from the gorgeous services of St. Peter's, to the traces of a simpler faith in the church of the catacombs'.[56]

The notion that the Church of these first centuries was purer than that after Constantine was also shared by Emma Dixon in her novel *Out of the mouth of the lion* (1875), which features a terrific(ally bad) frontispiece in which manly men look affronted as they support a swooning woman whilst a lion pounces a foot away (figure 13.2). In this story the *agape* or love feast held in the catacombs is, in good Protestant fashion, 'never more than a frugal meal, usually of bread, wine and a dish of herbs, which were spread on a small table in the centre'.[57] The theology behind such representations was developed by William Arthur, minister of the Wesleyan Chapel, Hinde Street, who addressed the YMCA in Exeter Hall in London in 1849–50. He noted that around 170,000 of the last lecture series had been sold, and we may use this figure to imply the wide dissemination of his own views. He attested that pagans cremated, but that Christians believed

---

[55] Kip (1854) 23.   [56] Kip (1854) 176 and 212.   [57] Dixon (1875) 80.

THE MARTYRS. [*See page* 19.

13.2 Emma Dixon [writing as Emma Leslie], *Out of the mouth of the lion*. London: Religious Tract Society, 1875, frontispiece.

in the Resurrection and thus they put their bodies to sleep as on a bed. The corpses in the catacombs were, therefore, not pagans.[58] He proceeded to show that these early Christians had carried out rituals in the same manner as under the reformed faith, as for instance in the Eucharist, in which they did not re-enact a sacrifice, but partook of a meal: 'they looked on spiritual worship – prayer and praise – as the only sacrifice that remains for us to offer'.[59] He appropriated the catacombs to reformed Christianity: 'looking at the spirit, the doctrine, the ministry, the rites of the Primitive and the Protestant Churches, a glow of fellowship with the first believers lights up our very soul. Antiquity is on our side. Church of the Catacombs! Thou art our Church.'[60] He saw the evidence as 'a standing protest and testimony, from the Martyr-Church of the first ages, against the corruption and idolatry that now, alas! reign all around'.[61]

The argument that the Church of the Catacombs was essentially Protestant rather than Catholic relied on vague assumptions as to when things had subsequently gone wrong. Much anger was focused on the figure of Constantine the Great, the man who had converted the Roman Empire's establishment to Christianity in the early fourth century. For many of those on

---

[58] Arthur (1850) 172.   [59] Arthur (1850) 193.   [60] Arthur (1850) 203.
[61] Arthur (1850) 204.

the Protestant side it was a disaster. It was the end of the early 'free' Church and the establishment of the fake rituals and luxurious debaucheries of Roman Catholicism. The first ecumenical council, held in 325, ended with a banquet given by Constantine: 'that assembly, which, after bartering away its freedom deemed it fitting to revel at the board of her ravisher, over the glory of its two-fold confession: – of Christ as the fellow of the Almighty, and of His Bride as the toy of the Caesar!'[62] Supposedly, he then 'stripped Pagan temples and altars of all their attributes, and . . . engrafted them upon the simple doctrine of the religion of Christ'.[63] The 'satellites of a Nero or a Diocletian' could no longer accuse Christianity of atheism, since 'they would see no difference, only in name, between the forms of old and modern Rome'.[64] The only pure response, for those of such opinions, was to turn to the intact and pre-Constantinian words of the Bible.[65] A fascinating modern study by Kee is heir to such anti-Constantine polemic, suggesting that Constantine was not Christian and that the whole thing was a cynical ploy with the aim of taking over the Church for his own ends, so replacing Christian with imperial ideology: nothing less than a 'betrayal of Christianity'.[66] Even Charles Maitland (1815–66), one of the more advanced English researchers of early medieval history of his time, was quite capable of some peculiarly muddled thinking on the process of Christianization, as when he assured his readers that the ancient church that evangelized England in the late sixth century at the behest of Pope Gregory the Great was not the (then) present one, which provided nothing but 'excommunication and the stake'.[67] Yet, clearly, this is inconsistent with his admission that 'the feeling of reverence for the dead soon became excessive. Sepulchres and remains, even in the fourth century, formed an object of veneration.'[68]

There was, finally, a group of Protestant visitors who, whilst, finding purity below ground and impurity above, also reflected elements of Gothic horror in experiencing the former as well as a delight in the 'Oriental' exuberance of the latter. The way in which moralizing merged seamlessly with entertainment can be seen in Selina Bunbury's *A visit to the catacombs, or the first Christian cemeteries at Rome; and a midnight trip to Vesuvius* (1849). She was strongly anti-Catholic and applauded the 'infant church in its underground cradle'.[69] In those confined spaces, she assured the reader, 'we find the religion of the New Testament, in its faith, patience and hope, tangibly brought before us'.[70] Joining the holy tourist route and led by the

---

[62] Cooper (1852) 373.   [63] Anon., 'A Quiet Looker-On' (1851) 14.
[64] Anon., 'A Quiet Looker-On' (1851) 14.   [65] Wood (1846) 120.
[66] Kee (1982) 168.   [67] Maitland (1846) 312.   [68] Maitland (1846) 64.
[69] Bunbury (1849) 3.   [70] Bunbury (1849) 5.

requisite monk, she entered the catacomb of St Sebastian. She relates that 'we descended into this vast house of the dead, which during the times of suffering and persecution, was made a dwelling place for the living. A more chilling receptacle for the former, a more frightful abode for the latter, can scarcely be imagined.'[71] These pious horrors contrasted with the splendour when, at Sta Maria Maggiore:

> enveloped in a cloud of incense, and partly screened by two immense fans made of ostrich feathers, came the pope [Pius IX], in the papal chair, borne on men's shoulders, to be set down in his place; his eyelids, as is customary, closely cast down; his fingers making the sacred sign. A god of man's device appeared to me revealed in this spectacle – one of the ancient gods of old Rome in real human flesh![72]

The pleasures of denunciation reach a high point in the hands of the aforementioned Hobart Seymour:

> I hope the ladies present will forgive me, I won't say anything very indelicate, but I was myself shown and I have handled a bottle of milk from her breasts as she suckled the child Jesus . . . and I did handle with my own hands – I hope the ladies will again forgive me for explaining myself in plain English – what seemed to be a red rag, but which was shown to me as piece of – what shall I call it? – the chemise of the Virgin Mary [which was used in former times to re-enact the birth of Jesus].[73]

These words are taken from a talk given at the Assembly Rooms in Bath, the city of bodily sanitation, in 1866. Rome itself became, in this cultural reading, a morally filthy ancient city of thrilling moral danger in which the Protestant hero could demonstrate his triumph over the seductions of the flesh. By battling through the physical and moral darkness he could reveal, for our pleasure and edification, both the inspirations of purity and the entertaining excesses of its dark opposite. The purpose of the seeming abomination of Rome in God's divine plan was, thus, ultimately, to affirm the truth of the Protestant faith and the superiority of its adherents both ancient and modern.

The semiotic terrors of the catacombs receded as they, and their contents, were categorized as primitive evidences of Christian devotion. The associated 'excesses' of contemporary Roman Catholic devotion, meanwhile, could be safely repackaged as moral entertainment, their dangers dissolved by mockery. Pollution, in Mary Douglas' terms, was contained by

---

[71] Bunbury (1849) 10.   [72] Bunbury (1849) v.   [73] Seymour (1866) 17–18.

such strategies if not entirely eliminated. It was the example of the Protestant martyrs of the Reformation that may have provided the inspiration for the semiotic cleansing of the early Christian martyrs from the dross of their medieval and contemporary relic cults. Hobart Seymour has the distinction of having produced an edition of Foxe's *Book of martyrs* that had been suitably expurgated of excessive violence and so was rendered suitable for family readings.[74] He, therefore, set himself up as the purifier not only of accounts of the early Church, but also of his own Protestant tradition. And, just as was the case with the 'ladies' in the audience, the Protestant Victorian public was enabled safely to enjoy the consumption of Rome and its religion as a titillating cultural product via the agency of the sanitizing interpretative technology of the guidebooks and tracts of Seymour and his fellow travellers. In order to purify one must of course have intimate contact with the impure. Whilst for Seymour, and his ilk, such contact was rendered harmless by their resolute Protestant faith, those whom they opposed must have had a very different reading of these affairs. For Roman Catholics, the Church of the catacombs and that of the popes were one and the same and the only vile bodies were not those of fake saints, or papal imposters, but of visiting Protestants who presumed to reinterpret the evidences of Roman piety with such peculiar and lip-smacking delight.

[74] Foxe (1838).

14 | Delinquency and pederasty: 'deviant' youngsters in the suburbs of Fascist Rome

MARTINA SALVANTE

On 15 November 1927 a boy of 15, named Otello, was conducted to a police station by two Fascists. He was accused of having led 'with the pretext to go and eat cardoons, a certain V. Marcello..., a 5-year-old child, to a meadow near the via Prenestina, and forced him to have sexual intercourse, while other boys were witnessing the disturbing scene from far off'. The accused had also involved a 10-year-old boy, Aldo, considered 'the unwitting instrument' of the older boy. Otello was charged with 'indecent behaviour, corruption, and carnal knowledge'. The 15-year-old boy was eventually brought to the juvenile section of the city prison in Via Giulia, while Aldo was sent to a house of correction.[1] These boys had been caught and brought to the police station of Porta Maggiore by two Fascist men who lived in the area and had quickly arrived on the scene.

The event had taken place in a mostly working-class suburb (*borgata*) of eastern Rome, in an area between Pigneto and Prenestina (Severino 2005).[2] The language in which this event was reported in the court proceedings showed some of the recurrent motives of anti-urban propaganda, denoting the city as a dangerous place for youngsters to grow up in. This included the moral connotation of the perpetrator (who was defined as 'wild'); the lack of parental control over the boys' behaviour; the sexual depravity of the uninhibited young males living in the city; and, in contrast, the exemplary role played by the two young Fascists in re-establishing order and reporting the culprits. Taking as a starting point this case of abusive sexual behaviour among teenage males living in the impoverished Roman suburbs in the

I am grateful to Mary Gibson for her helpful comments on an earlier draft of this chapter. I would also like to thank Mark Bradley and the anonymous reviewers for their constructive suggestions.

[1] Archivio di Stato di Roma (hereafter ASR), Tribunale Penale di Roma (hereafter TP), 1928, b. 103, no. 1114.
[2] The suburb of Pigneto was created at the end of the nineteenth century following the construction of an industrial estate along the Via Prenestina, where, thanks to some cooperatives set up by street-sweepers and tram and railway drivers, the first council houses were built. Therefore the workers' movement always had a strong presence in this suburb.

second half of the 1920s, I will aim to open up the discourse to the wider concepts of deviancy, obscenity and disgust, and to focus on the ways homosexuality was criminalized and often associated with delinquency. The stories here presented are part of a small sample of court cases of the late 1920s and early 1930s dealing with lewd or lascivious conduct, which can be traced in the Roman archives. All of them refer to young males – except the very last – and their 'deviant' sexual practices in a developing Rome. My aim is to focus on the connections between Fascist propaganda against 'corrupting' urbanism and 'abnormal' sexualities as perceived through the lens of moral pollution.

Definitions and perceptions

First of all, I would like to draw attention briefly to some *longue-durée* factors regarding same-sex sexuality. According to recent researches, the *vitio nefando* (the expression designated sodomy in early modern Italy) performed between male adults and minors was frequent in working-class areas of seventeenth-century Rome (Baldassari 2005). Usually, same-sex acts were an expression of asymmetrical power relations, determined by either the age disparity or the different social position of the participants; in fact, the male adult often had the active role and the adolescent the passive one. Nonetheless, youngsters had been perceived not only as victims of violence but also as perpetrators, or else as 'rent-boys' who offered their bodies in exchange for money or other items.[3]

At this point, it is necessary to give a definition of certain key terms used in this chapter, such as 'pederasty', 'deviancy', 'delinquency' and 'homosexuality', as they were perceived at that time. The word 'pederast' comes from the Greek *paiderastēs* (one who loves adolescents) and originally indicated a relationship between a male adult citizen and a freeborn adolescent male.[4] The term was more or less obsolete for centuries (the biblical 'sodomite', expressing moral condemnation, being preferred instead), but in the nineteenth century 'pederast' acquired the meaning of homosexual in general, without any age connotation. In turn, the expression 'homosexual' appeared

---

[3] For a discussion on this kind of sexual interactions in late nineteenth-century Rome, see Rizzo (forthcoming).

[4] The first significant work on Greek same-sex relations, and even now a landmark study, is Dover (1978). On homosexuality in the ancient Roman world, and its relationship to Greek behaviour, see Williams (2010).

only in the nineteenth century with the medicalization of sexual behaviour, and the attention that physiology, first, and psychoanalysis, later, gave to sexuality and its investigation.[5] Scientists questioned the possible causes of 'sexual inversion' (another expression often used in the 'scientific' field) and, in their approach, were divided between those who preferred the organic and constitutional explanation and those who, by contrast, ascribed homosexuality to an acquired, functional origin.

In the second half of the nineteenth century, the words 'prostitution', 'deviancy' and 'delinquency' came often to be associated, following Cesare Lombroso's theories, which have had a major influence on the evolution of criminological sciences and positivist law.[6] According to Lombroso (1906), homosexuality and criminality had in common a form of atavism and 'moral insanity', which originated from a faulty genetic heredity. Such theories strengthened the idea that homosexuality was a deviancy from a psychological and physical norm, as the 'born invert' was considered to be invariably amoral as well as to have distinctive bodily traits. But 'scientific' judgements eventually crossed over into the socio-political realm, often supporting policies of juridical repression or moral condemnation. In that context, disgust and the social rejection of same-sex relationships merged with attitudes to incest, nymphomania and sadism, which were all addressed as 'depravities', and therefore isolated, condemned and punished. Disgust is of course a complex emotion, often directed towards persons (or things) that are 'out of place' or do not fit into our accepted social classifications. As Mary Douglas demonstrated, disgust could be seen as a protective mechanism, which helps to trace boundaries between what fits the categories we create and what does not. In turn, Martha Nussbaum ((2004) 3) has illustrated how disgust plays an influential role in the development of law, by serving as a 'primary or even sole reason for making some acts illegal', homosexual relations being just one example. In her work she has analysed, for example, how frequently disgust is implicated in the definition and condemnation of obscene conduct, and points to the etymological origin of the adjective 'obscene', which Nussbaum gives as stemming from the Latin

---

[5] The term 'homosexual' was coined by Austro-Hungarian journalist Karl-Maria Kertbeny in the 1860s and later applied by sexologist Richard von Krafft-Ebing in his *Psychopathia sexualis* (1886).

[6] Cesare Lombroso's book *L'uomo delinquente* (*The criminal man*) was first published in 1876 and had four editions before the final version was published in four volumes in 1896–7. Cf. Lombroso and Ferrero (1893) on *La donna delinquente, la donna normale e la prostituta* (*The female offender*). On these, see Gibson (2002).

word *caenum* (dirt, filth, mud, mire). According to Nussbaum (and taking a different line from W. I. Miller (1998)), in liberal and democratic societies disgust cannot itself be considered a valid reason for the prohibition of particular acts.

Common taboos concerning sexuality in general, then, collided with the growing interest of science in transgressive behaviour at the end of the nineteenth century. Social and medical sciences proposed to manage 'deviants' and criminals through the investigation of body and mind, so that they would be able to identify and contain 'social dangers'. Homosexuality, for example, was pathologized and often, as happened in many Western countries, criminalized. Such reprobation had had a long tradition: from the Christian denunciation of sodomites as sinners and transgressors of God's principles, to the medieval condemnation of homosexuality as a crime against society, up to the modern identification of homosexuality with pathology (Leroy-Forgeot (1997) 42). Homosexuals were considered to have crossed specific social lines and transgressed the 'gender order'.[7] Robert Nye has explained 'the pathologization of the perversions' as 'a response to a perceived crisis in gender roles and widespread fears that "normal" sexual drives were being deflected from their rightful ends', that is to say, heterosexual, procreative intercourse (Nye (2004) 19).

What mostly frightened common sensibilities about homosexuality was, indeed, the 'confusion' between the strictly determined female and male features and functions. Homosexuals were considered to have crossed the strictly separated boundaries between femininity and masculinity. Starting around the seventeenth century, Western scientists had progressively described men's and women's bodies as decisively different but complementary, thus implying that the biologically determined 'two-sex' system also governed one's sexual destiny. According to Thomas Laqueur, 'to be a man or a woman was to hold a social rank, a place in society, to assume a cultural role, not to *be* organically one or the other of two incommensurable sexes. Sex before the seventeenth century, in other words, was still a sociological and not an ontological category' (Laqueur (1992) 8). This modern perception of sex and sexuality was truly dissimilar from what ancient Romans had understood of same-sex relations. Sex with other males was seen as largely unproblematic for Roman freeborn men – though in contexts carefully restricted by the status of the parties. Normally, Roman men did not categorize or judge sexual acts on the basis of whether only males or

---

[7] For a general overview of social structures based on gender, see Kimmel (2000). For a more historiographical discussion of sexualities, see Weeks (2000).

males and females were involved, but rather on the legal and social status of the persons involved.[8]

The advent of Fascism: fighting against urbanism and promoting the *uomo nuovo*

In spite of its early self-representation as an 'innovator and demolisher', later on Fascism made use of anti-modernist motifs, aimed at the restoration of a traditional, rural and patriarchal social order, whose essential ingredients were normative representations of masculinity and femininity. Such concepts converged in an exaltation of virility, as the means and expression of supremacy, which, in a reciprocal causal relation, was connected to the virtues of demographic growth and fertility (Falasca-Zamponi 1997; Bellassai 2005). In his Ascension Day speech, delivered on 26 May 1927 in front of the Chamber of Deputies, Benito Mussolini addressed the question of industrial urbanism by constructing the modern city as a pathological site of moral deviation and sterilization (Horn 1994). On these bases, added to the declared objective of the Italian population reaching 60 million by the middle of the century (Mussolini 1934), Mussolini planned the introduction of pro-natalist measures to increase the number of births in the country (Treves 2001). In 1926 a 'bachelor's tax', which aimed to promote early marriage by imposing a higher rate of tax on unmarried men aged between 25 and 65, had already been launched.

This tax, which discriminated against bachelors, recalled similar measures employed under the Roman emperor Augustus. Consequently, it was further assumed that classical *romanità* could provide an authentic model of virtues and values, offering a historical lesson to guard against the vices which corrupted late Roman civilization (Visser 1992). In particular, the age of Augustus was praised as the golden age of Roman values and virtues – such as the organic concept of the state, the imperial system and the return to traditional Roman family values – and was therefore portrayed as the obvious precursor of Mussolini's Italy. However, scholars of classical antiquity during the Fascist period appear not to have addressed the topic of same-sex sexual intercourse and its regulation in ancient Rome, but preferred instead to treat subjects deemed 'harmless' and in line with the themes of their day: imperialism, force, morality.

---

[8] For an interesting consideration of sexuality and its regulation in ancient Rome, see McGinn (1998).

Silence also characterized the contemporary legal approach to homosexual behaviours, since the Italian Kingdom's penal codes had not provided any rule against sexual acts involving same-sex individuals. That was the case for both the Zanardelli code of 1889 and the Rocco code of 1930 (from the names of the ministers of justice Giuseppe Zanardelli and Alfredo Rocco). One of the possible explanations for such an attitude by Italian lawmakers might be that only the Catholic Church was considered responsible for intervening within the sphere of morality and sexual habits, including the condemnation of sodomy (the connection between disgust and sex was always pivotal to the moral discourse of the Christian world: see Bradley, chapter 1, pp. 31–2, and Janes, both in this volume). The Rocco code did not introduce any explicit article against homosexuality either, although one had been provided in the elaboration phase (article number 528).[9] While compiling the new penal code in the 1920s, in fact, legislators debated on the opportunity to introduce a clause against homosexual acts, which was not present in the former liberal penal code of 1889: 'The president of the legislative commission, Giovanni Appiani, explained that the anti-homosexual law was "aimed at the protection of the moral health of the stock"' (Ebner (2004) 141). The exclusion of the article from the definitive text was motivated by the presumed scarcity of homosexuality in Italy, which (the argument went) would make the intervention of lawmakers redundant. As Michael R. Ebner has written, 'beneath the silence of the Fascist criminal and police codes thus lay an array of sanctions and institutions for the repression of homosexual behaviours' (Ebner (2004) 142). 'The concealment strategy' (Benadusi (2005) 114) on the one hand intended to spread a virile image of the dictatorship, free from the 'immoral debaucheries' of democracies, and on the other was part of a general policy of intervention to reform and normalize the sexual behaviours of Italians. In fact, the regime paid particular attention to issues surrounding sexuality and morality, both by intervening through the use of propaganda against 'deviant' and 'abnormal' behaviours and by applying strict social control through diverse repressive mechanisms.

As already observed, pederasts were considered to be potentially 'dangerous' individuals and were often associated with criminals. The advent of

---

[9] To cite the draft of the article: 'Article 528 – *Homosexual relations*: Whoever . . . accomplishes libidinous acts upon a same-sex person, or consents to those acts, is punished, should the fact rise to public scandal, with imprisonment from six months up to a maximum of three years. The following instances will be punished with detention from one to five years: 1) the offender is over 21 years and the victim under 18 years of age; 2) the act is executed on a customary or profit-seeking basis' (Ministero della Giustizia e degli Affari di Culto (1927) 206).

the Fascist regime, which aimed at perpetrating *bonifica umana* (a form of social engineering) by organizing and educating Italian people according to Fascist principles, complicated such perceptions. As Ruth Ben-Ghiat has observed: 'Through a combination of indoctrination, legislation, and punitive action, he [Mussolini] and his followers aimed to remould behaviours and bodies to combat domestic decadence and achieve international prestige' (Ben-Ghiat (2001) 3). In particular, Italy's youth were the favourite target of Fascist propaganda and indoctrination, as they represented the nation's future. For this reason, the dictatorship created organizations and institutions – like the Opera Nazionale Maternità e Infanzia, the Opera Nazionale Balilla and the juvenile court – purposefully aimed at assisting and organizing young people and at restraining any deviant inclinations they might have. The primary objective was to guard against any moral or social 'contagion' that might affect the minors, and to avoid contact with 'antisocial' and 'dangerous' environments or persons. Therefore, the government set out to monitor and keep records of any individuals who were considered 'anomalous' to the system.

Nevertheless, 'moral hygiene' – alongside physical hygiene – which Fascism boasted of promoting, was, according to many, undermined by urbanization. While the rural environment was praised as a model of traditional and hierarchical moral order, cities were viewed as filthy places of degeneration and perdition, which promoted sexual promiscuity and dangerous socialization between people of different ages, genders, classes and political faiths.[10] And the most exposed to such dangers were undoubtedly the city's youth, who were less prepared to recognize evil from good. These backward-looking elements of Fascist ideology, which valorized peasant culture and lifestyle, in some respects contradicted the dictatorship's aims of modernizing Italy.

'Deviant' youth in Rome: the legal situation

In another file deriving from Rome's tribunal court, I discovered a case involving twelve young boys, all born between 1910 and 1913. They were prosecuted in May 1929 for crimes described in article 79 (re-offending judged as one and the same crime), article 331 (rape) and article 336 (indictable offence because committed in a public place) of the Zanardelli penal code of 1889. They were accused of having perpetrated obscene acts

[10] See Mussolini (1928).

and of having 'had unions contrary to nature' with three minors younger than 12 over a period of two months (May–June 1927). The supposed offenders had been reported and arrested, 'so that such an immoral situation, which had so affected the population, might end'.[11] The victims' relatives had reported the crimes to Garbatella station's *Carabinieri* on 19 July 1927, horrified by 'indecent acts perpetrated by youngsters from the *rione* [district] of Garbatella'. In detail, the twelve boys were accused of having seduced, 'with the offer of bread, fruit, ice creams, and cigarettes and sometimes money', the three minors in order to force them to perform 'dirty libidinous acts', particularly 'masturbations and unions contrary to nature' in a Garbatella meadow. The charges were reinforced by the detailed statements of the three victims and by the partial confessions of some of the defendants, in relation to the 'type of indecent exposure' and the 'public nature of the crime scene'. However, during the preliminary proceeding the indicted were eventually freed by Garbatella's *Carabinieri*. Finally, the court, having taken into consideration the 'way the deeds took place', 'a certain predisposition of the young boys to accept the indecent behaviour', the young age of the accused and their lack of criminal records, changed the charge and sentenced them to five months imprisonment and to the payment of 100 lire fine each for the crime of corruption of minors.[12]

The above-mentioned crimes – rape (article 331), corruption of minors (article 335) and indictable offences (article 336) – were included in Title VIII of the Zanardelli penal code, dedicated to so-called 'Crimes against morality and the family'. These were all crimes that could be prosecuted only if reported, except for cases in which the action was perpetrated in a public space.[13] In actual fact, the penal code of the time was cautious in avoiding interference with the domestic or private sphere, which explains why violence, when occurring in families, was often not prosecuted as damage to personal integrity, but only as a cause of scandal.[14] As long as the offence took place in a private space, the law refrained from intervention to avoid making the offence public. It was rather when the action was already

---

[11] See ASR, TP, 1928, b. 103, no. 1116, *Processo verbale di arresto ad opera della Stazione territoriale dei Carabinieri Reali.*

[12] ASR, TP, Verdicts, no. 2632 (27 May 1929) and ASR, TP, 1928, b. 103, no. 1116.

[13] The other two exceptions were the death of the victim as a consequence of the action, and crimes originating from the misuse of paternal or tutorial authority (article 336). In Italian law some crimes can only be prosecuted following a request by the victim (*querela*). Such a request for prosecution is a prerequisite for criminal proceedings to be instituted and is generally envisaged for minor offences, while serious offences are prosecutable *ex officio.*

[14] Cf. Camera dei Deputati (1887) 213–14. See also Guarnieri (2003) and Rizzo (2004).

public – and constituted an offence to decency – that the juridical power intervened.

As stated in the introductory report, the Rocco penal code of 1930 aimed at punishing more harshly all public offences to decency, and therefore the expression 'a place open to the public' (which did not appear in the former Zanardelli code) was introduced to qualify the setting of crimes and misdemeanours. Hence, all offences perpetrated in a public space could be considered affronts to decency. As concerns sexual crimes, they were listed in Title ix of Book ii ('Of crimes against morality and decency', articles 519–44), while 'Crimes against the family' were listed together in Title xi.[15] As Nussbaum observes, 'Most behaviour penalized under the rubric of "public" lewdness takes place in seclusion – a washroom stall being the classic case, and secluded wooded areas' ((2004) 268–9). She regards the familiar distinction between the public and the private as misleading when thinking about the regulation of conduct. In fact, 'space that is "public", in the sense that it is part of publicly owned facilities and/or open to those who wish to enter, is not necessarily "public" in the sense that behaviour in it necessarily affects non-consenting parties' (Nussbaum (2004) 269). The American philosopher argues, therefore, that the 'private–public' boundary is vague as well as implicitly discriminatory, since it 'does not function symmetrically on both sides because it protects "normals" both in their choice to conceal and in their choice to make public, whereas "abnormals" are required to conceal' (Nussbaum (2004) 298). This latter observation captures perfectly the character of the cases under consideration here.

The scientific gaze

Scientists and authorities showed interest in dealing with the issue of homosexuality, in that it was a 'cause of crimes or acts which, if not strictly crimes, cause nonetheless a breach of the peace and an offence to decency and morality' (Falco (1935) 125). According to them, prostitution was among the crimes of which homosexuals were guilty. However, homosexuals were considered not only the perpetrators of crimes against property and persons (thefts, blackmails, homicides, etc.), but also the victims, in that they were forced to live out their sexuality in secret (Mieli 1926; Falco (1935)

---

[15] 'The pederast can be charged with crimes of rape or offence against decency, as outlined in articles 519 and 527, depending on whether the victim was a minor or an adult' (Salerno (1938) 701).

129–30; Musatti 1943). Shame and disgust, in fact, often caused them to hide and go underground, with the result that they might be exposed to any sort of danger.

A handful of criminologists conducted research focusing exclusively on male prostitutes, whose physical and psychological characteristics, as well as medical, familiar and personal histories, were carefully recorded (Salvante, forthcoming).[16] For example, Benigno Di Tullio, who served as a staff member of the Service of Clinical Anthropology in the Regina Coeli jail in Rome,[17] paid great attention to the examination of young inmates' biographies. In 1927 he published a study about sexual crimes committed by delinquent minors. The focus of his investigations was a group of youngsters who sold their bodies for money or food, and operated in a Roman *borgata* (see below) – forerunners of a sort of those *ragazzi (di vita)* whose stories were recounted by Pier Paolo Pasolini in the 1950s (1955; 1959).

Di Tullio was also professor of criminal anthropology at the Scuola di Polizia Scientifica directed by Salvatore Ottolenghi, a pupil of Cesare Lombroso.[18] The school's aim was to teach police investigators the innovative techniques of scientific investigation and recent criminological theories. In the light of new developments in criminology, a criminal records bureau was created, where the biographical files of arrested delinquents were collected, together with more detailed physiognomic and behavioural studies about prisoners, including presumed homosexuals. In the period 1927–39, Italy's elite police school went on to register over a thousand homosexuals among inmates. The bulletin of the Scuola di Polizia Scientifica published an annual report of all case files, including a section on 'homosexual offenders' (Ebner (2004) 144). Di Tullio paid particular attention to juvenile criminality, both by tracing the profiles of young delinquents and by investigating causes, as well as possible solutions (1928a; 1928b; 1928c).

The case of the twelve youngsters held in custody in 1927 took place in Garbatella, a suburb created under the Liberal governments, but significantly reshaped and enlarged under Fascism. Its development, like other

---

[16] An example can be found in the documentary evidence collected by Giuseppe Vidoni in the Genoa prison (1922; 1940).

[17] See Archivio Centrale dello Stato (hereafter ACS), Ministero di Grazia e Giustizia, Direzione Generale degli Istituti di Prevenzione e Pena, Archivio Generale, b. 801: *Roma. Riformatorio giudiziario. Personale sanitario. Anni 1921–25.*

[18] Benigno Di Tullio (1896–1979) practised as a medical doctor in the prisons of Rome from 1922 to 1928. Besides being the director of the centre for the observation of orphaned and delinquent minors in the period 1933–6, he became professor of criminal anthropology at the University of Rome. Moreover, he was among the founders of the International Society of Criminology based in Paris.

large-scale architectural projects in the twenty years of Mussolini's rule, was motivated by the project of redesigning the capital through a plan of 'clearance and slums' (Insolera (1993) 127–42), that is to say, the radical transformation of the old centre and the founding of new suburban districts. Moreover, at that time Rome was experiencing large-scale immigration of people who were aiming to find jobs and better conditions, and as a result its population increased from 691,661 in 1921 to 1,415,000 in 1941 (Insolera (1993) 143 n. 2). It was not an accident that of the twelve boys held by the police at Garbatella, only half were born in Rome, while the other half were immigrants from other parts of Italy, as well as further afield.

Rome under Fascism

Rome, with its glorious past and far-reaching symbolism, was the epicentre of all these developments – capital of Italy and the centre of Fascist ideology.

> For Mussolini, the glorious tradition of Rome represented a model of action, an inspiration for fascism's ideals of renewal . . . The Roman tradition constituted a firm and necessary point of reference in fascism's attempt to build Italy's identity as an aggressive and forward-looking country. (Falasca-Zamponi (1997) 92)

The city became the 'centerpiece of [the] "Fascist revolution"' and thus came to embody 'the values of the regime and its goal to change Italy through producing a new generation of Italians' (Painter (2005) xv). For this reason, Mussolini set out to redesign entire areas of the city. He commissioned archaeological excavations around ancient Roman remains and encouraged city engineers to develop spatial continuities between the buildings of antiquity and those of modernity. Furthermore, Mussolini decided to restructure the architecture of Rome by ordering the demolition of entire areas and developing new, more practical quarters. In this way, the capital would show its greatness through new, ostentatious buildings that were visually connected to its ancient heritage. Linking past and present primarily informed the massive public works projects that demolished old working-class neighbourhoods in order to 'liberate' ancient monuments, improve the flow of traffic, and create new sites for the Fascist spectacles that celebrated the regime's ongoing achievements (the district of EUR 42 south of the city centre was, and still is, one striking example of such grandiose urban planning). Moreover, Fascist authorities in many cases provided hygienic reasons for such drastic urban interventions in the capital, which were made in order to replace unhealthy populated areas with 'safer living spaces for obedient citizens' (Ghezzo (2010) 207).

Nonetheless, the demolitions and the new, imposing buildings also had significant effects on the social structure of the city, as the districts (*rioni* and *quartieri*) came increasingly to be organized on the basis of social class: middle-class districts were concentrated mostly on Rome's northern side, working-class areas on the southern (Bartolini (2002) 31). The lowest classes in particular were expelled from their homes in the city centre and relocated to suburban areas, in either renovated or newly built accommodation. In the process, new social boundaries between 'imperial' and suburban Rome were established. Discussing the *borgate*, Italo Insolera explains in his book *Roma moderna*:

> The word *borgata* was officially used for the first time in 1924 when Acilia was built 15 km distant from Rome in an area rife with malaria, to which were moved the people who had previously inhabited the areas around the Forums of Caesar and Trajan and the Via del Mare. There is something derogatory in this word, which comes from *borgo*, that is, a part of the city which is not sufficiently complete and organized to merit the definition of *quartiere*; or else, a rural agglomerate stunted in its growth by a feudal-type economy. *Borgata* is a subspecies of *borgo*, that is, a fragment of the city dropped in the middle of the countryside which is neither this nor that. (Insolera (1993) 135)

The Fascist authorities explained this relocation as being in the interests of public health and advocated the need to build new hygienic houses for everybody, in order to have a healthy and therefore strong population, in any social class. Therefore a building boom took place in the capital in the 1920s and 1930s, which on the one hand tried to address the housing shortage, and on the other redesigned the social aspect and equilibrium of Rome, exiling presumed 'dangerous' contingents and categories to isolated areas, where they could be monitored better. In the pages of *Capitolium*, the official journal of the Roman Fascist municipality, we read:

> We should consider moving agricultural, unskilled and unemployed workers, as well as families of irregular composition and of bad morality (which it would be unwise to mix with otherwise healthy environments), to lands owned by the municipality in the open countryside. These lands are not visible from the big highways, but are, however, accessible by public means of transport, and there we could let these people build their houses with discarded materials and others that the municipality may provide. (Ricci (1930) 148)

These statements demonstrate a vision of radically reshaping the composition of the *social body*, isolating and controlling the 'irregulars' or, as they could be described, 'infectious' limbs (Horn 1994).[19] Notwithstanding the attempt of the Fascist dictatorship to discourage immigration through the introduction of bans and forced displacements (Treves 1976), the capital still attracted masses of migrants looking for jobs in the Eternal City, which was the object of such praise in official propaganda. Suburbs came to represent an opportunity for immigrants to establish rural outposts close enough to the city, but within reach of the countryside. However, the suburban settlements, rising in the middle of nowhere and isolated by a lack of public transport, often compromised the already miserable economic conditions of those who lived there, because, as Insolera put it:

> The majority of deportees in the suburbs had lived previously by running small artisan businesses for the city they lived in. Transported outside it, they all of a sudden found themselves with no customers for their services and therefore no sources of profit. Nor was it possible to find a new clientele in the suburbs, where all the inhabitants were equally poor and their needs reduced to a minimum. (Insolera (1993) 136)[20]

Consequently, such settlements at the margins generally became symbols of social exclusion and extreme poverty. In the first decades of the twentieth century Rome was a rapidly growing city, as its population doubled and it established itself as the political centre of Italy and the seat of all government ministries. Attracted by the prospect of employment both within and for the government, a massive wave of immigration from the countryside hit the city. The ruling classes were therefore faced with the huge challenge of accommodating the new guests, and these social demands called for an adjustment in urban planning. In this context, complaints about uncontrollable areas, dangerous people and vicious temptations increased, demanding regular patrols and high moral standards. For the authorities, such a corrupting environment became the breeding ground of 'deviant' actors, who earned a living as thieves, pickpockets and male prostitutes.[21]

Giuseppe Vidoni, for example, maintained that male prostitution was often 'a substitute for crimes and that it usually constituted the most fertile

---

[19] The report continues: 'It would be possible to build, at minimum expense, real rural suburbs, with a population of between 1,000 and 1,500 people under the control of the *Carabinieri* or the Fascist militia and provided with health and social care' (Ricci (1930) 148).

[20] Insolera speaks of these people forcibly transported to the suburbs by the Fascist militia as 'deportees'.

[21] Similar associations were made in nineteenth- and twentieth-century Paris: see Peniston (2004) and Revenin (2005).

soil for the flourishing of the criminal underworld in the big cities' (Vidoni (1940) 401). From a moral point of view, the city was believed to be particularly dangerous for children and teenagers, as they were often neglected by their parents, both working outside the household. Young people were therefore exposed to unhealthy social influences, being less controlled by faraway relatives and neighbours.

Since scientific research had begun to define adolescence as an age of latent bisexuality and a period when a person's characteristics and sexual desires take shape, the existence of homosexual behaviour at this moment in life could be considered neither unusual nor worrying, provided that it did not persist beyond the teenage years. It was therefore considered imperative to intervene and correct delinquent minors by taking them away from 'abnormal' and sterile sexual practices and returning them to the norms of heterosexuality. Benigno Di Tullio wrote in 1925: 'I have evidence that among them [delinquent minors] active and passive pederasty is very common, being brought about by an age of physiological bisexuality, and by the total lack of moral and aesthetic restraints' (Di Tullio (1925) 187).

In addition, given that the strict social control on female sexuality made premarital relationships between men and women very difficult, men may have found same-sex relations the only way to give expression to their burgeoning sexual desires. In the case of homosexual relationships, importance was given to the distinction between active and passive roles. The man with the active role was not thought to have compromised his virility. Passivity, on the other hand, was considered a rejection of masculinity, in that it was a proper feminine characteristic (as Lombroso had argued). In fact, in contemporary police and juridical files, homosexuals who were held or sentenced for different crimes were often pejoratively described as 'passive pederasts'.[22]

Furthermore, the declared objective of 'creating' the Fascist *uomo nuovo* presupposed the encouragement of specific characteristics and categories considered appropriate to the 'virile' nature of the dictatorship.[23] Female and male sexualities were associated with concepts of passivity and activity, so that homosexuals were seen as men with feminine proclivities because they considered adopting a passive role in sexual intercourse. Their 'polluted' nature – not conforming to the heterosexual norm of the majority and to the virile image of the Fascist man – transfomed them into 'abject' beings.

---

[22] See, for example, the files of homosexuals sent to confinement sites stocked at ACS, Ministero dell'Interno, Direzione Generale di Polizia, Ufficio Confino Politico, Fascicoli Personali. Cf. also Goretti and Giartosio (2006).

[23] Cf. Spackman (1996).

As Julia Kristeva has demonstrated, it is not dirt and disease that instigate abjection, but rather anything that, by being intermediate and ambiguous, troubles any identity or order (1980; see also this volume, chapter 1, pp. 16–17).

The contrast between virility and femininity in men was also expressed in the dichotomy between the city and the country. The city with its pace, its economy and its social life was seen as a dangerous cause of 'de-virilization' of males. In this 'corrupting' environment that threatened virility, forms of criminality and deviancy associated with the urban criminal underworld were expected to flourish: 'The city, technology, comfort and the rhythms of modern life jeopardized virility because they denied man the benefits of a life in contact with nature, took his mind off the healthy, eternal struggle against obstacles and material and moral challenges, prohibited him from tempering his masculinity in an adventurous existence of continuous dangers and adversities' (Bellassai (2005) 319).

Even so, Fascist propaganda often deployed images of filthy and corrupting foreign metropoleis as metaphors for the *feminine*, while Rome enjoyed 'a *masculine* symbolic and semantic status that set her apart from the pathologization of the modern city' (Ghezzo (2010) 199–200). The 'deviancy' associated with youngsters was generated because they were not easily monitored and organized, in either social or sexual terms. The government's decision to create a juridical system exclusively for minors, with the remit of assessing them separately from adults, was of course an important step in the adaptation of Italian law to international trends, but it also demonstrates an aim of supervising the nation's youth more carefully. Institutionalization and discipline were considered the only reliable means of securing the moral redemption of delinquent minors, whose 'wild' behaviour was accordingly expected to conform to the Fascist ideal of social order. Younger generations in turn were supposed to comply with the standards of legitimate and productive heterosexuality which the dictatorship promoted in its pursuit of 'sexual normalization' (De Grazia (1992) 43).

Conclusions

'Out-of-place' sex was – and still is in many contexts – considered a threat to hetero-normative prolific unions. Sexuality in Fascist Rome, therefore, had to be severely controlled: brothels were regulated by the state, and independent street prostitution was combated through police raids. The same repressive attitude was applied to homosexuality, which, following

Lombroso's theories, was increasingly associated with prostitution, criminality and deviancy. The Fascist government, then, attempted to remove and isolate disorderly sexual presences from the urban environment through a substantial and highly symbolic reorganization of space.[24]

However, the urban development of Rome had increased the number of public places where 'illicit' dealings and meetings could occur. The introduction, with the 1930 penal code, of the extensive definition of crimes occurring in 'a place open to the public' testifies to broader aims of controlling and moulding diverse sexual practices and their distribution within the city. This approach is exemplified by the following tribunal court case documented in Rome's City Archives concerning an episode in February 1933. Although this document relates to adults instead of adolescents, it is telling to observe the enthusiasm expressed by a citizen in spying on and denouncing two men whom he saw having oral sex in a public urinal. As Sergio G. (a building contractor) was walking to work he noticed a man bowed in front of another one in a public toilet. Having some days previously noticed the same behaviour, 'he hid himself nearby and saw that the men, as soon as the street had emptied, went back to the urinal and resumed the same position'. At that point the builder stopped one of the two men – while the other escaped – and handed him over to a *Carabiniere*. Arturo R., the detainee, confessed to having met the other man by chance and asked for the episode not to be made public, as he was a family man. The *Carabinieri* were able to track down the second man, Carlo P. (unemployed). The two men were sentenced to five months in prison in November 1933 for the crime of offence against decency (article 527) by the lower court judge (*pretore*) of Rome. They appealed against the sentence in 1934. The judge deemed the behaviour illegal because it had happened in a public place, therefore causing offence to the moral senses of any passer-by who may have witnessed it, as the case documentation puts it: 'The lower court judge examined the quality of those acts and qualified them as obscene, due to the attitude the two men had in the urinal at the moment of their discovery'.[25] The judge of appeal ratified the former sentence and sentenced the defendants to the payment of procedural expenses, remitting, however, the prison sentence.

Urinals have been 'usual' meeting venues for homosexuals, especially in cities. They were dotted about in different areas of the city, freely accessible

---

[24] Maynard (1994) presents a stimulating piece of research, though relating to the city of Toronto, about the historical geography of sexuality within a single urban environment.

[25] ASR, TP, Verdicts, no. 3533 (25 October 1934).

and available for immediate sexual contact, providing a convenient context for partial nudity in male company. There are regrettably very few studies of public toilets in modern Italy and of their function both as a measure of urban hygiene and as a site for sexual encounters, as there have been (for example) for France.[26] As an example of 'filthy spaces', they offer historians of Italy a fascinating and revealing context for exploring modern approaches to dirt and pollution, both physical and moral.

[26] Revenin (2005).

Envoi. *Purity and danger*: its life and afterlife

JUDITH L. GOLDSTEIN

> I believe that ideas about separating, purifying, demarcating and
> punishing transgressions have as their main function to impose system
> on an inherently untidy experience.
>
> Douglas (1966) 5

The notion of picturesque dirt seems to have captured the imagination of
both nineteenth-century visitors to Rome and twenty-first-century analysts
of Rome's history. 'It was dirty, but it was Rome; and to anyone who has
long lived in Rome even its very dirt has a charm which the neatness of no
other place ever had', exulted William Story in the 1860s (Story (1871) 5).
The salience of the concept was rooted in the contradictory reactions to the
city recorded by many outside observers. The nineteenth-century devotion
to the picturesque qualities of Roman dirt was a way in which the 'filthy' and
the 'squalid' were made tolerable in the face of the city's other attractions
such as antiquity, art and climate (at least in winter). The joining of opposites
in the idea of dirt made picturesque provided the reassurance that Mary
Douglas and others have argued is the result of imposing categories on an
otherwise disturbing reality.

What would Mary Douglas' ideas about purity and danger look like today,
tested against the backdrop of contemporary Italy and its representations?
Where would we find enlightening contemporary parallels to the examples
she collected in *Purity and danger* (1966), which, in terms of geographical
regions, tacks most memorably between everyday domestic examples of
category disruption from England – the shoes left on the dining-room table –
and those further away in the comparatively exotic areas favoured by the
anthropological studies she cites – the Lele cult of the anomalous pangolin
or 'scaly anteater' (Douglas (1966) 44, 208)? To what domains would her
dominant concern with dirt and system lead us? What kind of 'thought
experiments' could we carry out to extend and apply her basic ideas and to
sound their strengths and weaknesses?

Another way of approaching these issues is to ask what would be the
twenty-first-century equivalent of the combination of the dangers and
pleasures of dirt that organized so much of the nineteenth-century

descriptions of Rome. I propose the current accounts of crime and criminal organizations as such a comparison: the depiction of Italy as lawless is a contemporary trope that operates as an equivalent background disturbance in need of a comforting set of categories through which to balance the negative with the positive. Therefore I look briefly in what follows at the Italy-based detective fiction and associated investigative writing concerned with the criminal clans, whose influence is described as pervasive throughout the country and beyond, as a kind of thought experiment. Like the nineteenth-century descriptions, this investigative writing (that of insiders and outsiders) combines description derived from 'being there' with the analysis of local mores.

Most relevant to our subject, in the light of Douglas' insistence that categories of moral pollution and physical dirt are interrelated, is a subset of that detective writing which focuses on the criminal organizations' domination of sanitation services. Robert Saviano went undercover in southern Italy for five years to research the economic activities of a criminal clan. In *Gomorrah*, he concluded, 'clan businesses have determined zoning regulations, infiltrated local sanitation services, purchased land immediately prior to its being zoned for building and then subcontracted the construction of shopping centres, and imposed patron saints' days festivities that depend on their multiservice companies, from catering to cleaning, from transportation to trash collection' (Saviano (2007) 46–7). The criminal organizations (in the above example, the Camorra) are held responsible (although distressingly unaccountable) for the heaping piles of garbage and the long-term environmental damage prevalent in Italy, especially in the South. The news media's images of rubbish bags piled high in the street of Naples haunts such accounts, locating them in a familiar everyday reality.

These fiction and non-fiction detective narratives, although also containing the traditional murder victims of the genre, present this dual criminal pollution – immoral and unhygienic – as harmful to the long-term health of both individual bodies and the body politic. Much of the detective fiction and investigative writing breaks with genre conventions by not concluding with a society returned to order after the disorder caused by crime. This investigative writing seeks to expose both the long-lasting effects of wrong-doing, and the difficulty of punishing the guilty.

In addition, the detective mystery can be understood (can be reframed for our purposes here) as a 'purity and danger' genre that offers ritual readings in which matter out of place, material and symbolic, is expelled. As W. H. Auden ((1962) 147) succinctly put it, 'The basic formula is this: a murder occurs; many are suspected; all but one suspect, who is the murderer, are eliminated; the murderer is arrested or dies.' The detective, police officer or amateur snoop occupies the role of the ritual specialist who presides over – and determines what constitutes – a return to the normal: overseeing the process of discovery, the action of expulsion and the renewed sense of security. 'The job of the detective is to restore the state of grace in which the aesthetic and the ethical are as one . . . The phantasy, then, which the detective story addict indulges is the phantasy of being restored . . . to a state of innocence' (Auden (1962) 154). Mary Douglas' work itself occupied a liminal space in 1970s anthropological writing, between structure and anti-structure, and between an earlier moment when social stability seemed to resemble a state of innocence, and a later one which increasingly interrogated the moral authority of the social order (of the status quo).[1]

The system at war with itself

> One fine day, a person 'worthy of respect' as you would say, comes to have a little talk . . . What he says might mean anything and nothing, allusive, blurred as the back of a piece of embroidery, a tangle of knots and threads with the pattern on the other. (Sciascia (2003) 20)

Like some of the most cited nineteenth-century visitors to Italy, best-selling mystery writer Donna Leon can point to a long residence in the country. Leon has lived in Venice for twenty-five years, and on her website, with its assorted interviews and reviews, she is portrayed as an insider/outsider so attached to her life in the city that she refuses to have her books translated into Italian because she does not want to become a celebrity to her neighbours. The reviews of her books point to their utility for readers seeking reliable information about the territory, saying that they are both useful to the tourist, and especially intriguing because they disclose things that would otherwise be hidden. Her book jackets note: 'Donna Leon takes readers . . . to a Venice that tourists rarely see'; 'If you're heading to Venice, take along a few of [Leon's] books to use for both entertainment and travel directions'; and even, 'Let Leon be your travel agent and tour guide to Venice.' The places mentioned in the books have been turned (like those of Andrea Camilleri which take place in Sicily) into a real guidebook that takes readers on walking tours of the bars, restaurants, monuments and institutions of the city.

The subject of *Through a glass, darkly* (2006) is the criminal poisoning of Venice's air and water for profit. Donna Leon's Inspector Guido Brunetti,

---

[1] For a review of recent anthropology see, for example, Ortner (1984); Comaroff (2010).

in the course of investigating a murder, discovers the dumping of toxic waste outside glass-making factories. His wife complains about a newspaper report on the situation. "'They prepare a document about this percolating industrial complex that's three kilometers from us probably filled with enough toxins and poisons to eliminate all of the northeast, and who do they ask for information about how dangerous those substances might be if not the very authorities who run the complex?'" (Leon (2006) 256). Brunetti will be able to see the murderer punished, but not the dumpers. As one of his officers concludes, "'So they get to pollute all they want and get away with it?'" (Leon (2006) 321).

In *About face* (2009) – published, significantly, after Robert Saviano's widely circulated and groundbreaking book *Gomorrah* – the scene of the crime, as it is in that book of investigative non-fiction, is significantly enlarged. Inspector Brunetti may solve local environmental crimes in Venice, but these are positioned against the globalization of material and moral pollution, and, in particular, the illegal dumping of the world's rubbish. *About face* thus references claims such as Saviano's that the reach of the criminal organizations goes from South to North and back again, manipulating the environmental codes of what was then the European Community, and establishing dominion over the high seas, in the pursuit of economic advantage as far away as China. As one of the policemen in Brunetti's Venetian office put it in the context of the local in *About face*, "'[D]on't we have enough trouble with our own garbage? Now they're bringing it in from other countries, too?'" (Leon (2009) 70).

The inability to punish the guilty is connected to the lack of shared values. This is one of the central problems in Leonardo Sciascia's 1961 *The day of the owl* (2003), his earliest 'Mafia novel' (Schneider and Schneider 1998), and a book considered to be historically important because it named the organization as such. In a much-cited passage, Inspector Bellodi, who is not Sicilian, interrogates a local informer after a Mafia killing: 'To the informer the law was not a rational thing born of reason, but something depending on a man, on the thoughts and the mood of this man here, on the cut he gave himself shaving or a good cup of coffee he has just drunk. To him the law was utterly irrational, created on the spot by those in command' (Sciascia (2003) 29). Sciascia continues,

> Captain Bellodi, on the contrary, an Emilian from Parma, was by family tradition and personal conviction, a republican, a soldier who followed what used to be called "the career of arms" in a police force, with the dedication of a man who has played his part in a revolution and has seen law created by it. This law, the law of the Republic, which safeguarded liberty and justice, he

> served and enforced . . . The informer would have been astounded to know
> [that Bellodi was] a man convinced that law rests on the idea of justice and
> that any action taken by the law should be governed by justice. (Sciascia
> (2003) 30)

Captain Bellodi, although he finds the murderer, cannot exact punishment
as he is undercut by the politicians who will not uphold the law, beholden
as they are to the criminals and their organization.

The sense of reassurance that should be the result of classic detective
stories – and perhaps of classic anthropological studies as well – falters
in the wake of such inconclusive endings. Saviano constantly searches for
ways to describe adequately the enormity and complexity of what he learns.
He uses metaphors of the body, as anthropologists do, but in despera-
tion: 'Rubbish has swollen the belly of southern Italy . . . to the point where
the body is ruined, the arteries are clogged, the lungs filled, the synapses
destroyed' (Saviano (2007) 281). Individual bodies also succumb to toxicity:
'Walking in the Campania hinterlands, one absorbs the odors of everything
that industry produces . . . Printer toner also fouls the land . . . Every time it
rained, a strong acid smell blossomed . . . Once inhaled, it lodges in the red
blood cells and hair and causes ulcers, respiratory and kidney problems and
lung cancer' (Saviano (2007) 285–6).

Those who dump waste even tolerate the environmental pollution of their
immediate neighbourhood. Saviano notes, 'The bosses have had no qualms
about saturating their towns with toxins . . . To flood an area with toxic waste
and circle one's city with poisonous mountain ranges is a problem only for
someone with a sense of social responsibility and a long-term concept of
power. In the here and now of business, there are no negatives, only a high
profit margin' (Saviano (2007) 284). Or, we might say, self-interest, variously
defined, has been decisively separated from concern for the collective.

By presenting a society so unlike those described in *Purity and danger*,
this investigative writing, fiction and non-fiction, forces us to question Mary
Douglas' underlying belief in the success of cleansing rituals. These rituals
do not always deliver up a society returned to stability. Most important,
this review of a literature in which designated persons are removed from
society through assassination, but often cannot be touched by the law, alerts
us to the social cost of discourses that too closely link cleaning and social
order. There are dangers not just in dirt made taboo, but also in the belief
of persons that propriety is the inevitable result of ritual and categorical
re-orderings.

At the heart of Mary Douglas' thinking about purity and pollution was
the received anthropological wisdom about the ritual process, borrowed

from Van Gennep (1960) and adapted by the colleagues in her discipline with whom her work was associated in the 1960s, 1970s and 1980s, especially Victor Turner and Edmund Leach. The focus on investigative writing allows us to pose the following questions to her powerful analysis of ritual, classification and system:

1. What if rituals of purification do not work (do not produce the desired outcome), or what if they work for a far shorter time than is, for example, the case with cyclical rituals that repeat on an annual basis? Van Gennep's ritual process is used to understand general ritual form, but it was put forth as a model for major rites of passage – becoming an adult, getting married, dying – in which the entire community participates or is involved on some level with an individual's change of status. Thus part of what is implicit in Mary Douglas' belief in the positive results of cleansing is this notion of visible, important and agreed-upon status change accomplished through the agency of ritual. Can we assume such agreement?
2. What if the constitution of the order or system – not the classification of dirt – is what is at issue? What if the background of an agreed-upon order, against which her optimistic definition of dirt is foregrounded, is absent?

To explore Mary Douglas' arguments further, it is useful to extrapolate from them their semiotic aspects, thinking explicitly about dirt and cleanliness as a system of communication, a language (see also Bradley, this volume, chapter 1, pp. 13–17). People talk with it. Our role as anthropologists and historians is to find its grammar, to see the role such distinctions play in discourses and in social actions which unite, for example, the aesthetic and the sanitary, the political and the religious, the personal and the public, the moral and the immoral, the legal and the criminal.

And it is at the juncture of theory and practice that the idealized methodologies of anthropology and official investigation most closely echo each other. In his comprehensive study *Excellent cadavers*, Alexander Stille showed that the success of the extraordinary anti-Mafia prosecutor Giovanni Falcone was based in part on his status as an insider/outsider, as a local who, speaking all the languages, real and symbolic, necessary to his work, consequently had a close working relationship with his 'informants'. Falcone said, 'Everything is a message, everything is full of meaning in the world of Cosa Nostra, no detail is too small to be overlooked . . . The interpretation of signs is one of the principal activities of a "man of honor" and consequently of the mafia-prosecutor' (Falcone, cited in Stille (1995) 7).

Falcone's statement references a science of signs and a semiotic anthropology.[2] As Clifford Geertz put it, 'The whole point of a semiotic approach to culture is, as I have said, to aid us in gaining access to the conceptual world in which our subjects live so that we can, in some extended sense of the term, converse with them' (Geertz (1973) 24). But as these conceptual worlds are increasingly at war with themselves – and are shattered, but perhaps not always renewed – we, as analysts perpetually in search of a method, will continue to face the difficulties of respecting the local while also embedding it in its ever-expanding larger contexts.

---

[2] See Eco and Sebeok (1983) for essays on semiotics and detective fiction.

Bibliography

Acciaiuoli, Z. (1518) *Oratio in laudem urbis Romae*. Rome: Jacopo Mazzocchi.

Achard, G. (1981) *Pratique rhétorique et idéologie politique dans les discours 'optimates' de Cicéron*. Leiden: Brill.

Adams, J. (1982) *The Latin sexual vocabulary*. Baltimore: Johns Hopkins University Press.

Adams, J. N. (1973) 'Two Latin words for "kill"', *Glotta* 51: 280–92.

Adams, W. H. D. (1877) *The catacombs of Rome: historical and descriptive, with a chapter on the symbolism of early Christian art*. London: Nelson.

Ager, S. L. (2005) 'Familiarity breeds: incest and the Ptolemaic dynasty', *Journal of Hellenic Studies* 125: 1–34.

Åkerman, S. (1991) *Queen Christina of Sweden and her circle: the transformation of a philosophical libertine*. Leiden: Brill.

Alberti, L. B. (1988) *On the art of building in ten books* (trans. J. Rykwert, N. Leach and R. Tavernor). Cambridge, MA: MIT Press.

Albertoni, M. (1983) 'La necropolis Esquilina arcaica e repubblicana', in G. R. Sartorio and L. Quilici (eds) *L'archeologia in Roma capitale tra sterro e scavo*. Venice: Marsilio, 140–55.

Aldrete, G. S. (2007) *Floods of the Tiber in ancient Rome*. Baltimore: Johns Hopkins University Press.

Allen, M. (ed.) (2007) *Cleansing the city: sanitary geographies in Victorian London*. Athens: Ohio University Press.

Amato, J. A. (2000) *Dust: a history of the small and the invisible*. Berkeley: University of California Press.

Ambühl, A. (1894) 'Oceanus', in A. F. v. Pauly, G. Wissowa, W. Kroll and K. Ziegler (eds) *Paulys Real-Encyclopädie der klassischen Altertumswissenschaft*. Stuttgart: J. B. Metzler, 1154.

Ammerman, A. J. (1990) 'On the origins of the Forum Romanum', *American Journal of Archaeology* 94: 627–45.

Ammerman, A. J. and Filippi, D. (2004) 'Dal Tevere all'Argileto: nuove osservazioni', *Bullettino della Commissione Archeologica Comunale di Roma* 105: 7–28.

Ammerman, A. J., Iliopoulos, I., Bondioli, F., Filippi, D., Hilditch, J., Manfredini, A., Pennisi, L. and Winter, N. A. (2008) 'The clay beds in the Velabrum and the earliest tiles in Rome', *Journal of Roman Archaeology* 21: 7–30.

Anderdon, W. H. (1852) *Two lectures on the catacombs of Rome*. London: Burns.

Anderson, A. (1993) *Tainted souls and painted faces: the rhetoric of fallenness in Victorian culture*. Ithaca, NY: Cornell University Press.

Anderson, W. (1995) 'Excremental colonialism: public health and the poetics of pollution', *Critical Inquiry* 21: 640–9.

André, J.-M. (1980) 'La notion de *pestilentia* à Rome: du tabu religieux à l'interprétation préscientifique', *Latomus* 39: 3–16.

Anon., 'A Quiet Looker-On' (1851) *The worship of saints, images, and relics* (Letters to a Romanist 5). Scarborough: Russell.

Anon., 'Adam Bede' (1860) *The natural history of Puseyism: with a short account of the Sunday opera at St. Paul's, Brighton*. Brighton: Smart.

Anon., 'Will o' the Wisp' (1843) *A paper lantern for Puseyites*. London: Smith Elder.

Ariès, P. (1974) *Western attitudes toward death: from the Middle Ages to the present* (trans. P. M. Ranum). Baltimore: Johns Hopkins University Press.

—— (1981) *The hour of our death* (trans. H. Weaver). New York: Knopf.

Arnaoutoglou, I. (1993) 'Pollution in the Athenian homicide law', *Revue Internationale des Droits de l'Antiquité* 40: 109–37.

Arrighi, V. (1996) 'Gli adepti del Savonarola: linee per un ricerca negli archivi familiari', in G. C. Garfagnini (ed.) *Savonarola e la Toscana: atti e documenti. Studi savonaroliani. Verso il v centenario*. Florence: SISMEL/Edizioni del Galluzzo, 41–6.

Arrizabalaga, J., Henderson, J. and French, R. (1997) *The great pox: the French disease in Renaissance Europe*. New Haven: Yale University Press.

Arthur, W. (1850) 'The Church in the catacombs', in *Twelve lectures delivered before the Young Men's Christian Association, in Exeter Hall, from November 1849, to February 1850*. London: Nisbet, 161–210.

Ashby, T. (1901) 'Recent excavations in Rome, Basilica Julia, Basilica Aemilia, Cloaca Maxima', *Classical Review* 15: 136–8.

Ashenburg, K. (2008) *Clean: an unsanitised history of washing*. London: Profile.

Assonitis, A. (2003) 'Art and Savonarolan reform at San Silvestro a Monte Cavallo in Rome: 1507–1540', *Archivum Fratrum Praedicatorum* 73: 205–88.

—— (2005) 'Fra Zanobi Acciaiuoli's *Oratio in laudem urbis Romae* (1518): antiquarianism and Savonarolism at the time of Raphael', in L. Catterson and M. Zucker (eds) *Watching art: writings in honor of James Beck/Studi in onore di James H. Beck*. Todi: Ediart, 55–63.

Astin, A. E. (1978) *Cato the Censor*. Oxford University Press.

Atti del Consiglio Comunale di Roma (1871). Rome: Salviucci.

Attridge, H. W. (2004) 'Pollution, sin, atonement, salvation', in S. I. Johnston (ed.) *Religions of the ancient world*. London: Belknap Press, 71–83.

Auden, W. H. (1962) 'The guilty vicarage', in W. H. Auden, *The dyer's hand, and other essays*. New York: Random House, 146–58.

Bacci, A. (1558) *Del Tevere: della natura et bontà dell'acque and delle inondationi libri II*. Rome: V. Luchino.

Baedeker, K. (1877) *Italy: handbook for travellers: second part, central Italy and Rome* (fifth edition, remodelled). Leipsig: Karl Baedeker.

(1881) *Italy: handbook for travellers: second part, central Italy and Rome* (seventh edition, revised). Leipsig: Karl Baedeker.

Bailey, C. (1932) *Phases in the religion of early Rome.* London: Oxford University Press.

Bajard, A. (1998) 'Quelques aspects de l'imaginaire romain de l'Océan de César aux Flaviens', *Revue des Etudes Latines* 76: 177–91.

Baldassari, M. (2005) *Bande giovanili e 'vizio nefando': violenza e sessualità nella Roma barocca.* Rome: Viella.

Baldwin, M. (1993a) 'Alchemy in the Society of Jesus in the seventeenth century: strange bedfellows?', *Ambix* 40: 41–64.

(1993b) 'Toads and plague: amulet therapy in seventeenth-century medicine', *Bulletin of the History of Medicine* 67: 227–47.

Barbara, A. and Perliss, A. (2006) *Invisible architecture: experiencing places through the sense of smell.* Milan: Skira.

Barber, M. (1981) 'Lepers, Jews, and Moslems: the plot to overthrow Christendom in 1321', *History* 66: 1–17.

Barbera, M. R., Pentiricci, M., Schingo, G., Asor Rosa, L. and Munzi, M. (2005) 'Ritrovamenti archeologici in piazza Vittorio Emanuele II', *Bullettino della Commissione Archeologica Comunale di Roma* 106: 302–37.

Barnes, D. S. (2006) *The great stink of Paris and the nineteenth-century struggle against filth and germs.* Baltimore: Johns Hopkins University Press.

Baroin, C. (2010) 'Intégrité du corps, maladie, mutilation et exclusion chez les magistrates et les sénateurs romain', in F. Collard and E. Samama (eds) *Handicaps et sociétés dans l'histoire: l'estrophié, l'aveugle et le paralytique de l'antiquité aux temps modernes.* Paris: L'Harmattan, 49–68.

Barrett, R. L. (2008) *Aghor medicine: pollution, death and healing in northern India.* Berkley: University of California Press.

Barry, F. (2011) 'The Mouth of Truth and the Forum Boarium: Oceanus, Hercules, and Hadrian', *Art Bulletin* 93.1: 7–37.

Barsani, L. (1987) 'Is the rectum a grave?', *October* 43: 197–222.

Bartoccini, F. (1985) *Roma nell'ottocento: il tramonto della 'città santa': nascita di una capitale.* Bologna: Cappelli Editore.

Bartolini, F. (2002) 'Condizioni di vita e identità sociali: nascita di una metropoli', in V. Vidotto (ed.) *Roma capitale.* Rome and Bari: Laterza, 3–36.

Bartosiewicz, L. (2003) '"There's something rotten in the state . . .": bad smells in antiquity', *European Journal of Archaeology* 6.2: 175–95.

Bashford, A. (1994) *Imperial hygiene: a critical history of colonialism, nationalism and public health.* Basingstoke: Palgrave Macmillan.

(1998) *Purity and pollution: gender, embodiment and Victorian medicine.* Basingstoke: Macmillan.

Bashford, A. and Hooker, C. (eds) (2001) *Contagion: historical and cultural studies.* London: Routledge.

Bauer, H. (1977) 'Kaiserfora und Ianustempel', *Mitteilungen des Deutschen Archäologischen Instituts, Römische Abteilung* 84.2: 301–29.

(1983) 'Il Foro Transitorio e il Tempio di Giano', *Atti della Pontificia Accademia Romana di Archeologia, Rendiconti* 49: 1976–7.

(1989) 'Die Cloaca Maxima in Rom', *Mitteilungen des Leichtweiss-Institut für Wasserbau der Technischen Universität Braunschweig* 103: 45–67.

(1993) 'Cloaca, Cloaca Maxima', in E. M. Steinby (ed.) *Lexicon Topographicum Urbis Romae i*. Rome: Quasar, 288–90.

Bauman, R. (1996) *Crime and punishment in ancient Rome*. London: Routledge.

Bayet, J. (1973) *Histoire politique et psychologique de la religion Romaine*. Paris: Payot.

Beard, M. (1980) 'The sexual status of Vestal Virgins', *Journal of Roman Studies* 70: 12–27.

(1995) 'Re-reading (Vestal) virginity', in R. Hawley and B. Levick (eds) *Women in antiquity: new assessments*. London and New York: Routledge, 166–77.

(2007) *The Roman triumph*. Cambridge, MA, and London: Harvard University Press.

Beard, M., North, J. and Price, S. (1998) *Religions of Rome* (2 volumes). Cambridge University Press.

Beck, R. (2004) 'Sin, pollution and purity: Rome', in S. I. Johnston (ed.) *Religions of the ancient world*. London: Belknap Press, 509–11.

Bellasis, E. (1851) *The Anglican bishops versus the Catholic hierarchy: a demurrer to further proceedings*. London: Toovey.

Bellassai, S. (2005) 'The masculine mystique: antimodernism and virility in Fascist Italy', *Journal of Modern Italian Studies* 10: 314–35.

Benadusi, L. (2005) *Il nemico dell'uomo nuovo: l'omosessualità nell'esperimento totalitario fascista*. Milan: Feltrinelli.

Bendlin, A. (1998) 'Reinheit/Unreinheit', in H. Cancik, B. Gladigow and K. H. Kohl (eds) *Handbuch religionswissenschaftlicher Grundbegriffe* (5 volumes). Stuttgart: Kohlhammer, iv.412–15.

(2007) 'Purity and pollution', in D. Ogden (ed.) *A companion to Greek religion*. Oxford: Wiley-Blackwell, 178–89.

Benetelli, L. M. (1703, reprint 2001) *Le saette di Gionata scagliate a favor degli Ebrei*. Venice: Antonio Bortoli.

Ben-Ghiat, R. (2001) *Fascist modernities: Italy, 1922–1945*. Berkeley: University of California Press.

Berlinguer, G. and Della Seta, P. (1960) *Borgate di Roma*. Rome: Editori Riuniti.

Berner, B. (1998) 'The meaning of cleaning: the creation of harmony and hygiene in the home', *History and Technology* 14: 312–52.

Bernstein, A. (1993) *The formation of hell: death and retribution in the ancient and early Christian worlds*. Ithaca, NY: Cornell University Press.

Bersezio, V. (*c.* 1872) *Roma: la capitale d'Italia*. [publisher unknown]

Bertolaso, B. (1969) 'La peste romana del 1656–1657 dalle lettere inedite di S. Gregorio Barbarigo', *Fonti e Ricerche di Storia Ecclesiastica Padovana* 2: 217–69.

Bettini, M. (1995) '*In vino stuprum*', in O. Murray and M. Tecuşan (eds) *In vino veritas*. London: British School at Rome, 224–35.

Bignami Odier, J. and Partini, A. M. (1983) 'Cristina di Svezia e le scienze occulte', *Physis* 25: 251–78.

Blacktin, S. C. (1934) *Dust.* London: Chapman and Hall.

Blake, M. E. (1947) *Ancient Roman construction in Italy from the prehistoric period to Augustus.* Washington, DC: Carnegie Institution of Washington.

Blake, M. E. and Van Deman, E. B. (1947) *Ancient Roman construction in Italy from the prehistoric period to Augustus: a chronological study based in part upon the material accumulated by Esther Boise Van Deman.* Washington, DC: Carnegie Institution.

Blastenbrei, P. (2006) 'Violence, arms and criminal justice in papal Rome, 1560–1600', *Renaissance Studies* 20: 68–87.

Bleeker, C. (1966) 'Guilt and purification in ancient Egypt', *Numen* 13.2: 81–7.

Blonski, M. (2008) 'Les *sordes* dans la vie politique romaine: la saleté comme tenue de travail?', *Mètis* n.s. 6: 41–56.

Boatwright, M. T. (1998) 'Luxuriant gardens and extravagant women: the *horti* of Rome between Republic and Empire', in M. Cima and E. La Rocca (eds) *Horti romani: atti del convegno internazionale: Roma, 4–6 maggio 1995.* Rome: 'L'Erma' di Bretschneider, 71–82.

Bodel, J. (1994) *Graveyards and groves: a study of the Lex Lucerina (American Journal of Ancient History 11).* Cambridge, MA: American Journal of Ancient History.

(1999) 'Death on display: looking at Roman funerals', in B. Bergmann and C. Kondoleon (eds) *The art of ancient spectacle.* Washington, DC: Yale University Press, 258–81.

(2000) 'Dealing with the dead: undertakers, executioners and potter's fields in ancient Rome', in V. Hope and E. Marshall (eds) *Death and disease in the ancient city.* London: Routledge, 128–51.

Boiteux, M. (2006) 'Le bouclage: Rome en temps de peste (1656–1657)', *Roma Moderna e Contemporanea* 14: 175–201.

Boltanski, L. (2004) *Distant suffering: morality, media and politics.* Cambridge University Press.

Bömer, F. (1957) 'Interpretationen zu den *Fasti* des Ovid', *Gymnasium* 64: 112–35.

Bonsegale, G. and Biagi, M. C. (eds) (1993) *Riletture del vero: gli acquarelli di Ettore Roesler Franz.* Rome: Fratelli Palombi Editori.

Borca, F. (1997) '"*Palus omni modo vitanda*": a liminal space in ancient Roman culture', *Classical Bulletin* 73: 3–12.

Boswell, J. (1989) 'Jews, bicycle riders, and gay people: the determination of social consensus and its impact on minorities', *Yale Journal of Law and Humanities* 1: 205–28.

Bouma, J. W. (1996) *Religio votiva: the archaeology of Latial votive religion* (3 volumes). Drachten: University of Groningen.

Boym, S. (2001) *The future of nostalgia.* New York: Basic Books.

Bradley, M. (2002) 'It all comes out in the wash: looking harder at the Roman *fullonica*', *Journal of Roman Archaeology* 15: 21–44.

(2006) 'Thinking with dirt: Roman sewers and the politics of cleanliness', *Omnibus* 51: 3–5.

(2011) 'Obesity, corpulence and emaciation in Roman art', *Papers of the British School at Rome* 79: 1–40.

(forthcoming a) 'Pollution, Greece and Rome', in R. Bagnall, K. Brodersen, C. Champion, A. Erskine and S. Huebner (eds) *The Blackwell encyclopedia of ancient history*. Oxford: Wiley-Blackwell.

(forthcoming b) 'Purification, Roman', in R. Bagnall, K. Brodersen, C. Champion, A. Erskine and S. Huebner (eds) *The Blackwell encyclopedia of ancient history*. Oxford: Wiley-Blackwell.

(forthcoming c) *Smell* (Senses in Antiquity series, vol. II). Durham: Acumen.

Braithwaite, J. (1989) *Crime, shame, and reintegration*. Cambridge University Press.

Brakke, D. (1995) *Athanasius and the politics of asceticism*. Oxford University Press.

Bremmer, J. (1983) 'Scapegoat rituals in ancient Greece', *Harvard Studies in Classical Philology* 87: 299–320.

(2007) 'Human sacrifice: a brief introduction', in J. Bremmer (ed.) *The strange world of human sacrifice*. Leuven: Peeters, 1–8.

Briquel, D. (1980) 'Sur le mode d'exécution en cas de parricide et en cas de *perduellio*', *Mélanges de l'Ecole Française de Rome. Antiquité* 92: 87–107.

Briscoe, J. (2008) *A commentary on Livy, books 38–40*. Oxford University Press.

Brizzi, B. (1978) *Roma fine secolo nelle fotografie di Ettore Roesler Franz*. Rome: Edizioni Quasar.

Broughton, T. R. S. (1951–86) *The magistrates of the Roman Republic I–III*. Atlanta: Scholars Press.

Brown, K. (2008) *Foul bodies: cleanliness in early America*. London: Yale University Press.

Brown, P. (1988) *The body and society: men, women, and sexual renunciation in early Christianity*. New York: Columbia University Press.

(1998) 'Asceticism pagan and Christian', in A. Cameron and P. Garnsey (eds) *The Cambridge ancient history XIII: the late Empire, AD 337–425*. Cambridge University Press, 601–31.

Brundage, J. (1987) *Law, sex and Christian society in medieval Europe*. Chicago University Press.

Brunt, P. A. and Moore, J. M. (1967) *Res Gestae Divi Augusti: the achievements of the Divine Augustus*. Oxford University Press.

Bruschi, A. (1966) 'La "riforma artistica" di Girolamo Savonarola e la crisi dell'umanesimo rinascimentale', *Quaderni dell'Istituto di Storia dell'Architettura* 13: 1–7.

Bufalini, L. (1551, reprint 1560) *Roma*. Rome: Antonio Treviso.

Bullard, M. M. (1976) '*Mercatores Florentini Romanam curiam sequentes* in the early sixteenth century', *Journal of Medieval and Renaissance Studies* 6: 51–71.

Bunbury, S. (1849) *A visit to the catacombs, or the first Christian cemeteries at Rome; and a midnight trip to Vesuvius*. London: Robinson.

Burke, J. (2000) 'Patronage and identity in Renaissance Florence: the case of Santa Maria a Lecceto', in M. Rogers (ed.) *Fashioning identities in Renaissance art*. Aldershot: Ashgate, 51–62.

Burkert, W. (1983) *Homo necans: the anthropology of ancient Greek sacrificial ritual and myth* (trans. P. Bing). Berkeley: University of California Press.

(1996) *Creation of the sacred: tracks of biology in early religions*. Cambridge, MA: Harvard University Press.

Burriss, E. (1925) 'Cato, *De agri cultura*, 83', *Classical Journal* 21.3: 221.

(1929) 'The nature of taboo and its survival in Roman life', *Classical Philology* 24.2: 142–63.

(1931) *Taboo, magic, spirits*. New York: Macmillan.

Bushman, R. and Bushman, C. (1988) 'The early history of cleanliness in America', *Journal of American History* 74.4: 1213–38.

Byrne, J. P. (2004) *The black death*. Westport, CT: Greenwood Press.

Cahn, H. A. (1981) 'Oceanus', in J. Boardman (ed.) *Lexicon iconographicum mythologiae classicae* VIII. Zurich: Artemis, 599–607.

Calabi, D. (2006) *Foreigners and the city: an historiographical exploration of the early modern period*. Milan: Fondazione Eni Enrico Mattei.

Cambiaso, D. (1908) 'La peste in Val Polcevera negli anni 1579–1580', *Giornale Storico e Letterario della Liguria* IX: 211–12.

Camera dei Deputati (1887) *Progetto del Codice penale per il Regno d'Italia e disegno di legge che ne autorizza la pubblicazione*. Rome: Stamperia Reale.

Campkin, B. and Cox, R. (eds) (2007) *Dirt: new geographies of cleanliness and contamination*. London: I. B. Tauris.

Camporesi, P. (1990) *La miniera del mondo: artieri, inventori, impostori*. Milan: Mondadori.

Cancik, C., Gladigow, B. and Kohl, K. H. (1998) *Handbuch religionswissenschaftlicher Grundbegriffe* (5 volumes). Stuttgart: Kohlhammer.

Cannon, J. (1983) 'The detective fiction of Leonardo Sciascia', *Modern Fiction Studies* 29.4: 282–91.

Capparoni, P. (1935) 'La difesa di Roma contro la peste del 1656–57 come risulta dall'opera del cardinale Gastaldi "Tractatus de avertenda et profliganda peste"', *Atti e Memorie dell'Accademia di Storia dell'Arte Sanitaria* 34: 1–12.

Carandini, A. and Papi, E. (1999) *Palatium e Sacra Via* II. Rome: Istituto Poligrafico e Zecca dello Stato.

Carawan, E. (1993) 'The *Tetralogies* and Athenian homicide trials', *American Journal of Philology* 114.2: 235–70.

Carmichael, A. (1986) *Plague and the poor in Renaissance Florence*. Cambridge University Press.

(1993) 'History of public health and sanitation in the West before 1700', in K. Kiple (ed.) *The Cambridge world history of human disease*. Cambridge University Press, 192–200.

Carson, A. (1990) 'Putting her in her place: women, dirt and desire', in D. M. Halperin, J. J. Winkler and F. I. Zeitlin (eds) *Before sexuality: the construction of erotic experience in the ancient world.* Princeton University Press, 135–69.

Cassio, A. (1756–7) *Corso dell'acque antiche portate sopra xiv. aquidotti da lontane contrade nelle xiv. regioni dentro Roma; delle moderne, e di altre in essa nascenti; con l'illustrazione di molte antichità da scrittori, e antiquari non conosciute, ne nominate (2 volumes).* Rome.

Ceccarelli, G. (1938) *La 'Spina' dei Borghi.* Rome: Danesi.

Celani, E. (1890) 'La venuta di Borso d'Este in Roma, 1471', *Archivio della Società Romana di Storia Patria* 13: 361–450.

Cerasoli, F. (1900) 'Notizie circa la sistemazione di molte strade di Roma', *Bullettino della Commissione Archeologica Comunale di Roma* 27: 342–62.

Cerretani, B. (1994) *Storia fiorentina.* Florence: Leo S. Olschki.

Ceselli, M. (1873) 'Le acque potabili di Roma, e loro varie applicazioni agli usi domestici ed industriali', *Buonarroti* 8.2: 102–14.

Chabod, F. (1951) *Storia della politica estera italiana dal 1870 al 1896* (reprinted 1990). Bari: Laterza.

Chaniotis, A. (1997) 'Reinheit des Körpers: Reinheit des Sinnes in den griechischen Kultgesetzen', in J. Assmann and T. Sundermeier (eds) *Schuld, Gewissen und Person: Studien zur Geschichte des inneren Menschen.* Gütersloh: Gütersloher, 142–79.

Charlier, P. (2009) *Male mort: morts violentes dans l'antiquité.* Paris: Fayard.

Cherubelli, P. (1940) *M. Mariano da Genazzano O.S.A. saggio bibliografico.* Florence: Monografie Storiche Agostiniane.

Chiappini, L. (1967) *Gli estensi.* Varese: Dall'Oglio.

Choay, F. (1965) *La città: utopie e realtà* (reprinted 2000). Turin: Einaudi.

Ciappelli, G. (1997a) *Carnevale e quaresima.* Rome: Edizioni di Storia e Letteratura.

(1997b) 'I bruciamenti delle vanità e la transizione verso un nuovo ordine carnavalesco', in A. Fontes, J.-L. Fournel and M. Plaisance (eds) *Savonarole: enjeux, débats, questions* (Actes du Colloque International, Paris, 25–7 January 1996). Paris: Université de la Sorbonne Nouvelle, 133–47.

(1999) 'Il rogo della cultura: i bruciamenti delle vanità', in *Girolamo Savonarola, l'uomo e il frate* (Atti del xxxv Convegno Storico Internazionale, Todi, 11–14 October 1998). Spoleto: Centro Italiano di Studi sull'Alto Medioevo, 261–95.

Cibin, L. (2003) *Selciato Romano: il sampietrino.* Rome: Gangemi.

Cichorius, C. (1922) *Römische Studien.* Leipzig: Teubner.

Cifani, G. (2008) *Architettura romana arcaica: edilizia e società tra monarchia e repubblica.* Rome: 'L'Erma' di Bretschneider.

Cinozzi, P. (1898) 'Estratto d'una epistola de vita e moribus Reverendis Patris Hieronimi Savonarole de Ferraria', in P. Villari and E. Casanova (eds) *Scelta di prediche e scritti di Fra Girolamo Savonarola.* Florence: Sansoni, 1–28.

Cipolla, C. (1981) *Fighting the plague in seventeenth-century Italy.* Madison: University of Wisconsin Press.

(1992) *Miasmas and disease: public health and the environment in the pre-industrial age.* New Haven: Yale University Press.

Claasen, J. (1999) *Displaced persons: the literature of exile from Cicero to Boethius.* London: Duckworth.

Clark, G. (1998) 'Bodies and blood: late antique debates on martyrdom, virginity and resurrection', in D. Montserrat (ed.) *Changing bodies, changing meanings: studies on the human body in antiquity.* London: Routledge, 99–115.

Clarke, J. R. (2005) 'Augustan domestic interiors: propaganda or fashion?', in K. Galinsky (ed.) *The Cambridge companion to the age of Augustus.* Cambridge University Press, 264–80.

Classen, C. (1993) *Worlds of sense: exploring the senses in history and across cultures.* London: Routledge.

Classen, C., Howes, D. and Synnott, A. (1994) *Aroma: the cultural history of smell.* London: Routledge.

Cleere, E. (2004) 'Victorian dust traps', in W. A. Cohen and R. Johnson (eds) *Filth: dirt, disgust, and modern life.* Minneapolis: University of Minnesota Press, 133–54.

Cloud, J. (1971) '*Parricidium* from the *lex Numae* to the *lex Pompeia de parricidiis*', *Zeitschrift der Savigny-Stiftung für Rechtsgeschichte, romanistische Abteilung* 88: 1–66.

Coarelli, F. (1968) 'La Porta Trionfale e la Via dei Trionfi', *Dialoghi di Archeologia* 2: 55–103.

(1983) *Il Foro Romano: periodo arcaico.* Rome: Quasar.

(1985) *Foro Romano II: periodo Repubblicano e Augusteo.* Rome: Quasar.

(1988) *Il Foro Boario: dalle origini alla fine della Repubblica.* Rome: Quasar.

(1993a) 'Campus Sceleratus', in E. M. Steinby (ed.) *Lexicon topographicum urbis Romae I.* Rome: Quasar, 225.

(1993b) 'Clivus Publicius', in E. M. Steinby (ed.) *Lexicon topographicum urbis Romae I.* Rome: Quasar, 284.

(1993c) 'Carcer', in E. M. Steinby (ed.) *Lexicon topographicum urbis Romae I.* Rome: Quasar, 236–7.

(1997) *Il Campo Marzio: dalle origini alla fine della Repubblica.* Rome: Quasar.

Coarelli, F. and Battaglini, G. (2004) *Gli scavi di Roma, 1878–1921.* Rome: Quasar.

Cohen, E. S. (1998) 'Seen and known: prostitutes in the cityscape of late-sixteenth-century Rome', *Renaissance Studies* 12.3: 392–409.

Cohen, W. (2004) 'Introduction: locating filth', in W. A. Cohen and R. Johnson (eds) *Filth: dirt, disgust, and modern life.* Minneapolis: University of Minnesota Press, vii–xxxvii.

Cohen, W. and Johnson, R. (eds) (2004) *Filth: dirt, disgust and modern life.* Minnesota: University of Minnesota Press.

Comaroff, J. (2010) 'The end of anthropology again: on the future of an in/discipline', *American Anthropologist* 112.4: 524–38.

Comella, A. (1981) 'Tipologia e diffusione dei complessi votivi in Italia in epoca medio- e tardo-Repubblicana', *Mélanges de l'Ecole Française de Rome. Antiquité* 93: 717–803.

Conrad, L. and Wujastyk, D. (eds) (1999) *Contagion: perspectives from pre-modern societies.* Aldershot: Ashgate.

Cooley, M. G. L. (ed.) (2003) *The age of Augustus* (LACTOR 17). Kingston upon Thames: London Association of Classical Teachers.

Cooper, B. H. (1852) *The free Church of ancient Christendom, and its subjugation under Constantine.* London: Cockshaw.

Corazza, A. and Lombardi, L. (1995) 'Idrogeologia dell'area del centro storico di Roma', in R. Funicello (ed.) *Memorie descrittive della carta geologica d'Italia: La geologia di Roma, il centro storico I.* Rome: Istituto Poligrafico e Zecca dello Stato, 179–211.

Corbeill, A. (2004) *Nature embodied: gesture in ancient Rome.* Princeton University Press.

Corbin, A. (1982) *Le miasme et la jonquille: l'odorat et l'imaginaire social XVIIIe–XIXe siècles.* Paris: Aubier Montaigne.

   (1986) *The foul and the fragrant: odor and the French social imagination.* Cambridge, MA: Harvard University Press.

Cordero, F. (1986) *Savonarola: voce calamitosa, 1452–1494.* Rome: Laterza.

Cornell, T. (1981) 'Some observations on the *crimen incesti*', in M. Gras (ed.) *Le délit religieux dans la cité antique* (Collection de l'Ecole Française de Rome 48). Rome: Ecole Française de Rome, 26–37.

Corradi, A. (1973) *Annali delle epidemie occorse in Italia dalle prime memorie fino al 1850* (5 volumes). Bologna: Forni.

Corsini, A. (1922) *Medici ciarlatani e ciarlatani medici.* Bologna: Zanichelli.

Cox, R., George, R., Horne, R. H., Nagle, R., Pisani, E., Ralph, B. and Smith, V. (eds) (2011) *Dirt: the filthy reality of everyday life.* London: Profile Books.

Crawford, R. (1914) *Plague and pestilence in literature and art.* Oxford University Press.

Crescimbeni, G. M. (1715) *L'istoria della basilica diaconale, collegiata, e parrocchiale di S. Maria in Cosmedin di Roma.* Rome: Antonio di Rossi.

Cressedi, G. (1984) 'Il Foro Boario e il Velabro', *Bullettino della Commissione Archeoligica Comunale di Roma* 89.2: 249–96.

Creytens, R. (1970) 'Les Actes Capitulaires de la congregation Toscano-Romaine O.P. (1496–1530)', *Archivum Fratrum Praedicatorum* 40: 125–230.

Curtis, V. (1998) *The dangers of dirt: household hygiene and health.* Wageningen: Grafisch Service Centrum.

   (2007) 'Dirt, disgust and disease: a natural history of hygiene', *Journal of Epidemiology and Community Health* 61: 660–4.

Curtis, V. and Biran, A. (2001) 'Dirt, disgust, and disease: is hygiene in our genes?', *Perspectives in Biology and Medicine* 44.1: 17–31.

D'Onofrio, C. (1986) *Le fontane di Roma.* Rome: Staderini.

Dall'Aglio, S. (2002) 'Un breve scritto savonaroliano ritrovato: i quesiti rivolti a Zanobi Acciaiuoli', *Archivio Storico Italiano* 160: 113–28.

(2010) *Savonarola and Savonarolism*. Toronto: Centre for Reformation and Renaissance Studies.

Danckert, W. (1963) *Unehrliche Leute: die verfemten Berufe*. Bern: Francke.

Dante Alighieri (1886) *The Divine Comedy: Paradiso* (trans. Henry Wadsworth Longfellow). London: Routledge.

David, J.-M. (1984) 'Du comitium à la roche Tarpéienne ... sur certains rituels d'exécution capitale sous la République, le règne d'Auguste et de Tibère', in J. L. Voisin (ed.) *Du châtiment dans la cité: supplices corporels et peine de mort dans le monde antique* (Collection de l'Ecole Française de Rome 79). Rome: Ecole Française de Rome, 131–76.

Davies, J. (1999) *Death, burial and rebirth in the religions of antiquity*. London and New York: Routledge.

Davies, P. J. E. (2000) *Death and the emperor: Roman imperial funerary monuments from Augustus to Marcus Aurelius*. Cambridge University Press.

(2006) 'Exploring the international arena: the Tarquins' aspirations for the Temple of Jupiter Optimus Maximus', in C. Mattusch and A. Donohue (eds) *Common ground: the proceedings of the xvith international congress of classical archaeology, Boston 2003*. Woodbridge: Oxbow, 186–9.

(2007) 'The personal and the political', in S. E. Alcock and R. Osborne (eds) *Blackwell guides to archaeology: classical archaeology*. Oxford: Wiley-Blackwell, 307–34.

(forthcoming) *Architecture, art and politics in republican Rome*. Cambridge University Press.

de Bonis, F. A. (1879) *I vandali a Roma*. Rome: Tipografia Forense.

De Grazia, V. (1992) *How Fascism ruled women: Italy, 1922–1945*. Berkeley: University of California Press.

de Kleijn, G. (2001) *The water supply of ancient Rome: city area, water, and population*. Amsterdam: Gieben.

De Maio, R. (1969) *Savonarola e la Curia Romana*. Rome: Edizioni di Storia e Letteratura.

Del Monte, A. (1989) *Ratto della Signora Anna del Monte trattenuta a' catecumini tredici giorni dalli 6 fino alli 19 maggio anno 1749* (ed. G. Sermoneta). Rome: Carucci.

Del Prete, F. (2002) *Il fondo fotografico del Piano Regolatore di Roma 1883: la visione trasformata*. Rome: Gangemi Editore.

DeLaine, J. (2002) 'The Temple of Hadrian at Cyzicus and Roman attitudes to exceptional construction', *Papers of the British School at Rome* 70: 205–30.

Delumeau, J. (1957–9) *Vie économique et sociale de Rome dans la seconde moitié du xvie siècle* (2 volumes). Paris: De Boccard.

Di Agresti, D. (1976) *Aspetti di vita pratese nel cinquecento*. Florence: Leo S. Olschki.

(1980) *Sviluppi della riforma monastica savonaroliana*. Florence: Leo S. Olschki.

Di Palma, W. (ed.) (1990) *Cristina di Svezia: scienza e alchimia nella Roma barocca.* Bari: Dedalo.

Di Teodoro, F. P. (1994) *Raffaello, Baldassar Castiglione, e la lettera a Leone X.* Bologna: Nuova Alfa.

Di Tullio, B. (1925) 'Spunti sulla vita e sulla patologia sessuale nelle carceri', *Rassegna di Studi Sessuali e di Eugenia* 5: 176–88.

(1927) 'Nella delinquenza minorile: reati sessuali e prostituzione maschile', *Bollettino dell'Amministrazione Carceraria* 4: 2–8.

(1928a) 'Sulla profilassi e terapia della criminalità minorile', *Maternità ed Infanzia* 3: 415–21.

(1928b) 'Sulla tendenza istintiva al delitto nei fanciulli', *Maternità ed Infanzia* 3: 545–8.

(1928c) 'I centri di osservazione per i minorenni delinquenti nei rapporti dell'igiene sociale e della difesa della razza', *Maternità ed Infanzia* 3: 960–3.

Dickens, C. (1850) 'A December vision', *Household Words* 2: 265–7.

Dickie, J. (1999) *Darkest Italy: the nation and stereotypes of the Mezzogiorno, 1860–1900.* New York: Palgrave Macmillan.

Dixon, E. [writing as Emma Leslie] (1875) *Out of the mouth of the lion.* London: Religious Tract Society.

Dixon, S. (1988) *The Roman mother.* London: Croom Helm.

Domaszewski, A. (1975) *Abhandlungen zur römischen Religion.* New York: Arno Press.

Donnini, M. (1999) 'Sul latino del Savonarola: problemi di stile', in *Girolamo Savonarola, l'uomo e il frate* (Atti del xxxv Convegno Storico Internazionale, Todi, 11–14 October 1998). Spoleto: Centro Italiano di Studi sull'Alto Medioevo, 75–120.

Dosio, G. A. (1561) *Roma.* Rome: Barptoloemei Phaletij.

Douglas, M. (1966, reprinted 2002 with additional preface) *Purity and danger: an analysis of concepts of pollution and taboo.* London: Routledge.

(1975) *Implicit meanings: essays in anthropology* (second edition). London: Routledge.

(1991) 'Witchcraft and leprosy: two strategies of exclusion', *Man* 26: 723–36.

(1998) 'La pureté du corps', *Terrain: Carnets du Patrimoine Ethnologique* 31: 5–12.

(1999) *Leviticus as literature.* Oxford University Press.

(2000) Review of Valeri (1999), *Science* 289.5488: 2288.

Douglas, M. and Wildavsky, A. (1982) *Risk and culture.* Berkeley: University of California Press.

Dover, K. J. (1978) *Greek homosexuality.* Cambridge, MA: Harvard University Press.

Drake, H. (2006) *Violence in late antiquity: perceptions and practices.* Aldershot: Ashgate.

Drobnick, J. (ed.) (2006) *The smell culture reader.* Oxford: Berg.

Dugger, C. W. (2006) 'Toilets underused to fight disease: U.N. study finds', *New York Times*, 10 November.

Dumézil, G. (1966) *Archaic Roman religion* (2 volumes) (trans. P. Krapp, 1970). University of Chicago Press.

Du Pérac, E. (1575) *Vestigi dell'antichità di Roma.* Rome: L. della Vacheria.

Dupré Raventós, X. and Remolà, J. A. (eds) (2000) *Sordes urbis: la eliminación de residuos en la ciudad romana.* Rome: 'L'Erma' di Bretschneider.

Durand, J. (1989) 'Greek animals: toward a typology of edible bodies', in M. Detienne and J. Vernant (eds) *The cuisine of sacrifice among the Greeks* (trans. P. Wissing). University of Chicago Press, 87–118.

Durkheim, E. (1912) *Les formes élémentaires de la vie religieuse: le système totémique en Australie.* Paris: Librairie Félix Alcan.

Dyck, A. R. (2004) *A commentary on Cicero, De legibus.* Ann Arbor: University of Michigan Press.

   (2008) *Cicero: Catilinarians.* Cambridge University Press.

Dyer, R. (1969) 'The evidence for Apolline purification rituals at Delphi and Athens', *Journal of Hellenic Studies* 89: 38–56.

Ebner, M. R. (2004) 'The persecution of homosexual men under Fascism', in P. Willson (ed.) *Gender, family and sexuality: the private sphere in Italy, 1860–1945.* Basingstoke: Palgrave Macmillan, 139–56.

Eckstein, A. M. (1982) 'Human sacrifice and fear of military disaster in Republican Rome', *American Journal of Ancient History* 7: 69–95.

Eco, U. and Sebeok, T. (eds) (1983) *The sign of three: Dupin, Holmes, Peirce.* Bloomington: Indiana University Press.

Edlund-Berry, I. (2006) 'Hot, cold, or smelly: the power of sacred water in Roman religion, 400–100 BCE', *Yale Classical Studies* 33: 162–80.

Edwards, C. (1993) *The politics of immorality in ancient Rome.* Cambridge University Press.

   (1996) *Writing Rome: textual approaches to the city.* Cambridge University Press.

   (1997) 'Unspeakable professions: public performance and prostitution in ancient Rome', in J. Hallett and M. Skinner (eds) *Roman sexualities.* Princeton University Press, 66–95.

   (2007) *Death in ancient Rome.* New Haven: Yale University Press.

Eliade, M. (1957) *Le sacré et le profane.* Paris: Gallimard.

Ellis, H. T. (1871) *A visit to the catacombs of Paris and other subterranean experiences.* Durham: Walker.

Engels, D. (2007) *Das Römische Vorzeichenwesen (753–27 v. Chr.).* Stuttgart: Franz Steiner.

Engen, T. (1982) *The perception of odors.* New York: Academic.

   (1991) *Odor sensation and memory.* New York: Praeger.

Erasmo, M. (2008) *Reading death in ancient Rome.* Columbus: Ohio State University Press.

Eschebach, H. (1983) 'Die innerstädtische Gebrauchwasserversorgung dargestellt am Beispiel Pompejis', in J.-P. Boucher (ed.) *Journées d'Etudes sur les Aqueducs Romains.* Paris: Les Belles Lettres, 79–132.

Evans, E. C. (1935) 'Roman descriptions of personal appearance in history and biography', *Harvard Studies in Classical Philology* 46: 43–84.

Evans, H. B. (1994) *Water distribution in ancient Rome: the evidence of Frontinus.* Ann Arbor: University of Michigan Press.

Evelyn, J. (1955) *Diary* (6 volumes) (ed. E. S. de Beer). Oxford University Press.

Fagan, G. G. (2002) *Bathing in public in the Roman world.* Ann Arbor: University of Michigan Press.

Fahlbusch, H. (1977) 'The development of the Pergamon water supply between 200 BC and 300 AD', in *Proceedings of the 17th Congress of the International Association of Hydraulic Research.* Baden-Baden: International Association of Hydraulic Research, 758–62.

Falasca-Zamponi, S. (1997) *Fascist spectacle: the aesthetics of power in Mussolini's Italy.* Berkeley: University of California Press.

Falco, G. (1935) *La sessuologia nel codice penale italiano.* Milan: Società Palermitana Editrice Medica.

Falletti, F. (1988) 'Le origini del convento di S. Maria Maddalena in Pian di Mugnone: indagine storica-documentaria', *Rivista d'Arte* 40: 63–124.

Fantham, E. (1991) '*Stuprum*: public attitudes and penalties for sexual offences in Republican Rome', *Echos du Monde Classique* 35: 267–91.

(ed.) (1998) *Ovid Fasti IV.* Cambridge University Press.

Fardon, R. (1987) 'The faithful disciple: on Mary Douglas and Durkheim', *Anthropology Today* 3.5: 4–6.

(1999) *Mary Douglas: an intellectual biography.* London: Routledge.

Favro, D. (1992) '*Pater urbis*: Augustus as city father of Rome', *Society of Architectural Historians* 51.1: 61–84.

Favro, D. G. (1996) *The urban image of Augustan Rome.* New York: Cambridge University Press.

(2005) 'Making Rome a world city', in K. Galinsky (ed.) *The Cambridge companion to the age of Augustus.* Cambridge University Press, 234–63.

Fea, C. (1832) *Storia delle acque antiche sorgenti in Roma, perdute, e modo di ristabilire II: dei condotti antico-moderni delle acque, Vergine, Felice e Paola, e loro autori.* Rome: Stamperia della RCA.

Feci, S. (1998) 'Tra il tribunale e il ghetto: le magistrature, la comunità e gli individui di fronte ai reati degli ebrei romani nel seicento', *Quaderni Storici* 33: 568–85.

Fehrle, E. (1910) *Die kultische Keuschheit im Altertum* (Religionsgeschichtliche Versuche und Vorarbeiten 6). Giessen: A. Töpelmann.

Fenelli, M. (1975) 'Contributo per lo studio del votivo anatomico: i votivi antomici di Lavinio', *Archeologia Classica* 27: 206–52.

Filarete, F. (*c.* 1465) *Libro architettonico.* Milan: Codex Magliabechiano.

Filipepi, S. (1898) 'Estratto della cronaca di Simone Filipepi', in P. Villari and E. Casanova (eds) *Scelta di prediche e scritti di Fra Girolamo Savonarola.* Florence: Sansoni, 453–518.

Finn, R. (2009) *Asceticism in the Graeco-Roman world.* Cambridge University Press.

Fiorani, L. (1978) 'Astrologi, superstiziosi e devoti nella società romana del seicento', *Ricerche per la Storia Religiosa di Roma* 2: 99–112.

Flemming, R. (1999) '*Quae corpora quaestum facit*: the sexual economy of female prostitution in the Roman Empire', *Journal of Roman Studies* 89: 38–62.

(2000) *Medicine and the making of Roman women*. Oxford University Press.

Foa, A. (1990) 'The old and the new: the growth of syphilis (1494–1530)', in E. Muir and G. Ruggiero (eds) *Sex and gender in historical perspective*. Baltimore and London: Johns Hopkins University Press, 26–45.

Fögen, T. and Lee, M. (eds) (2009) *Bodies and boundaries in Graeco-Roman antiquity*. Berlin and New York: De Gruyter.

Foot, J. (1999) 'The dead *Duce*', *History Today* 49.8: 14–15.

Fordyce, C. (ed.) (1977) *P. Vergili Maronis Aeneidos, libri VII–VIII*. Oxford University Press.

Foxe, J. (1838) *The acts and monuments of the Church; containing the history and sufferings of the martyrs* (2 volumes) (ed. M. Hobart Seymour). London: Scott, Webster and Geary.

Frank, T. (1924a) 'The Tullianum and Sallust's *Catiline*', *Classical Journal* 19: 495–8.

(1924b) *Roman buildings of the Republic: an attempt to date them from their materials*. Rome: American Academy in Rome.

Fraschetti, A. (1981) 'Le sepolture rituali del Foro Boario', in M. Gras (ed.) *Le délit religieux dans la cité antique* (Collection de l'Ecole Française de Rome 48). Rome: Ecole Française de Rome, 51–115.

Frazer, J. (1890) *The golden bough: a study in comparative religion*. London: Macmillan.

(1929) *The Fasti of Ovid* (5 volumes). London: Macmillan.

Frischer, B. (1983) '*Monumenta et arae honoris virtutisque causa*: evidence of memorials for Roman civic heroes', *Bullettino della Commissione Archeologica Comunale di Roma* 88: 51–86.

Frosini, P. (1977) *Il tevere: le inondazioni di Roma e i provvedimenti presi dal governo italiano per evitarle*. Rome: Accademia Nazionale dei Lincei.

Gabriele, M. (1986) *Il giardino di Hermes: Massimiliano Palombara alchimista e rosacroce nella Roma del seicento*. Rome: Ianua.

Gade, K. (1986) 'Homosexuality and rape of males in Old Norse law and literature', *Scandinavian Studies* 58: 124–41.

(1989) 'Penile puns: personal names and phallic symbols in Skaldric poetry', in J. Friedman and P. Hollahan (eds) *Essays in medieval studies: proceedings of the Illinois Medieval Association*. Chicago: Illinois Medieval Association, 57–67.

Galasso, G. (1982) *Napoli spagnola dopo Masaniello: politica, cultura, società* (2 volumes). Florence: Saggiatore.

Galinsky, G. K. (1992) 'Venus, polysemy and the *Ara Pacis Augustae*', *American Journal of Archaeology* 96: 457–75.

(1996) *Augustan culture: an interpretive introduction*. Princeton University Press.

Gandy, M. (1999) 'The Paris sewers and the rationalization of urban space', *Transactions of the Institute of British Geographers* 24.1: 23–44.

Garbrecht, G. (1979) 'L'alimentation en eau de Pergame', *Dossiers d'Archéologie* 38: 26–33.

   (1987) 'Die Wasserversorgung des antiken Pergamon', in G. Garbrecht (ed.) *Die Wasserversorgung antiker Städte II*. Mainz: Philipp von Zabern, 13–47.

Garfagnini, G. C. (1988) 'Savonarola e la profezia: tra mito e storia', *Studi Medievali* 29: 175–201.

Garland, R. (1995) *The eye of the beholder: deformity and disability in the Graeco-Roman world*. Ithaca, NY: Cornell University Press.

Gaston, R. W. (1983) 'British travellers and scholars in the Roman catacombs, 1450–1900', *Journal of the Warburg and Courtauld Institutes* 46: 144–65.

Gatrell, V. A. C. (1994) *The hanging tree: execution and the English people 1770–1868*. Oxford University Press.

Geertz, C. (1973) *The interpretation of cultures*. New York: Basic Books.

Gentilcore, D. (2006) *Medical charlatanism in early modern Italy*. Oxford University Press.

Gerlini, E. (1943) *Piazza Navona*. Rome: Tipografia delle Terme.

Gherardi, A. (1887) *Nuovi documenti e studi intorno a Girolamo Savonarola*. Florence: Sansoni.

Ghezzo, F. (2010) 'Topographies of disease and desire: mapping the city in Fascist Italy', *Modern Language Notes* 125: 195–222.

Giannelli, G. (1993) 'Arx', in E. M. Steinby (ed.) *Lexicon topographicum urbis Romae I*. Rome: Quasar, 127–9.

Gibson, M. (2002) *Born to crime: Cesare Lombroso and the origins of biological criminology*. Westport, CT: Praeger.

Gilbert, P. (2004) *Mapping the Victorian social body*. Albany: State University of New York Press.

Gilbert, P. K. (2004) 'Medical mapping: the Thames, the body, and *Our Mutual Friend*', in W. A. Cohen and R. Johnson (eds) *Filth: dirt, disgust, and modern life*. Minneapolis: University of Minnesota Press, 78–102.

Ginori Conti, P. (1937) *La vita del beato Ieronimo Savonarola, scritta da un anonimo del sec. XVI e già attribuita a Fra Pacifico Burlamacchi*. Florence: Leo S. Olschki.

Ginouvès, R. (1962) *Balaneutiké: recherches sur le bain dans l'antiquité grecque* (BEFAR 200). Paris: De Boccard.

Girard, R. (1977) *Violence and the sacred* (trans. P. Gregory). Baltimore: Johns Hopkins University Press.

Glare, P. G. W. (ed.) (1982) *The Oxford Latin dictionary*. Oxford University Press.

Gleason, K. L. (1990) 'The garden portico of Pompey the Great: an ancient public park preserved in the layers of Rome', *Expedition* 32.2: 4–13.

   (1994) 'Porticus Pompeiana: a new perspective on the first public park of ancient Rome', *Journal of Garden History* 14: 13–27.

Glinister, F. (2000) 'Sacred rubbish', in E. Bispham and C. Smith (eds) *Religion in archaic and Republican Rome and Italy: evidence and experience*. Edinburgh University Press, 54–70.

Goldschmidt, W. (1973) 'Guilt and pollution in Sebei mortuary ritual', *Ethos* 1.1: 75–105.

Gorer, G. (1965) *Death, grief and mourning in contemporary Britain*. London: Cresset Press.

Goretti, G. and Giartosio, T. (2006) *La città e l'isola: omosessuali al confino nell'Italia fascista*. Rome: Donzelli.

Goubert, J.-P. (1989) *The conquest of water: the advent of health in the Industrial Age*. Cambridge: Polity.

Gowers, E. (1993) *The loaded table: representations of food in Roman literature*. Oxford: Clarendon Press.

(1995) 'The anatomy of Rome from Capitol to Cloaca', *Journal of Roman Studies* 85: 23–32.

Gowing, A. M. (2005) *Empire and memory: the representation of the Roman Republic in imperial culture*. Cambridge University Press.

Gradel, I. (2002) 'Jupiter Latiaris and human blood: fact or fiction?', *Classica et Mediaevalia* 53: 235–53.

Graf, F. (2000) 'The rite of the Argei', *Museum Helveticum* 57.2: 94–103.

Grainger, R. (1998) *The social symbolism of grief and mourning*. London: Jessica Kingsley.

Green, M. (2001) *Dying for the gods: human sacrifice in Iron Age and Roman Europe*. Stroud: Tempus.

Greenblatt, S. (1990) 'Filthy rites', in S. Greenblatt (ed.) *Learning to curse: essays in early modern culture*. London and New York: Routledge, 59–79.

Greenhalgh, M. (n.d.) 'Dates and deeds in Renaissance planning of Rome' (http://rubens.anu.edu.au/htdocs/bycountry/italy/rome/popolo/themes/heemskerck/new/planning.new.html, accessed January 2012).

Gregorovius, F. (1948) *The ghetto and the Jews of Rome* (trans. M. Hadas). New York: Schocken.

Grimal, P. (1943) *Les jardins romains à la fin de la République et aux deux premiers siècles de l'Empire: essai sur le naturalisme romain*. Paris: De Boccard.

Grosz, E. (1992) 'Bodies-cities' in B. Colomina (ed.) *Sexuality and space*. Princeton Architectural Press, 241–54.

Grueber, H. A. (1910) *Coins of the Roman Republic in the British Museum*. London: British Museum.

Gruen, E. S. (1974) *The last generation of the Roman Republic*. Berkeley: University of California Press.

Guarnieri, P. (2003) 'L'incesto scandaloso: legge e mentalità nell'Italia unita', *Passato e Presente* 21: 45–68.

Gundersheimer, W. L. (1988) *Ferrara estense: lo stile del potere*. Modena: Panini.

Haggerty, G. E. (2006) *Queer Gothic.* Urbana: University of Illinois Press.

Hallam, E., Hockey, J. and Howarth, G. (eds) (1999) *Beyond the body: death and social identity.* London: Routledge.

Halliday, S. (1999) *The great stink of London: Sir Joseph Bazalgette and the cleansing of the Victorian metropolis.* Stroud: Sutton.

Halperin, D. M., Winklerl, J. J. and Zeitlin, F. I. (eds) (1990) *Before sexuality: the construction of erotic experience in the ancient world.* Princeton University Press.

Harries, J. (2007) *Law and crime in the Roman world.* Cambridge University Press.

Harrington, H. (2004) *The purity texts.* London: T & T Clark.

Hart, E. (1884) *London, old and new: a sanitary contrast.* London: Allman.

Harvey, P. B. and Schultz, C. (2006) *Religion in republican Italy.* Cambridge University Press.

Haslam, D. and Haslam, F. (2009) *Fat, gluttony and sloth: obesity in literature, art and medicine.* Liverpool University Press.

Hawthorne, N. (2002) *The marble faun* [first published 1859]. Oxford University Press.

Heinze, R. (ed.) (1897) *T. Lucretius carus de rerum natura: Buch iii.* Leipzig: Teubner.

Henriques, F. (1966) *Prostitution and society.* New York: Grove Press.

Hershman, P. (1974) 'Hair, sex and dirt', *Man* 9.2: 274–98.

Hertz, R. (1960) *Death and the right hand* (trans. R. Needham and C. Needham). London: Cohen and West.

Heskel, J. (1994) 'Cicero as evidence for attitudes to dress in the late Republic', in J. L. Sebesta and L. Bonfante (eds) *The world of Roman costume.* Madison: University of Wisconsin Press, 133–45.

   (2001) 'Cicero as evidence for attitudes to dress in the late Republic', in J. L. Sebesta and L. Bonfante (eds) *The world of Roman costume.* Madison: University of Wisconsin Press, 133–45.

Heubner, S. R. (2007) '"Brother–sister" marriage in Roman Egypt: a curiosity of humankind or a widespread family strategy?', *Journal of Roman Studies* 97: 21–49.

Hilton, B. (1988) *The age of atonement: the influence of evangelicalism on social and economic thought, 1795–1865.* Oxford University Press.

Hinard, F. (ed.) (1987) *La mort, les morts et l'au delà dans le monde romain.* Centre de Publications de l'Université; de Caen.

   (ed.) (1995) *La mort au quotidien dans le monde Romain.* Paris: De Boccard.

Hinard, F. and Dumont, J.-C. (eds) (2003) *Libitina: pompes funèbres et supplices en Campanie à l'époque d'Auguste.* Paris: De Boccard.

Hobson, B. (2009) *Latrinae et foricae: toilets in the Roman world.* London: Duckworth.

Hockey, J., Katz, J. and Small, N. (eds) (2001) *Grief, mourning and death ritual.* Buckingham: Open University Press.

Hodge, T. (1992) *Roman aqueducts and water supply.* London: Duckworth.

Holland, L. A. (1961) *Janus and the bridge* (Papers and Monographs of the American Academy in Rome 21). Rome: American Academy in Rome.

Holliday, P. J. (1990) 'Time, history and ritual in the *Ara Pacis Augustae*', *Art Bulletin* 72: 542–57.

Hood, W. (1993) *Fra Angelico at San Marco*. New Haven: Yale University Press.

Hope, V. (2007) *Death in ancient Rome: a sourcebook*. London: Routledge.

(2009) *Roman death: the dying and the dead in ancient Rome*. London: Continuum.

Hope, V. and Marshall, E. (2000) *Death and disease in the ancient city*. London: Routledge.

Hopkins, A. and Wyke, M. (eds) (2005) *Roman bodies: antiquity to the eighteenth century*. London: British School at Rome.

Hopkins, J. N. (2007) 'The Cloaca Maxima and the monumental manipulation of water in archaic Rome', *Waters of Rome* 4: 1–12. www3.iath.virginia.edu/waters/Journal4Hopkins.pdf, accessed July 2011.

(2010) *The topographical manipulation of archaic Rome: a new interpretation of architecture and geography in the early city*. PhD dissertation: University of Texas at Austin.

(forthcoming) 'The creation of the Forum and the making of monumental Rome', in J. Becker and E. C. Robinson (eds) *Studies in Italian urbanism: the first millennium* BCE (Journal of Roman Archaeology supplement). Portsmouth, RI: Journal of Roman Archaeology.

Hopkins, K. (1980) 'Brother–sister marriage in Roman Egypt', *Comparative Studies in Society and History* 22.3: 303–54.

(1983) *Death and renewal*. Cambridge University Press.

Horn, D. G. (1991) 'Constructing the sterile city: pronatalism and social sciences in interwar Italy', *American Ethnologies* 18: 581–601.

(1994) *Social bodies: science, reproduction, and Italian modernity*. Princeton University Press.

Houlbrouk, M. (2005) *Queer London: perils and pleasures in the sexual metropolis, 1918–1957*. University of Chicago Press.

Hoy, S. (1996) *Chasing dirt: the American pursuit of cleanliness*. New York and London: Oxford University Press.

Hubbard, P. (1999) *Sex and the city: geographies of prostitution in the urban West*. Aldershot: Ashgate.

Hubert, H. and Mauss, M. (1964) *Sacrifice: its nature and function* (trans. W. D. Halls). Chicago: Cohen and West.

Hughes, D. D. (1991) *Human sacrifice in ancient Greece*. London: Routledge.

Hughes, J. D. (1994) *Pan's travail: environmental problems of the ancient Greeks and Romans*. Baltimore: Johns Hopkins University Press.

Hülsen, C. (1902) 'Jahresbericht über neue Funde und Forschungen zur Topographie der Stadt Rom: Neue Reihe. I. Die Ausgrabung auf dem Forum Romanum 1898–1902', *Mitteilungen des Deutschen Archäologischen Instituts, Römische Abteilung* 17: 22–57.

Hülsen, C. and Egger, H. (1911) *Die römischen Skizzenbücher von Marten van Heemskerck im Königlichen Kupferstichkabinett zu Berlin, mit Unterstützung der Generalverwaltung der Königlichen Museen zu Berlin.* Vienna: F. Wolfrum.

Humm, M. (1996) 'Appius Claudius Caecus et la construction de la via Appia', *Mélanges de l'Ecole Française de Rome. Antiquité* 108: 693–749.

Ignatieff, M. (1995) 'The seductiveness of moral disgust', *Social Research* 62: 77–97.

Improta, M. C. (1989–90) 'Gli affreschi del beato Angelico', in T. S. Centi *et al.* (eds) *La chiesa e il convento di San Marco II.* Florence: Cassa di Risparmio di Firenze, 106–7.

Infessura, S. (1890) *Diaria rerum Romanarum* (ed. O. Tommasini). Rome: Istituto Storico Italiano.

Ingholt, H. (1969) 'The Prima Porta statue of Augustus', *Archaeology* 22: 177–87, 304–18.

Ingram, G., Bouthilette, A.-M. and Retter, Y. (eds) (1997) *Queers in space: communities, public places, sites of resistance.* Seattle: Bay Press.

Insolera, I. (1993) *Roma moderna: un secolo di storia urbanistica 1870–1970.* Turin: Einaudi.

(2001) *Roma moderna: un secolo di storia urbanistica 1870–1970* (ninth edition) (first published 1962). Turin: Einaudi.

Jackson, M. and Marra, F. (2006) 'Roman stone masonry: volcanic foundations of the ancient city', *American Journal of Archaeology* 110.3: 403–36.

Jackson, R. (1988) *Doctors and diseases in the Roman empire.* London: British Museum.

(1990) 'Waters and spas in the classical world', in R. Porter (ed.) *Medical history of waters and spas.* London: Wellcome Institute for the History of Medicine, 1–13.

James, E. O. (1962) *Sacrifice and sacrament.* London: Thames and Hudson.

James, H. (1987) *Italian hours* [first published 1909]. New York: Ecco Press.

Janes, D. (2009a) 'Dickens and the Catholic corpse', in M. Hollington and F. Orestano (eds) *Dickens and Italy: Little Dorrit and Pictures from Italy.* Newcastle: Cambridge Scholars Press, 170–85.

(2009b) *Victorian reformation: the fight over idolatry in the Church of England, 1840–1860.* Oxford University Press.

Jansen, G. (1991) 'Water systems and sanitation in the houses of Herculaneum', *Mededeelingen van het Nederlands Historisch Instituut te Rome* 50: 145–65.

Jarcho, S. (1986) *Italian broadsides concerning public health.* New York: Futura.

Jeffrey, S. (1997) *The idea of prostitution.* Melbourne: Spinifex.

Jenkins, P. (1998) *Moral panic: changing concepts of the child molester in modern America.* New Haven: Yale University Press.

Jenks, C. (2003) *Transgression.* London: Routledge.

Jenner, M. (1995) 'The politics of London air: John Evelyn's "Fumifugium" and the Restoration', *Historical Journal* 38.3: 535–51.

(1997) 'Overground, underground: pollution and place in urban history', *Journal of Urban History* 24.1: 97–110.

(1998) 'Bathing and baptism: Sir John Floyer and the politics of cold bathing', in K. Sharpe and S. Zwicker (eds) *Refiguring revolutions: aesthetics and politics from the English Revolution to the Romantic Revolution.* Berkeley and Los Angeles: University of California Press, 197–216.

(2000) 'From conduit community to commercial network? Water in London 1500–1725', in P. Griffiths and M. Jenner (eds) *Londinopolis: essays in the social and cultural history of early modern London.* Manchester University Press, 250–72.

(2002) 'The roasting of the rump: scatology and the body politic in Restoration England', *Past and Present* 177: 84–120.

Jennison, G. (1937) *Animals for show and pleasure in ancient Rome.* Manchester University Press.

Johnston, S. I. (ed.) (2004) *Religions of the ancient world.* Harvard: Belknap Press.

Jońca, M. (2004) 'Blood-revenge and murder trial in the early Roman law', *Eos* 91: 44–51.

Jones, C. (1987) '*Stigma*: tattooing and branding in Graeco-Roman antiquity', *Journal of Roman Studies* 77: 139–55.

(2000) 'Stigma and tattoo', in J. Caplan (ed.) *Written on the body.* London: Reaktion, 1–16.

Karmon, D. (2005) 'The Renaissance of the Acqua Vergine' (*Waters of Rome* Occasional Papers 3). www.iath.virginia.edu/waters/karmon.html, accessed July 2011.

Kaster, R. (1997) 'The shame of the Romans', *Transactions of the American Philological Association* 131: 143–89.

Kazen, T. (2002) *Jesus and purity Halakhah: was Jesus indifferent to purity?* Stockholm: Almqvist and Wiksell.

Keaveney, A. (2005) *Sulla: the last Republican.* London and New York: Routledge.

Kee, A. (1982) *Constantine versus Christ: the triumph of ideology.* London: SCM.

Kelley, M. (2004) *The uncanny.* Cologne: Walther König.

Kellum, B. (1985) 'Sculptural programs and propaganda in Augustan Rome: the Temple of Apollo on the Palatine', in R. Winkes (ed.) *The age of Augustus: interdisciplinary conference held at Brown University, April 30–May 2, 1982. Publications d'Histoire de l'Art et d'Archéologie de l'Université Catholique de Louvain.* Providence, RI, and Louvain-la-Neuve: Center for Old World Archaeology and Art and Institut Supérieur d'Archéologie et d'Histoire de l'Art, Collège Erasme, 170–6.

Kimmel, M. S. (2000) *The gendered society.* Oxford University Press.

King, H. (ed.) (2005) *Health in antiquity.* Oxford and New York: Routledge.

Kip, W. I. (1854) *The catacombs of Rome as illustrating the Church of the first three centuries.* New York: Redfield.

Kirkpatrick, J. (2003) 'Purity and pollution' (online essay published as a response to a graduate seminar, New College, Oxford, January 2003). www.classics.ox. ac.uk/faculty/princeton/KirkpatrickPurity0310.rtf, accessed July 2011.

Klawans, J. (2000) *Impurity and sin in ancient Judaism*. Oxford University Press.

Kleiner, D. E. E. (1978) 'The great friezes of the *Ara Pacis Augustae*: Greek sources, Roman derivatives and Augustan social policy', *Mélanges de l'Ecole Française de Rome. Antiquité* 90: 753–85.

(2005) 'Semblance and storytelling in Augustan Rome', in K. Galinsky (ed.) *Cambridge companion to the age of Augustus*. Cambridge University Press, 197–233.

Knapp, P. (1967) 'Purging and curbing: an inquiry into disgust, satiety and shame', *Journal of Nervous and Mental Disease* 144: 514–34.

Koloski-Ostrow, A. O. (1996) 'Finding social meaning in the public latrines of Pompeii', in N. de Haan and G. C. M. Jansen (eds) *Cura aquarum in Campania: proceedings of the Ninth International Congress on the History of Water Management and Hydraulic Engineering in the Mediterranean Region*. Leiden: Stichting Babesch, 79–86.

(forthcoming) *The archaeology of sanitation in Roman Italy: waters, sewers and toilets*. London: University of North Carolina Press.

Kövecses, Z. (2002) *Metaphor: a practical introduction*. Oxford University Press.

Koven, S. (2001) 'Toward a geography of sexuality', *Journal of Urban History* 27: 497–504.

(2004) *Slumming: sexual and social politics in Victorian London*. Princeton University Press.

Koves-Zulauf, T. (1972) *Reden und Schweigen; Religion bei Plinius Minor*. Munich: Fink.

Krafft-Ebing, R. von (1886) *Psychopathia sexualis: Eine klinische-forensische Studie*. Stuttgart: Enke.

Kristeva, J. (1980) *Pouvoirs de l'horreur: essai sur l'abjection*. Paris: Editions du Seuil.

(1982) *Powers of horror: an essay on abjection* (trans. L. Roudiez). New York: Columbia University Press.

Kyle, D. (1995) 'Animal spectacles in ancient Rome: meat and meaning', *Nikephoros* 7: 181–205.

(1998) *Spectacles of death in ancient Rome*. London: Routledge.

Lachmann, R. (1988–9) 'Bakhtin and Carnival: culture as counter-culture', *Cultural Critique* 11: 115–52.

Lagerspetz, O. (2008) *Lika: Kirja maailmasta, kodistamme*. Helsinki: Multikustannus.

Lahiji, N. and Friedman, D. (1997) 'At the sink: architecture in abjection', in N. Lahiji and D. Friedman (eds) *Plumbing: sounding modern architecture*. New York: Princeton Architectural Press, 35–61.

Lakoff, G. and Turner, M. (1980) *Metaphors we live by*. University of Chicago Press.

Lancaster, L. C. (2005) *Concrete vaulted construction in imperial Rome: innovations in context.* Cambridge University Press.

Lanciani, R. (1874) 'Puticoli', *Bullettino della Commissione Archeologica Comunale di Roma* 2: 42–53.

  (1889) 'Il foro di Augusto', *Bullettino della Commissione Archaeologica Comunale di Roma* 17: 30–1.

  (1890) 'Ricerche sulle xiv regioni urbane, Cloaca Maxima', *Bullettino della Commissione Archeologica Comunale di Roma* 18: 95–102.

  (1891) *Ancient Rome in the light of recent discoveries.* Boston: Houghton.

  (1897) *The ruins and excavations of ancient Rome: a companion book for students and travelers.* Boston: Houghton.

  (1988) *Notes from Rome.* Rome: British School at Rome.

  (1989–94) *Storia degli scavi di Roma e notizie intorno le collezioni Romane di antichità.* Rome: Quasar.

Landucci, L. (1883) *Diario fiorentino: dal 1450 al 1516: continuato da un anonimo fino al 1542* (ed. I. Della Badia). Florence: Sansoni.

Lankewish, V. A. (2000) 'Love among the ruins: the catacombs, the closet and the Victorian "early Christian" novel', *Victorian Literature and Culture* 28: 239–73.

Laporte, D. (1978) *Histoire de la merde: prologue.* Paris: C. Bourgeois.

Laqueur, T. (1983) 'Bodies, death, and pauper funerals', *Representations* 1: 109–31.

  (1992) *Making sex: body and gender from the Greeks to Freud.* Cambridge, MA: Harvard University Press.

Lassen, E. M. (1992) 'The ultimate crime: *parricidium* and the concept of the family in the late Roman Republic and early Empire', *Classica et Mediaevalia* 43: 147–61.

Latte, K. (1960) *Römische Religionsgeschichte.* Munich: Beck.

Laughran, M. (1998) *The body, public health and social control in sixteenth-century Venice.* PhD dissertation: University of Connecticut.

Laurence, R. (1997) 'Writing the Roman metropolis', in H. M. Parkins (ed.) *Roman urbanism: beyond the consumer city.* London and New York: Routledge, 1–20.

Le Gall, J. (1953) *Recherches sur le culte du Tibre.* Paris: Presses Universitaires de France.

Lee, Dr (1843) 'Tractarianism: on Dr Pusey's late sermon', *Pulpit* 44: 525–6.

Leguilloux, M. (2004) 'L'identification des tanneries antiques par l'archéologie et l'archéozoologie', in E. C. De Sena and H. Dessales (eds) *Metodi e approcci archeologici: l'industria e il commercio nell'Italia antica/Archaeological methods and approaches: industry and commerce in ancient Italy.* Oxford: Archaeopress, 38–48.

Lennon, J. (2010a) 'Pollution and ritual impurity in Cicero's *De domo sua*', *Classical Quarterly* 60.2: 427–45.

  (2010b) 'Jupiter Latiaris and the Taurobolium: inversions of cleansing in Christian polemic', *Historia* 59.3: 381–4.

(2010c) 'Menstrual blood in ancient Rome: an unspeakable impurity', *Classica et Mediaevalia* 61: 71–87.

Leon, D. (2006) *Through a glass, darkly*. London: Heinemann.

(2009) *About face*. London: Heinemann.

Leroy-Forgeot, F. (1997) *Histoire juridique de l'homosexualité en Europe*. Paris: Presses Universitaires de France.

Levillain, P. and O'Malley, J. (eds) (2002) *The papacy: an encyclopedia II*. London: Routledge.

Levine, E. B. (1971) *Hippocrates*. New York: Twayne.

Leyerle, B. (2009) 'Refuse, filth, and excrement in the *Homilies* of John Chrysostom', *Journal of Late Antiquity* 2: 337–56.

Leyser, C. (1998) 'Masculinity in flux: nocturnal emission and the limits of celibacy in the early Middle Ages', in D. Hadley (ed.) *Masculinity in medieval Europe*. London: Londman, 103–19.

Liebeschuetz, J. H. W. G. (1979) *Continuity and change in Roman religion*. Oxford University Press.

Liedman, S.-E. (1997) *I skuggan av framtiden: modernitetens idéhistoria*. Stockholm: Bonnier Alba.

Lilja, S. (1972) *The treatment of odours in the poetry of antiquity*. Helsinki: Societas Scientiarum Fennica.

Lindenlauf, A. (2004) 'The sea as a place of no return in ancient Greece', *World Archaeology* 35.3: 416–33.

Lindsay, H. (1998) 'Eating with the dead: the Roman funerary banquet', in I. Nielsen and H. S. Nielsen (eds) *Meals in a social context*. Oxford: Aarhus University Press, 67–81.

(2000) 'Death-pollution and funerals in the city of Rome', in V. Hope and E. Marshall (eds) *Death and disease in the ancient city*. London: Routledge, 152–73.

Lintott, A. (1999a) *The constitution of the Roman Republic*. Oxford University Press.

(1999b) *Violence in Republican Rome* (second edition). Oxford University Press.

Liotta, S. (1963) *Michele Savonarola: la sua epoca e la sua opera*. Rome: Istituto di Storia della Medicina dell'Università.

Lippini, P. (1990) *La vita quotidiana di un convento medievale: gli ambienti, le regole, l'orario e le mansioni dei Frati Domenicani del tredicesimo secolo*. Bologna: Edizioni Studio Domenicano.

Lloyd, G. (2003) *In the grip of disease: studies in the Greek imagination*. Oxford University Press.

Logan, W. (1995) *Dirt: the ecstatic skin of the earth*. New York: Riverhead.

Lomas, K. (1997) 'The idea of a city: elite ideology and the evolution of urban form in Italy, 200 BC–AD 100', in H. M. Parkins (ed.) *Roman urbanism: beyond the consumer city*. London and New York: Routledge, 21–41.

Lombroso, C. (1896–7) *L'uomo delinquente: in rapporto all'antropologia, alla giurisprudenza ed alle discipline carcerie* (4 volumes). Turin: Fratelli Bocca.

(1906) 'Du parallelisme entre l'homosexualité et la criminalité innée', *Archivio di Psichiatria* 27: 378–81.

Lombroso, C. and Ferrero, G. (1893) *La donna delinquente, la donna normale e la prostituta.* Milan: Fratelli Bocca.

Long, P. O. (2008) 'Hydraulic engineering and the study of antiquity: Rome, 1557–1570', *Renaissance Quarterly* 61: 1098–138.

Longhurst, R. (2001) *Bodies: exploring fluid boundaries.* London: Routledge.

Lora, E. and Simionati, R. (eds) (1994) *Enchiridion delle encicliche* (8 volumes). Bologna: Dehoniane.

Loreto, R. and Cervantes, F. J. (eds) (1994) *Limpiar y obedecer: la basur, el agua y la muerte en la puebla de Los Angeles, 1650–1925.* Universidad de Puebla.

Lott, J. B. (2004) *The neighborhoods of Augustan Rome.* Cambridge University Press.

Louz, F. and Richard, P. (1978) *Sagesses du corps.* Paris: Maisonneuve et Larose.

Lovisi, C. (1998) 'Vestale, *incestus* et juridiction pontificale sous la République Romaine', *Mélanges de l'Ecole Française de Rome. Antiquité* 110: 699–735.

Lowenthal, D. (1986) *The past is a foreign country.* Cambridge University Press.

Luckman, H. and Kulzer, L. (1999) *Purity of heart in early ascetic and monastic literature.* Collegeville: Liturgical Press.

Lugones, M. (1994) 'Purity, impurity and separation', *Signs* 19: 458–79.

Lupu, E. (2004) *Greek sacred law: a collection of new documents.* Leiden: Brill.

Luschino, B. (2002) *Vulnera diligentis* (ed. S. Dall'Aglio) (Savonarola e la Toscana 17). Florence: SISMI Edizioni del Galluzzo.

MacBain, B. (1980) 'Appius Claudius Caecus and the Via Appia', *Classical Quarterly* 30: 356–72.

(1982) *Prodigy and expiation: a study in religion and politics in Republican Rome.* Brussels: Latomus.

Maccoby, H. (1982) *The sacred executioner: human sacrifice and the legacy of guilt.* London: Thames and Hudson.

Maccormac, H. (1852) *On the connexion of atmospheric impurity with disease.* Belfast: Belfast Social Inquiry Society.

MacFarlane, C. (1852) *The catacombs of Rome.* London: Routledge.

MacKinnon, M. (2004) *Production and consumption of animals in Roman Italy: integrating the zooarchaeological and textual evidence.* Portsmouth, RI: Journal of Roman Archaeology.

Maddoli, G. (1971) 'Il rito degli Argei e il culto di Hera a Roma', *Parola del Passato* 26: 153–66.

Magdelaine, C. (2003) 'Ville, déchets et pollution urbaine chez les médecins grecs', in P. Ballet, P. Cordier and N. Dieudonné-Glad (eds) *La ville et ses déchets dans le monde romain: rebuts et recyclages.* Montagnac: Monique Mergoil, 27–36.

Maglen, K. (2002) '"First line of defence": British quarantine and the port sanitary authorities in the nineteenth century', *Social History of Medicine* 15: 413–28.

Maischberger, M. (1999) 'Transtiberim', in E. M. Steinby (ed.) *Lexicon topographicum urbis Romae v.* Rome: Quasar, 77–82.

Maitland, C. (1846) *The Church in the catacombs: a description of the primitive Church of Rome, illustrated by its sepulchral remains.* London: Longman.

Mandich, G. (1936) 'Le privative industriali veneziane 1450–1550', *Rivista del Diritto Commerciale* 34: 511–47.

(1958) 'Primi riconoscimenti veneziani di un diritto di privativa agli inventori', *Rivista di Diritto Industriale* 7: 101–55.

Manfroni, G. (1920) *Sulla soglia del Vaticano 1870–1901: dalle memorie di Giuseppe Manfroni a cura del figlio Camillo I.* Bologna: Nicola Zanichelli.

Martin, P. M. (1982) *L'idée de royauté à Rome.* Clermont-Ferrand: Adosa.

Martini, F. di Giorgio (*c.* 1478–81) *Trattato di architettura, ingegneria e arte militare* (Ashburnham 361). Urbino: Biblioteca Medicea Laurenzina, ms 282.

Martini, M. C. (2004) *Le Vestali: un sacerdozio funzionale al 'cosmo' romano* (Collection Latomus 282). Brussels: Latomus.

Maynard, J. (1993) *Victorian discourses on sexuality and religion.* Cambridge University Press.

Maynard, S. (1994) 'Through a hole in the lavatory wall: homosexual subcultures, police surveillance, and the dialectics of discovery, Toronto 1890–1930', *Journal of the History of Sexuality* 5: 207–42.

McCarthy, D. J. (1969) 'The symbolism of blood and sacrifice', *Journal of Biblical Literature* 88.2: 166–76.

(1973) 'Further notes on the symbolism of blood and sacrifice', *Journal of Biblical Literature* 92.2: 205–10.

McCartney, E. S. (1952) 'Speaking eyes', *Classical Journal* 47.5: 187–8.

McClary, A. (1980) 'Germs are everywhere: the germ threat as seen in magazine articles 1890–1920', *Journal of American Culture* 3: 33–46.

McClintock, A. (1995) *Imperial leather: race, gender and sexuality in the colonial contest.* London: Routledge.

McGinn, T. (1998) *Prostitution, sexuality and the law in ancient Rome.* New York: Oxford University Press.

McGushin, P. (ed.) (1980) *Sallust, Bellum Catilinae.* Bristol Classical Press.

McLaughlin, M. (ed.) (2000) *Britain and Italy from romanticism to modernism: a festschrift for Peter Brand.* Oxford: Legenda.

McLaughlin, T. (1971) *Coprophilia, or a peck of dirt.* London: Cassell.

McManners, J. (1981) *Death and the Enlightenment: changing attitudes to death among Christians and unbelievers in eighteenth-century France.* Oxford University Press.

Meersseman, G. G. (1946) 'L'architecture domenicaine au XIIIe siècle: législation et pratique', *Archivum Fratrum Praedicatorum* 16: 136–90.

Mehl, V. and Brulé, P. (eds) (2008) *Le sacrifice antique: vestiges, procédures et stratégies.* Rennes: Press Universitaire de Rennes.

Meier, C. (1982) *Caesar: a biography.* New York: Basic Books.

Meigs, A. (1978) 'A Papuan perspective on pollution', *Man* 13.2: 304–18.

(1984) *Food, sex and pollution: a New Guinea religion.* Brunswick, NJ: Rutgers University Press.

Mekacher, N. (2006) *Die Vestalischen Jungfrauen in der römischen Kaiserzeit.* Wiesbaden: Reichert.

Melosi, M. (1981) *Garbage in the city: refuse, reform and the environment, 1880–1980.* Chicago: Dorsey Press.

(2000) *The sanitary city: urban infrastructure in America from colonial times to the present.* Baltimore: Johns Hopkins University Press.

Menninghaus, W. (2003) *Disgust: theory and history of a strong sensation* (trans. H. Eiland and J. Golb). Albany: State University of New York Press.

Mercurio, S. (1645) *De gli errori popolari d'Italia, libri sette.* Padua: Francesco Bolzetta.

Metcalf, P. and Huntington, R. (1991) *Celebrations of death: the anthropology of mortuary ritual.* Cambridge University Press.

Metropolitan Sanitary Association (1850) *The public health, a public question: first report of the Metropolitan Sanitary Association on the chief evils affecting the sanitary condition of the metropolis, with suggestions for their removal and containing the proceedings of the public meeting held at Freemasons Hall, Feb. 6th, 1850.* London: Metropolitan Sanitary Association.

Meyer, M. (2005) *Thicker than water: the origins of blood as ritual and symbol.* Oxford: Routledge.

Meyer, W. W. (2010) 'A tale of two cities: John Henry Newman and the church of the catacombs', *Journal of Ecclesiastical History* 61: 746–63.

Mieli, A. (1926) 'Per la lotta contro la delinquenza collegata a manifestazioni sessuali', *Rassegna di Studi Sessuali e di Eugenia* 5: 256–61.

Miglio, M. (2001) 'Savonarola di fronte ad Alessandro VI e alla Curia', in G. C. Garfagnini (ed.) *Una città e il suo profeta: Firenze di fronte al Savonarola.* Florence: SISMEL/Edizioni del Galluzzo, 109–18.

Milgrom, J. (1976) 'Israel's sanctuary: the priestly "Picture of Dorian Gray"', *Revue Biblique* 83: 390–9.

(1992) *Leviticus 1–16: a new translation with introduction and commentary.* New York: Doubleday.

Milizia, F. (1787) *Roma delle belle arti del disegno. Parte prima: dell'architettura civile.* Bassano: [Giuseppe Remondini].

Miller, P. (1999) 'The bodily grotesque in Roman satire: images of sterility', *Arethusa* 32: 257–83.

Miller, W. I. (1993) *Humiliation: and other essays on honor, social discomfort, and violence.* Ithaca, NY: Cornell University Press.

(1998) *The anatomy of disgust.* London: Harvard University Press.

Millon, H. A. and Smyth, C. H. (1996) 'The project for the Castel Sant'Angelo in the Dyson Perrins codex', in C. L. Striker (ed.) *Architectural studies in memory of Richard Krautheimer.* Mainz: Philipp Von Zabern, 111–17.

Miner, H. (1956) 'Body ritual among the Nacirema', *American Anthropologist* 58: 503–7.

Ministero della Giustizia e degli Affari di Culto (1927) *Progetto preliminare di un nuovo codice penale.* Rome: Tipografia delle Mantellate.

Mittarelli, B. and Costadoni, A. (1755–73) *Annales camaldulenses ordinis Sancti Benedicti quibus plura interserentur tum ceteras italico-monasticas res, tum historiam ecclesiasticam remque diplomaticam illustrantia* (9 volumes). Venice: Giovanni Battista Pasquali.

Moccheggiani Carpano, C. (1984) 'Le Cloache dell'antica Roma', in R. Luciani (ed.) *Roma sotterranea.* Rome: Fratelli Palombi, 164–78.

Modigliani, A. (1998) 'Roma e Firenze, *Tuscus et Remi*: due modelli in opposizione?', *Studi Romani* 46: 5–28.

Modio, G. B. (1556) *Il Tevere . . . dove si ragiona in generale della natura di tutte le acque, e in particolare di quella del fiume di Roma.* Rome: Vincenzo Luchini.

Moe, N. (2002) *The view from Vesuvius: Italian culture and the Southern question.* Berkeley, Los Angeles and London: University of California Press.

Moehring, H. R. (1959) 'The persecution of Jews and the adherents of the Isis cult at Rome AD 19', *Novum Testamentum* 3.4: 293–304.

Molteni, M. (1995) *Ercole de' Roberti.* Cinisello Balsamo: Cassa di Risparmio di Ferrara.

Montagnes, B. (1974) 'L'attitude des prêcheurs à l'égard des oeuvres d'art', in *La naissance et l'essor du gothique méridional au xiiie siècle* (Cahiers de Fanjeaux 9). Toulouse: Cahiers de Fanjeaux, 87–100.

Montserrat, D. (ed.) (1998) *Changing bodies, changing meanings: studies on the human body in antiquity.* London: Routledge.

Moore, J. (1997) 'Mary Douglas: symbols and structures, pollution and purity', in J. Moore (ed.) *Visions of culture: an introduction to anthropological theories and theorists.* London: AltaMira Press, 248–60.

Moore, R. I. (1987) *The formation of a persecuting society: power and deviance in Western Europe, 950–1250.* New York: Blackwell.

Moreau, P. (2002) *Incestus et prohibitae nuptiae.* Paris: Belles Lettres.

Morisi Guerra, A. (1991) 'Sulle orme del Savonarola: la riscoperta degli apologisti greci antipagani', *Rivista di Storia della Chiesa in Italia* 45: 89–109.

Morley, N. (2005) 'The salubriousness of the Roman city', in H. King (ed.) *Health in antiquity.* London and New York: Routledge, 192–204.

Morrison, A. (1986) *The culture of shame.* London: Aronson.

Morrison, S. (2008) *Excrement in the late Middle Ages: sacred filth and Chaucer's fecopoetics.* New York: Palgrave Macmillan.

Morselli, C. (1995) 'Forum Iulium', in E. M. Steinby (ed.) *Lexicon topographicum urbis Romae* ii. Rome: Quasar, 299–306.

Morselli, C., Tortorici, E. and Alvaro, C. (1989) *Curia, Forum Iulium, Forum Transitorium* i. Rome: De Luca, Soprintendenza Archeologica di Roma.

Moulinier, L. (1952) *Le pur et l'impur dans la pensée des Grecs d'Homère à Aristote* (Etudes et Commentaires 11). Paris: C. Klincksieck.

Mullin, A. (1996) 'Purity and pollution: resisting the rehabilitation of a virtue', *Journal of the History of Ideas* 57.3: 509–24.

Mumford, L. (1961) *The city in history.* New York: Harcourt, Brace and World.

Murray, O. and Tecuşan, M. (eds) (1995) *In vino veritas.* London: British School at Rome.

Musatti, C. (1943) 'Omosessualità', in E. Florian, A. Niceforo and N. Pende (eds) *Dizionario di criminologia II.* Florence: Vallardi, 602–3.

Mussolini, B. (1928) 'Prefazione', in R. Korherr (ed.) *Regresso delle nascite: morte dei popoli.* Rome: Libreria del Littorio, 7–23.

(1934) 'Discorso dell'Ascensione', in B. Mussolini, *Scritti e discorsi di Benito Mussolini VI.* Milan: Hoepli, 37–77.

Mustakallio, K. (1992) 'The *crimen incesti* of the Vestal Virgins and the prodigious pestilence', in T. Viljama, A. Timonen and C. Krötzl (eds) *Crudelitas: the politics of cruelty in the ancient and medieval world.* Krems: Medium Aevum Quotidianum, 56–62.

(1994) *Death and disgrace: capital penalties with post-mortem sanctions in early Roman historiography* (Annales Academiae Scientarum Fennicae Dissertationes Humanarum Litterarum 72). Helsinki: Suomalainen Tiedeakatemia.

Nagy, B. (1985) 'The Argei puzzle', *American Journal of Ancient History* 10: 1–27.

Nardi, C. (1991) 'Una pagina "umanistica" di Teodoreto di Ciro e un'interpretazione di Zanobi Acciaiuoli', *Atti e Memorie dell'Accademia Toscana di Scienze e Lettere: La Colombaria* 66: 9–64.

Narducci, P. (1889) *Sulle fognatura della città di Roma.* Rome: Forzani.

Nash, E. (1968) *Pictorial dictionary of ancient Rome.* New York: Praeger.

Nasta, M. (2001) '*Scelus fraternae necis*: Rómulo, del asesinato a la apoteosis', *Cuadernos de Filología Clásica Estudios Latinos* 20: 67–82.

Neudecker, R. (1994) *Der Pracht der Latrine: zum Wandel öffentlicher Bedürfnisanstalten in der kaiserzeitlichen Stadt.* Munich: Friedrich Pfeil.

Neusner, J. (1973) *The idea of purity in ancient Judaism.* Leiden: Brill.

Neves, L. (2004) 'Cleanliness, pollution and disgust in modern industrial societies: the Brazilian case', *Journal of Consumer Culture* 4.3: 385–405.

Niccoli, O. (1996) 'I bambini del Savonarola', in G. C. Garfagnini (ed.) *Savonarola e la Toscana: atti e documenti. Studi savonaroliani. Verso il V centenario.* Florence: SISMEL/Edizioni del Galluzzo, 279–88.

Nicolet, C. (1991) *Space, geography and politics in the early Roman Empire.* Ann Arbor: University of Michigan Press.

Nielsen, I. and Nielsen, H. S. (eds) (1998) *Meals in a social context.* Oxford: Aarhus University Press.

Northcote, J. S. (1857) *The Roman catacombs; or, some account of the burial-places of the early Christians in Rome.* London: Charles Dolman.

Nummedal, T. (2007) *Alchemy and authority in the Holy Roman Empire.* University of Chicago Press.

Nussbaum, M. (1999) '"Secret sewers of vice": disgust, bodies, and the law', in S. Bandes (ed.) *The passions of law.* New York: New York University Press, 19–62.

(2004) *Hiding from humanity: disgust, shame and the law.* Princeton University Press.

Nussdorfer, L. (1997) 'The politics of space in early-modern Rome', *Memoirs of the American Academy in Rome* 42: 161–86.

Nutton, V. (1983) 'The seeds of disease: an explanation of contagion and infection from the Greeks to the Renaissance', *Medical History* 27: 1–34.

(2004) *Ancient medicine.* London: Routledge.

Nye, R. (2004) 'Sexuality', in T. A. Meade and M. E. Wiesner-Hanks (eds) *A companion to gender history.* Oxford: Blackwell, 11–25.

O'Connell Davidson, J. (1998) *Prostitution, power and freedom.* London: Polity.

O'Connor, E. (2001) *Raw material: producing pathology in Victorian culture.* Durham, NC: Duke University Press.

Ogden, D. (ed.) (2007) *A companion to Greek religion.* Oxford: Wiley-Blackwell.

Ogilvie, R. (1961) '*Lustrum condere*', *Journal of Roman Studies* 51: 31–9.

(1969) *The Romans and their gods.* London: Chatto and Windus.

Olyan, S. and Nussbaum, C. (eds) (1998) *Sexual orientation and human rights in American religious discourse.* Oxford University Press.

Orlin, E. M. (1997) *Temples, religion and politics in the Roman Republic.* Leiden: Brill.

Ortner, S. (1984) 'Theory in anthropology since the sixties', *Comparative Studies in Society and History* 26.1: 126–66.

Orwell, G. (1933) *Down and out in Paris and London.* New York: Harper.

Orwell, S. and Angus, I. (eds) (1968) *The collected essays, journalism and letters of George Orwell 1: an age like this.* London: Secker and Warburg.

Östenberg, I. (2009) *Staging the world: spoils, captives, and representations in the Roman triumphal procession* (Oxford Studies in Ancient Culture and Representation). Oxford University Press.

Otter, C. (2004) 'Cleansing and clarifying: technology and perception in nineteenth-century London', *Journal of British Studies* 43: 40–65.

Pailler, J.-M. (1988) *Bacchanalia: la répression de 186 av. J.-C. à Rome et en Italie: vestiges, images, tradition.* Paris: De Boccard.

Painter, B. W. (2005) *Mussolini's Rome: rebuilding the eternal city.* New York: Palgrave Macmillan.

Palmer, P. (1989) *Domesticity and dirt: housewives and domestic servants in the United States, 1920–1945.* Philadelphia: Temple University Press.

Palmer, R. (1990) '"In this our lightye and learned tyme": Italian baths in the era of the Renaissance', in R. Porter (ed.) *Medical history of waters and spas.* London: Wellcome Institute for the History of Medicine, 14–22.

Pancino, C. (2001) '"I medicamenti sono di tre sorti": magia, scienza e religione ne "Gli errori popolari d'Italia" di Scipione Mercurio (1603)', in A. Prosperi (ed.) *Il piacere del testo: saggi e studi per Albano Biondi 1*. Rome: Bulzoni, 385–421.

Parker, H. (2004) 'Why were the Vestals virgins? Or the chastity of women and the safety of the state', *American Journal of Philology* 125: 563–601.

Parker, R. (1983) *Miasma: pollution and purification in early Greek religion*. Oxford University Press.

Parker, S. (1996) 'Full brother–sister marriage in Roman Egypt: another look', *Cultural Anthropology* 11.3: 362–76.

Parry, J. (1982) 'Sacrificial death and the necrophagous ascetic', in M. Bloch and J. Parry (eds) *Death and the regeneration of life*. Cambridge University Press, 74–111.

Pascal, C. B. (1988) 'Tibullus and the *Ambarvalia*', *American Journal of Ancient History* 109.4: 523–36.

Pasolini, P. P. (1955) *Ragazzi di vita*. Milan: Garzanti.

(1959) *Una vita violenta*. Milan: Garzanti.

Pasquali, S. (2002) 'Roma antica: memorie materiali, storia e mito', in G. Ciucci (ed.) *Roma moderna: storia di Roma dall'antichità a oggi*. Rome and Bari: Laterza, 323–47.

Pastor, L. von (1938–61) *The history of the popes from the close of the Middle Ages* (40 volumes). London: Routledge and Kegan Paul.

Pastore, A. (1988) 'Tra giustizia e politica: il governo della peste a Genova e Roma nel 1656/7', *Rivista Storica Italiana* 100: 126–54.

Patterson, J. R. (2000) 'On the margins of the city of Rome', in V. Hope and E. Marshall (eds) *Death and disease in the ancient city*. London: Routledge, 85–103.

Payer, P. (1993) *The bridling of desire: views of sex in the later Middle Ages*. University of Toronto Press.

Peachin, M. (2004) *Frontinus and the curae of the Curator Aquarum*. Stuttgart: Steiner.

Pecchiai, P. (1948) *Roma nel cinquecento*. Bologna: Capelli.

Peniston, W. A. (2004) *Pederasts and others: urban culture and sexual identity in nineteenth-century Paris*. New York: Harrington Park Press.

Pensabene, P. and Panella, C. (1994–5) 'Reimpiego e progettazione architettonica nei monumenti tardoantichi di Roma 2: Arco quadrifronte (Giano) del Foro Boario', *Atti della Pontificia Accademia Romana di Archeologia. Rendiconti* 67: 25–67.

Perini, D. A. (1917) *Un emulo di Fra Girolamo Savonarola, Fra Mariano da Genazzano*. Rome: Unione Editrice.

Perkins, C. (1990) 'Vitellius the *spectaculum*: a note on *Histories* 3.84.5', *Classical Bulletin* 66.1–2: 47–9.

Petronio, A. T. (1552) *Ad Julium III Pont. Opt. Max. de Aqua Tiberina: opus quidem novum sed ut omnibus qui hac aqua utuntur utile, ita et necessarium.* Rome: Apud Valerium et Aloisium Doricos Fratres Brixienses.

Phillips, E. D. (1973) *Greek medicine.* London: Thames and Hudson.

Pianciani, L. (1874) *Diciotto mesi di amministrazione municipale.* Rome: Tipografia del don Pirloncino.

Pico della Mirandola, G. (1998) *Vita di Hieronimo Savonarola (Volgarizzamento anonimo)* (ed. R. Castagnola). Florence: SISMEL/Edizioni del Galluzzo.

Picozzi, S. (1975) 'L'esplorazione della Cloaca Massima', *Capitolium* 50: 2–9.

Pike, D. (2005a) *Subterranean cities: the world beneath Paris and London, 1800–1945.* Ithaca, NY: Cornell University Press.

  (2005b) 'Sewage treatments: vertical space and waste in nineteenth-century Paris and London', in W. A. Cohen and R. Johnson (eds) *Dirt, disgust, and modern life.* Minneapolis: University of Minnesota Press, 51–77.

Pile, S. (1996) *The body and the city: psychoanalysis, space and subjectivity.* London: Routledge.

Pisani Sartorio, G. and Colini, A. M. (1986) 'Portus Tiberinus', *Archeologia Laziale* 7.2: 157–97.

Pitman, E. R. (1869) *Vestina's martyrdom: a story of the catacombs.* London: Partridge.

Platner, S. and Ashby, T. (1929a) 'Cloaca Maxima', in S. Platner and T. Ashby, *A topographical dictionary of ancient Rome.* London: Oxford University Press, 126–7.

  (1929b) 'Mundus', in S. Platner and T. Ashby, *A topographical dictionary of ancient Rome.* London: Oxford University Press, 346–8.

Polizzotto, L. (1994) *The elect nation: the Savonarolan movement in Florence 1494–1545.* Oxford University Press.

  (2009) 'L'eredità dell'osservanza domenicana in Savonarola e nei savonaroliani', *Memorie Domenicane* 40: 175–88.

Pollini, J. (1995) 'The Augustus of Prima Porta and the transformation of the Polykleitan heroic ideal: the rhetoric of art', in W. G. Moon (ed.) *Polykleitos, the Doryphoros, and tradition.* Madison: University of Wisconsin Press, 262–82.

Poorthius, M. and Schwartz, J. (eds) (1999) *Purity and holiness: the heritage of Leviticus.* Leiden: Brill.

Porter, D. (1998) *Health, civilization and the state.* London: Routledge.

Porter, R. (ed.) (1990) *Medical history of waters and spas.* London: Wellcome Institute for the History of Medicine.

Pötscher, W. (1998) 'Der Funktion der Argei', *Acta Classica Universitatis Scientarium Debreceniensis* 34.5: 225–34.

Potter, D. (1999) 'Odor and power in the Roman empire', in J. Porter (ed.) *Constructions of the classical body.* Ann Arbor: University of Michigan Press, 169–89.

Preto, P. (1978) *Peste e società a Venezia nel 1576.* Vicenza: Neri Pozza.

Prieur, J. (1986) *La mort dans l'antiquité Romaine.* Rennes: Ouest-France.

Purcell, N. (2001) 'Dialectical gardening', *Journal of Roman Archaeology* 14: 546–56.

Quilici Gigli, S. (1995) 'Attenzione e disattenzione ai principi igienici nella pianificazione romana: contributi dall'archeologia', *Medicina nei Secoli* 7: 551–9.

Radford, F. (1994) '"Cloacal obsession": Hugo, Joyce, and the sewer museum of Paris', *Mattoid* 48: 66–85.

Rajberti, G. (1857) *Il viaggio di un ignorante ossia: ricetta per gli ipocondriaci composta dal dottore Giovanni Rajberti*. Milan: Bernardoni.

Re, C. (1880) *Statuti della città di Roma*. Rome: Tipografia della Pace.

Re, E. (1920) 'I maestre di strade', *Archivio della Societa Romana di Storia Patria* 43: 5–102.

Redigonda, A. (1960) 'Acciaiuoli, Zanobi', *Dizionario Biografico degli Italiani* I: 93–94.

Reeder, J. C. (1992) 'Typology and ideology in the Mausoleum of Augustus: *tumulus* and *tholos*', *Classical Antiquity* 11: 265–307.

Rehak, P. and Younger, J. G. (2006) *Imperium and cosmos: Augustus and the northern Campus Martius*. Madison: University of Wisconsin Press.

Reid, D. (1991) *Paris sewers and sewermen: realities and representations*. Cambridge, MA: Harvard University Press.

Reimers, P. (1989) '"*Opus omnium dictu maximum*": literary sources for the knowledge of Roman city drainage', *Opuscula Romana* 17: 137–41.

(1991) 'Roman sewers and sewage networks: neglected areas of study', in A.-M. Leander Touati, E. Rystedt and Ö. Wikander (eds) *Munuscula Romana*. Stockholm: Svenska Institutet i Rom, 111–16.

Remijsen, S. and Clarysse, W. (2008) 'Incest or adoption? Brother–sister marriage in Roman Egypt revisited', *Journal of Roman Studies* 98: 53–61.

Revenin, R. (2005) *Homosexualité et prostitution masculines à Paris, 1870–1918*. Paris: Harmattan.

Reynolds, R. (1943) *Cleanliness and godliness*. London: Allen and Unwin.

Ricci, R. (1930) 'Baracche e sbaraccamenti: relazione per il 1929 a S.E. il principe Francesco Boncompagni Ludovisi, Governatore di Roma, del delegato ai servizi assistenziali del Governatorato, Raffaello Ricci', *Capitolium* 5: 142–9.

Richards, J. (1991) *Sex, dissidence and damnation: minority groups in the Middle Ages*. London: Routledge.

Richardson, L. (1992a) 'Cloaca Maxima', in L. Richardson, *A new topographical dictionary of ancient Rome*. Baltimore: Johns Hopkins University Press, 91–2.

(1992b) 'Carcer', in L. Richardson, *A new topographical dictionary of Rome*. Baltimore: Johns Hopkins University Press, 71.

Richardson, R. (1988) *Death, dissection and the destitute*. London: Penguin.

Richlin, A. (1983) *The garden of Priapus: sexuality and aggression in Roman humor*. New Haven: Yale University Press.

Richter, O. (1889) 'Cloaca Maxima in Rom', *Antike Denkmäler* 1: 25.

Ridolfi, R. (1997) *Vita di Girolamo Savonarola*. Florence: Lettere.

Rindisbacher, H. J. (2005) 'A cultural history of disgust', *KulturPoetik* 5.1: 119–27.

Rinne, K. (2001–2) 'The landscape of laundry in late cinquecento Rome', *Studies in the Decorative Arts* 9.1: 34–60.

(2010) *The waters of Rome: aqueducts, fountains, and the birth of the baroque city.* New Haven and London: Yale University Press.

Rives, J. (1995) 'Human sacrifice among pagans and Christians', *Journal of Roman Studies* 85: 65–85.

(2007) *Religion in the Roman Empire.* Oxford: Wiley-Blackwell.

Rizzo, D. (2004) *Gli spazi della morale.* Rome: Biblink.

(forthcoming) '"Per due soldi": carriere sessuali di bambini e adolescenti nella Roma di fine ottocento', in L. Guidi and M. Rosaria Pelizzari (eds) *Le nuove frontiere della storia di genere II.* Salerno: Collana di Studi e Ricerche dell'Università di Salerno.

Robertson, A. (1996) *The grotesque interface: deformity, debasement, dissolution.* Frankfurt: Iberoamericana.

Robertson Smith, W. (1927) *Lectures on the religion of the Semites: the fundamental institutions* (third edition, revised). London: A & C Black.

Robinson, J. (2000) 'Feminism and the spaces of transformation', *Transactions of the Institute of British Geographers* 25.3: 285–301.

Robinson, O. F. (1984) 'Baths: an aspect of Roman local government law', in V. Giuffrè (ed.) *Sodalitas: scritti in onore di A. Guarino.* Naples: Biblioteca di Labeo, 1065–82.

(1992) *Ancient Rome: city planning and administration.* London and New York: Routledge.

(2007) *Penal practice and penal policy in ancient Rome.* London and New York: Routledge.

Rocciolo, D. (2006) '"*Cum suspicione morbid contagiosi obierunt*": società, religione e peste a Roma nel 1656–1657', *Roma Moderna e Contemporanea* 14: 111–34.

Rodocanachi, E. (1906) *The Roman Capitol in ancient and modern times.* New York: Heinemann.

Rogers, R. S. (1932) 'Fulvia Paulina C. Sentii Saturnini', *American Journal of Philology* 53.3: 252–6.

Rohde, E. (1894) *Psyche: Seelencult und Unsterblichkeitsglaube der Griechen.* Leipzig: Akademische Verlagsbuchhandlung von J. C. B. Mohr.

Rose, H. J. (1924) *The Roman questions.* Oxford: Clarendon Press.

Rosselli, D. (1996) '"*Tamquam bruta animalia*": l'immagine dei vagabondi a Roma tra cinque e seicento', *Quaderni Storici* 31: 363–404.

Rossetti, G. (1894) *Disquisitions on the antipapal spirit which produced the reformation* (trans. C. Ward). London: Smith, Elder.

Rousseau, P. (2010) *Ascetics, authority, and the Church in the age of Jerome and Cassian* (second edition). University of Notre Dame Press.

Roux, J.-P. (1988) *Le sang: mythes, symboles et réalités.* Paris: Fayard.

Rüpke, J. (1990) *Domi militiae: die religiöse Konstruktion des Krieges in Rom.* Stuttgart: Franz Steiner.

(2007) *Religion of the Romans* (trans. R. Gordon). Cambridge: Polity.

Sala, G. A. (1869) *Rome and Venice: with other wanderings in Italy, in 1866–7.* London: Tinsley Brothers.

Salerno, L. (1938) 'Omosessualità', in L. Salerno, *Enciclopedia di polizia.* Milan: Bocca, 655–6.

Salvante, M. (forthcoming) 'La prostituzione maschile nel discorso scientifico della prima metà del novecento in Italia', in L. Guidi and M. Rosaria Pelizzari (eds) *Le nuove frontiere della storia di genere II.* Salerno: Collana di Studi e Ricerche dell'Università di Salerno.

Samaritani, A. (1976) *Michele Savonarola: riformatore cattolico nella Corte Estense a metà del sec. XV.* Ferrara: Deputazione Provinciale Ferrarese di Storia Patria.

San Juan, R.-M. (2001) *Rome, a city out of print.* Minneapolis: University of Minnesota Press.

Sansa, R. (2002a) 'L'odore del contagio: ambiente urbano e prevenzione delle epidemie nella prima età moderna', *Medicina e Storia* 2: 83–108.

(2002b) 'Strategie di prevenzione a confronto: l'igiene urbana durante la peste romana del 1656–1657', *Roma Moderna e Contemporanea* 14: 93–110.

Santoro Bianchi, S. (2001) 'L'iconografia musiva di Oceano e le sue corrispondenze letterarie', in D. Paunier and C. Schmidt (eds) *La mosaïque Gréco-Romaine: actes du VIIIème Colloque International pour l'Etude de la Mosaïque Antique et Médiévale, Lausanne 6–11 octobre 1997.* Lausanne: Cahiers d'Archéologie Romande, II.84–95.

Saviano, R. (2007) *Gomorrah* (trans. V. Jewiss). New York: Picador.

Savio, P. (1972) 'Ricerche sulla peste di Roma degli anni 1656–1657', *Archivio della Società Romana di Storia Patria* 26: 113–42.

Savonarola, G. (1898) 'Ut quid Deus repulisti in finem', in P. Villari and E. Casanova (eds) *Scelta di prediche e scritti di Fra Girolamo Savonarola.* Florence: Sansoni, 35–52.

(1955a) *Prediche sopra Ezechiele* (edited by R. Ridolfi in two volumes). Rome: Angelo Belardetti Editore.

(1955b) *Prediche sopra l'Esodo* (2 volumes) (ed. P. G. Ricci). Rome: Angelo Belardetti Editore.

(1959a) *Compendio di rivelazioni e Dialogus de veritate prophetica* (ed. A. Crucitti). Rome: Angelo Belardetti Editore.

(1959b) *De simplicitate Christianae vitae* (ed. P. G. Ricci). Rome: Angelo Belardetti Editore.

(1959c) *Lettere e scritti apologetici* (eds R. Ridolfi, V. Romano and A. Verde). Rome: Angelo Belardetti Editore.

(1961) *Triumphus Crucis* (ed. M. Ferrara). Rome: Angelo Belardetti Editore.

(1962) *Prediche sopra Ruth e Michea* (2 volumes) (ed. V. Romano). Rome: Angelo Belardetti Editore.

(1968) *Poesie* (ed. M. Martelli). Rome: Angelo Belardetti Editore.

(1971–2) *Prediche sopra Amos e Zaccaria* (3 volumes) (ed. P. Ghiglieri). Rome: Angelo Belardetti Editore.

(1982) *Scritti filosofici* (2 volumes) (eds E. Garin and G. C. Garfagnini). Rome: Angelo Belardetti Editore.

(1989) *Sermones in primam divi Ioannis epistolam* (eds A. Verde and E. Giaconi). Florence: SISMEL/Edizioni del Galluzzo.

Scanlan, J. (2005) *On garbage*. London: Reaktion.

Schaub, B., Lauener, R. and von Mutius, E. (2006) 'The many faces of the hygiene hypothesis', *Journal of Allergy and Clinical Immunology* 117: 969–77.

Scheid, J. (2003) *An introduction to Roman religion*. Edinburgh University Press.

Scheidel, W. (2003) 'Germs for Rome' in C. Edwards and G. Woolf (eds) *Rome the cosmopolis*. Cambridge University Press, 158–78.

Schlör, J. (1998) *Nights in the big city: Paris, Berlin, London 1840–1930*. London: Reaktion.

Schnall, S., Benton, J. and Harvey, S. (2008) 'With a clean conscience: cleanliness reduces the severity of moral judgments', *Psychological Science* 19: 1219–22.

Schnapp, J. T. (2003) 'Trionfo e invisibilità dell'asfalto', in M. Zardini (ed.) *Asfalto: il carattere della città*. Milan: Mondadori Electa, 140–5.

Schneider, J. and Schneider, P. (1998) 'Il Caso Sciascia: dilemmas of the antimafia movement in Sicily', in J. Schneider (ed.) *Italy's 'Southern question'*. Oxford: Berg, 245–61.

Schnitzer J. (1910) *Savonarola nach den Aufzeichnungen des Florentiners Piero Parenti* (Quellen und Forschungen zur Geschichte Savonarolas 4). Leipzig: Von Duncker und Humblot.

(1931) *Savonarola* (2 volumes). Milan: Fratelli Treves.

Schult, H. (2002) *H. A. Schult: art is action: actions are experienced pictures*. Tübingen and Berlin: Wasmuth.

Schultz, C. E. (2006) *Women's religious activity in the Roman Republic*. Chapel Hill: University of North Carolina Press.

(2010) 'The Romans and ritual murder', *Journal of the American Academy of Religion* 78: 516–41.

Sciascia, L. (2003) *The day of the owl* (trans. A. Colquhoun and A. Oliver). New York: Review Books.

Scobie, A. (1986) 'Slums, sanitation, and mortality in the Roman world', *Klio* 68: 399–433.

Seager, R. (1979) *Pompey the Great: a political biography*. Oxford: Blackwell.

Sealey, R. (2006) 'Aristotle, *Athenaion Politeia* 57.4: trial of animals and inanimate objects for homicide', *Classical Quarterly* 56.2: 475–85.

Segarizzi, A. (1900) *Della vita e delle opere di Michele Savonarola medico padovano del secolo xv*. Padua: Fratelli Gallina.

Sennett, R. (1996) *The uses of disorder*. London: Faber and Faber.

Severino, C. G. (2005) *Roma mosaico urbano: il pigneto fuori Porta Maggiore*. Rome: Gangemi.

Seymour, M. H. (1843) 'The English Communion contrasted with the Roman Mass, preached by Rev. M. Hobart Seymour, St Georges Church, Southwark, Sunday eve Nov 5th, 1843', *Pulpit* 44: 375–83.

(1850) *Mornings among the Jesuits at Rome* (third edition, considerably enlarged). London: Seeleys.

(1866) *The Jubilee at Rome: a lecture delivered at the Assembly Rooms, Bath, April, 16th, 1866*. Bath: R. E. Peach.

Sforza Pallavicino, P. (1837) *Descrizione del contagio che da Napoli si comunicò a Roma nell'anno 1656 e de' saggi provvedimenti ordinate allora da Alessandro VII*. Rome: Collegio Urbano.

Shackleton Bailey, D. R. (ed.) (1965–71) *Cicero's Letters to Atticus* (7 volumes). Cambridge University Press.

Shaw, B. D. (1992) 'Explaining incest: brother–sister marriage in Graeco-Roman Egypt', *Man* 27.2: 267–99.

(1996) 'Seasons of death: aspects of mortality in imperial Rome', *Journal of Roman Studies* 86: 100–38.

Shearman, J. (2003) *Raphael in early modern sources (1483–1602)* (2 volumes). New Haven: Yale University Press.

Shields, R. (1991) *Places on the margin: alternative geographies of modernity*. London: Routledge.

Shildrick, M. (1997) *Leaky bodies and boundaries: feminism, postmodernism and (bio)ethics*. London: Routledge.

Shilling, C. (1993) *The body and social theory*. London: Sage.

Shonfield, K. (2001) 'Dirt is matter out of place', in J. Hill (ed.) *Architecture: the subject is matter*. London: Routledge, 29–44.

Shove, E. (2003) *Comfort, cleanliness and convenience: the social organization of normality*. Oxford: Berg.

Shweder, R. (1991) 'Menstrual pollution, soul loss, and the comparative study of emotions', in R. Shweder (ed.) *Thinking through cultures: expeditions in cultural psychology*. Cambridge, MA: Harvard University Press, 241–65.

Sibley, D. (1995) *Geographies of exclusion: society and difference in the West*. London: Routledge.

Sinatra, M. (2006) *La Garbatella a Roma 1920–1940*. Milan: Franco Angeli.

Sivulka, J. (2001) *Stronger than dirt: a cultural history of advertising personal hygiene in America, 1875 to 1940*. Amherst, NY: Humanity Books.

Sjöblad, A. (2009) *Metaphors Cicero lived by: the role of metaphor and simile in De senectute* (Studia Graeca et Latina Lundensia 16). Centre for Languages and Literature, Lund University.

Slack, P. (1985) *The impact of plague in Tudor and Stuart England*. London: Routledge and Kegan Paul.

(1988) 'Responses to plague in early modern Europe: the implications of public health', *Social Research* 55: 433–53.

Small, J. (1982) *Cacus and Marsyas in Etrusco-Roman legend.* Princeton University Press.

Smith, M. M. (2007) *Sensory history.* Oxford and New York: Berg.

Smith, V. (2007) *Clean: a history of personal hygiene and purity.* Oxford University Press.

Soja, E. W. (1996) *Thirdspace: journeys to Los Angeles and other real-and-imagined places.* Oxford: Blackwell.

Sonnino, E. (2006) 'Cronache della peste di Roma: notizie dal Ghetto e lettere di Girolamo Gastaldi (1656–7)', *Roma Moderna e Contemporanea* 14: 35–74.

Sonnino, E. and Traina, R. (1982) 'La peste del 1656–57 a Roma: organizzazione sanitaria e mortalità', in Società Italiana di Demografia Storica, *La demografia storica delle città italiane.* Bologna: CLUEB, 433–52.

Sontag, S. (1979) *Illness as metaphor.* London: Allen Lane.

Sourvinou Inwood, C. (1995) *'Reading' Greek death to the end of the classical period.* Oxford: Clarendon Press.

Spackman, B. (1996) *Fascist virilities: rhetoric, ideology, and social fantasy in Italy.* Minneapolis: University of Minnesota Press.

Staccioli, R. A. (2003) *The roads of the Romans.* Los Angeles: J. Paul Getty Museum.

Stallybrass, P. and White, A. (1986) *The politics and poetics of transgression.* London: Methuen.

Staples, A. (1998) *From Good Goddess to Vestal Virgins: sex and category in Roman religion.* London: Routledge.

Stille, A. (1995) *Excellent cadavers: the Mafia and the death of the first Italian Republic.* London: Cape.

Storey, T. (2008) *Carnal commerce in counter-reformation Rome.* Cambridge University Press.

Story, W. W. (1871) *Roba di Roma* (sixth edition, with additions). London: Chapman and Hall.

Stouck, M.-A. (1999) *Medieval saints.* New York: Broadview Press.

Stow, K. (1987) 'The Jewish family in the Rhineland: form and function', *American Historical Review* 92: 1085–110.

(1992) *Alienated minority: the Jews of medieval Latin Europe.* Cambridge, MA, and London: Harvard University Press.

(1995 and 1997) *The Jews in Rome* (2 volumes). Leiden: Brill.

(2001) *Theater of acculturation: the Roman ghetto in the sixteenth century.* Seattle: University of Washington Press.

(2006) *Jewish dogs: an image and its interpreters.* Stanford University Press.

Strange, C. and Bashford, A. (eds) (2003) *Isolation: places and practices of exclusion.* London: Routledge.

Strocchia, S. (1992) *Death and ritual in Renaissance Florence.* Baltimore: Johns Hopkins University Press.

Stuart, K. (1999) *Defiled trades and social outcasts: honour and ritual pollution in early modern Germany*. Cambridge University Press.

Sundt, R. A. (1987) '"*Mediocres domos et humiles habeant fratres nostri*": Dominican legislation on architecture and architectural decoration in the 13th century', *Journal of the Society of Architectural Historians* 46: 349–407.

Sussman, L. (ed.) (1994) *The declamations of Calpurnius Flaccus*. Leiden: Brill.

Syrjämaa, T. (2006) *Constructing unity, living in diversity: a Roman decade*. Helsinki: Finnish Academy of Science and Letters.

(2007a) *Edistyksen luvattu maailma/Edistysusko maailmannäyttelyissä 1851–1915*. Helsinki: Suomalaisen Kirjallisuuden Seura.

(2007b) 'Talking about beggars: nineteenth-century perspectives on Roman poverty', in B. Althammer (ed.) *Bettler in der europäischen Stadt der Moderne*. Frankfurt and Oxford: Peter Lang, 133–50.

(2008) 'Gli spazi del passato in Roma Capitale: visioni, dibattiti e pratiche 1870–1881'. Institutum Romanum Finlandiae. www.irfrome.org/ei/images/stories/syrjamaa.pdf, accessed July 2011.

Tatum, W. (1993) 'Ritual and morality in Roman religion', *Syllecta Classica* 4: 13–20.

Thein, A. G. (2002) *Sulla's public image and the politics of civic renewal*. PhD dissertation: University of Pennsylvania.

Thomas, K. (1994) 'Cleanliness and godliness in early modern England', in A. Fletcher and P. Roberts (eds) *Religion, culture and society in early modern Britain: essays in honour of Patrick Collinson*. Cambridge University Press, 56–83.

Thomas, P. (2001) *Cleaning yourself to death: how safe is your home?* Dublin: Newleaf.

Thome, G. (1992) 'Crime and punishment, guilt and expiation: Roman thought and vocabulary', *Acta Classica* 35: 73–98.

Thompson, E. (2006) 'Noise and noise abatement in the modern city', in M. Zardini (ed.) *Sense of the city: an alternate approach to urbanism*. Montreal: Canadian Centre for Architecture, 190–9.

Thompson, M. (1979) *Rubbish theory*. Oxford University Press.

Thorndike, L. (1928) 'Sanitation, baths and street-cleaning in the Middle Ages and Renaissance', *Speculum* 3: 192–203.

Thornton, M. K. and Thornton, R. L. (1989) *Julio-Claudian building programs: a quantitative study in political management*. Wauconda, IL: Bolchazy-Carducci.

Thorsheim, P. (2006) *Inventing pollution: coal, smoke, and culture in Britain since 1800*. Athens: Ohio University Press.

Tobia, B. (1991) *Una patria per gli Italiani: spazi, itinerari, monumenti nell'Italia unita (1870–1900)*. Rome and Bari: Laterza.

Tomes, N. (1998) *The gospel of germs: men, women and the microbe in American life*. Cambridge, MA, and London: Harvard University Press.

Torrioli, E. (1992) 'Ciao, Augustin' [first published 1883], in A.-C. Faitrop Porta (ed.) *Novelle della Roma umbertina*. Rome: Salerno Editrice, 194–205.

Towner, J. (1996) *An historical geography of recreation and tourism in the Western world 1540–1940*. Chicester: Wiley.

Toynbee, J. (1971) *Death and burial in the Roman world*. London: Thames and Hudson.

Treves, A. (1976) *Le migrazioni interne nell'Italia fascista: politica e realtà demografica*. Turin: Einaudi.

(2001) *Le nascite e la politica nell'Italia del novecento*. Milan: LED.

Trexler, R. C. (1974) 'Ritual in Florence: adolescence and salvation in the Renaissance', in C. Trinkhaus (ed.) *The pursuit of holiness in late medieval and Renaissance religion*. Leiden: Brill, 200–64.

Troncarelli, F. (ed.) (1985) *La città dei segreti: magia, astrologia e cultura esoterica a Roma (xv–xviii)*. Milan: Franco Angeli.

Turcan, R. (1996) *The gods of ancient Rome* (trans. A. Nevill). Edinburgh University Press.

Turner, V. (1969) *The ritual process: structure and anti-structure*. New York: Aldine.

(1979) 'Betwixt and between: the liminal period in *Rites de passage*', in W. Lessa, E. Vogt and J. Watanabe (eds) *Reader in comparative religion: an anthropological approach*. New York: Harper and Row, 234–43.

Vaglieri, D. (1900) 'Venus Cloacina', *Bullettino della Commissione Archeologica Comunale di Roma* 28: 61–2.

(1903) 'Gli scavi recenti del Foro Romano', *Bullettino della Commissione Archeoligica Comunale di Roma* 31: 97–9.

Valeri, V. (1999) *The forest of taboos: morality, hunting, and identity among the Huaulu of the Moluccas*. Madison: University of Wisconsin Press.

van Driel-Murray, C. (2002) 'Ancient skin processing and the impact of Roman tanning technology', in F. Audoin-Rouzeau and S. Beyries (eds) *Le travail du cuir de la préhistoire à nos jours: xxiie Rencontres Internationals d'Archéologie et d'Histoire d'Antibes*. Antibes: Editions Association pour la Promotion et la Diffusion des Connaissances Archéologiques, 251–65.

van Essen, C. C. (1956) 'Venus Cloacina', *Mnemosyne* 4.9: 137–44.

Van Gennep, A. (1960) *Rites of passage* (trans. M. Vizedom and G. Caffee, Intro. S. Kimball). London: Routledge and Kegan Paul.

Varner, E. (2004) *Mutilation and transformation: damnatio memoriae and Roman imperial portraiture*. Leiden: Brill.

Vasaly, A. (1993) *Representations: images of the world in Ciceronian oratory*. Berkeley: University of California Press.

Vasoli, C. (1962) 'L'attesa della nuova era in ambienti e gruppi fiorentini del quattrocento', in *L'attesa dell'età nuova nella spiritualità della fine del medioevo* (Convegni del Centro di Studi sulla Spiritualità Medioevale iii, 16–19 October 1962). Todi: Accademia Tudertina, 370–432.

Verde, A. (1983) 'La congregazione di S. Marco dell'Ordine dei Frati Predicatori: il "reale" della predicazione savonaroliana', *Memorie Domenicane* 14: 151–237.

(1988) 'Le lezioni o i sermoni sull'*Apocalisse* di Girolamo Savonarola (1490): "*Nova dicere et novo modo*"', *Memorie Domenicane* 19: 5–109.

(1994) 'Il movimento spirituale savonaroliano fra Lucca, Bologna, Ferrara, Pistoia, Perugia, Prato, Firenze: il volgarizzamento delle prediche sullo Spirito Santo di Fra Girolamo Savonarola', *Memorie Domenicane* 26: 5–206.

Verde, A. and Giaconi, E. (1992) 'Epistolario di Fra Vincenzo Mainardi da San Gimignano domenicano 1481–1527', *Memorie Domenicane* 22: vii–727.

Vidler, A. (1992) *The architectural uncanny: essays in the modern unhomely*. Cambridge, MA: MIT Press.

Vidoni G. (1922) 'Per lo studio della prostituzione maschile', *Manicomio* 37: 225–46.

(1940) 'Ancora per lo studio della prostituzione maschile', *Giustizia Penale* 46: 386–401.

Vidotto, V. (ed.) (2002) *Roma capitale* (Storia di Roma dall'Antichità a Oggi). Rome and Bari: Laterza.

Vigarello, G. (1988) *Concepts of cleanliness: changing attitudes in France since the Middle Ages* (trans. J. Birrell). Cambridge University Press.

Viljama, T., Timonen, A. and Krötzl, C. (eds) (1992) *Crudelitas: the politics of cruelty in the ancient and medieval world*. Krems: Medium Aevum Quotidianum.

Villari, P. (1930) *La storia di Girolamo Savonarola e de' suoi tempi* (2 volumes). Florence: Felice Le Monnier.

Ville, G. (1981) *La gladiature en Occident des origines à la morte de Domitien*. Rome: Ecole française de Rome.

Villoresi, L. (2006) 'Roma sotterranea', *National Geographic Italia* 18.1: 2–25.

Vinikas, V. (1992) *Soft soap, hard sell: American hygiene in an age of advertisement*. Iowa State University Press.

Viscogliosi, A. (2009) 'Il Foro Transitorio', in F. Coarelli (ed.) *Divus Vespasianus: il bimillenario dei Flavi*. Milan: Electa, 202–9.

Visino, S. (1994) *I pittori del grand tour: viaggio a Roma alla ricerca delle aure*. Latina: Argonauta.

Visscher, F. de (1963) *Le droix des tombeaux Romains*. Milan: Giuffrè.

Visser, R. (1992) 'Fascist doctrine and the cult of *romanità*', *Journal of Contemporary History* 27: 5–22.

Viviani, A. (1873) *Relazione intorno al piano regolatore della città di Roma*. Rome: L. Cecchini.

Voisin, J. L. (ed.) (1984) *Du châtiment dans la cité: supplices corporels et peine de mort dans le monde antique*. Rome: Ecole Française de Rome.

von Hesberg, H. (1992) *Römische Grabbauten*. Darmstadt: Wissenschaftliche Buchgesellschaft.

Vout, C. (forthcoming) *The hills of Rome: signature of an eternal city*. Cambridge University Press.

Voute, P. (1972) 'Notes sur l'iconographie d'Océan: à propos d'une fontaine à mosaïques découverte à Nole (Campanie)', *Mélanges de l'Ecole Française de Rome. Antiquité* 84: 639–73.

Waddington, K. (2006) *The bovine scourge: meat, tuberculosis and the public's health, 1860s–1914.* Woodbridge: Boydell and Brewer.

Walkowitz, J. (1980) *Prostitution and Victorian society: women, class and the state.* Cambridge University Press.

Wallace, E. (ed.) (1869) *The writings of Cyprian, Bishop of Carthage II.* Edinburgh: T & T Clark.

Wallace-Hadrill, A. (1982) 'The golden age and sin in Augustan ideology', *Past and Present* 95: 19–36.

(1998) '*Horti* and Hellenization', in M. Cima and E. La Rocca (eds) *Horti romani: atti del convegno internazionale: Roma, 4–6 maggio 1995.* Rome: "L'Erma" di Bretschneider, 1–12.

Warde, A. (1997) *Consumption, food and taste: culinary antinomies and commodity culture.* London: Sage.

Warde-Fowler, W. (1911a) 'The original meaning of the word *sacer*', *Journal of Roman Studies* 1: 57–63.

(1911b) *The religious experience of the Roman people from the earliest times to the age of Augustus.* London: Macmillan.

(1912) 'Mundus patet', *Journal of Roman Studies* 2: 25–33.

Wasserfall, R. R. (ed.) (1999) *Women and water: menstruation in Jewish life and law.* Hanover: Brandeis University Press.

Watson, A. (1992) *The state, law, and religion: pagan Rome.* Athens: University of Georgia Press.

Watson, J. (1982) 'Of flesh and bones: the management of death pollution in Cantonese society', in M. Bloch and J. Parry (eds) *Death and the regeneration of life.* Cambridge University Press, 155–86.

(1988) 'Funeral specialists in Cantonese society: pollution, performance, and social hierarchy', in J. Watson and E. Rawski (eds) *Death ritual in late imperial and modern China.* Berkeley: University of California Press, 109–34.

Weeks, J. (2000) *Making sexual history.* Cambridge: Polity.

Weinstein, D. (1970) *Savonorola and Florence: prophecy and patriotism in the Renaissance.* Princeton University Press.

Weinstock, S. (1971) *Divus Julius.* Oxford University Press.

Welch, T. S. (2005) *The elegiac cityscape: Propertius and the meaning of Roman monuments.* Columbus: Ohio State University Press.

Wey, F. (1879) *Roma: descrizione e ricordi.* Milan: Fratelli Treves.

Wheatley, T. and Haidt, J. (2005) 'Hypnotically induced disgust makes moral judgments more severe', *Psychological Science* 16: 780–4.

Wheeler, M. (2006). *The old enemies: Catholic and Protestant in nineteenth-century culture.* Cambridge University Press.

Whittaker, C. (1993) 'The poor', in A. Giardina (ed.) *The Romans*. University of Chicago Press, 272–99.

Widding Isaken, L. (2002) 'Masculine dignity and the dirty body', *Nordic Journal of Feminist and Gender Studies* 10.3: 137–46.

Wildfang, R. L. (2006) *Rome's Vestal Virgins: a study of Rome's Vestal priestesses in the late Republic and early Empire*. London: Routledge.

Wilkie, J. (1986) 'A submerged sensuality: technology and perceptions of bathing', *Journal of Social History* 19: 649–64.

Williams, C. A. (2010) *Roman homosexuality* (second edition). Oxford University Press.

Williams, J. H. C. (1999) 'Septimius Severus and Sol, Carausius and Oceanus: two new Roman acquisitions at the British Museum', *Numismatic Chronicle* 159: 307–13.

Williams, M. (1991) *Washing 'the great unwashed': public baths in urban America, 1840–1920*. Columbus: Ohio State University Press.

Wilton-Ely, J. (2004) '"Classic ground": Britain, Italy and the grand tour', *Eighteenth-Century Life* 28: 136–65.

Winkler, M. (1991) 'Satire and the grotesque in Juvenal, Arcimboldo and Goya', *Antike und Abendland* 37: 22–42.

Winter, F. E. (2006) *Studies in Hellenistic architecture*. University of Toronto Press.

Winter, N. A., Iliopoulos, I. and Ammerman, A. J. (2009) 'New light on the production of decorated roofs of the 6th c. BC at sites in and around Rome', *Journal of Roman Archaeology* 22: 6–28.

Winterbottom, M. (ed.) (1974) *The elder Seneca: Declamations 1: Controversiae books 1–6* (Loeb Classical Library). Cambridge, MA, and London: Harvard University Press.

Wiseman, N. (1850) *The social and intellectual state of England, compared with its moral condition: a sermon delivered in St. Johns Catholic Church, Salford, on Sunday, July 28th, 1850*. London: Richardson.

(1855) *Fabiola; or, the Church of the Catacombs*. London: Burns and Lambert.

Wiseman, T. P. (1979) 'Topography and rhetoric: the trial of Manlius', *Historia* 28.1: 32–50.

(1990) 'The central area of the Roman Forum', *Journal of Roman Archaeology* 3: 245–7.

(1993) 'Rome and the resplendent Aemilii', in H. D. Jocelyn and H. Hunt (eds) *Tria lustra: essays and notes presented to John Pinsent*. Liverpool Classical Monthly, 181–92.

(1998a) '*Conspicui postes tectaque digna deo*', in T. P. Wiseman, *Historiography and imagination: eight essays on Roman culture*. University of Exeter Press, 98–115.

(1998b) 'A stroll on the ramparts', in M. Cima and E. La Rocca (eds) *Horti romani: atti del convegno internazionale: Roma, 4–6 maggio 1995*. Rome: "L'Erma" di Bretschneider, 13–22.

(1999) 'Saxum Tarpeium', in E. M. Steinby (ed.) *Lexicon topographicum urbis romae IV*. Rome: Quasar, 237–8.

Wissowa, G. (1912) *Religion und Kultus der Römer*. Munich: Beck.

(1923–4) 'Vestalinnenfrevel', *Archiv für Religionswissenschaft* 22: 201–14.

Wissowa, G., Kroll, W., Mittelhaus, K. and Ziegler, K. (eds) (1894–1978) *Paulys Real-encyclopädie der klassischen Altertumswissenschaft*. Stuttgart: J. B. Metzler.

Wood, T. (1846) *The origin, learning, religion, and customs of the ancient Britons: with an account of the introduction of Christianity into Britain, and the idolatry and conversion of the Saxon: remarks on the errors and progress of popery: considerations on the Christian Church, its foundation, superstructure, and beauty*. London: William Brown.

Woodburn, J. (1982) 'Social dimensions of death in four African hunting and gathering societies', in M. Bloch and J. Parry (eds) *Death and the regeneration of life*. Cambridge University Press, 187–211.

Wright, L. (1960) *Clean and decent: the fascinating history of the bathroom and the water closet, and of the sundry habits, fashions and accessories of the toilet, principally in Great Britain, France, and America*. London: Routledge and Kegan Paul.

Wrigley, R. (2007) '"It was dirty, but it was Rome": dirt, digression, and the picturesque', in R. Wrigley (ed.) *Regarding romantic Rome*. Bern: Peter Lang, 157–79.

(forthcoming) 'Making sense of Rome', *Journal for Eighteenth-Century Studies* 35.

Wülker, L. (1903) *Die geschichtliche Entwicklung des Prodigienwesens bei den Römern: Studien zur Geschichte und Überlieferung der Staatsprodigien*. Leipzig: E. Glausch.

Wuthnow, J., Hunter, D., Bergesen, A. and Kurzweil, E. (1984) *Cultural analysis: the work of Peter L. Berger, Mary Douglas, Michel Foucault, and Jürgen Habermas*. London: Routledge and Kegan Paul.

Yegül, F. (1992) *Baths and bathing in classical antiquity*. Cambridge, MA: MIT Press.

(2010) *Bathing in the Roman world*. Cambridge University Press.

Younger, S., Arnold, R. and Schapiro, R. (eds) (1999) *The definition of death: contemporary controversies*. Baltimore: Johns Hopkins University Press.

Zanazzo, G. (1908) *Usi, costume e pregiudizi del popolo di Roma: tradizioni popolari romane*. Turin: Società Tipografico-Editrice Nazionale.

Zanker, P. (1990) *The power of images in the age of Augustus*. Ann Arbor: University of Michigan Press.

Zardini, M. (2006) 'The ground of the modern city', in M. Zardini (ed.) *Sense of the city: an alternate approach to urbanism*. Montreal: Canadian Centre for Architecture, 239–45.

Zhong, C. and Liljenquist, K. (2006) 'Washing away your sins: threatened morality and physical cleansing', *Science* 313: 1451–2.

Zimmermann, M. (ed.) (2009) *Extreme Formen von Gewalt in Bild und Text des Altertums: Münchner Studien zur Alten Welt.* Munich: Herbert Utz.

Ziolkowski, A. (1998–9) 'Ritual cleaning-up of the city: from the Lupercalia to the Argei', *Ancient Society* 29: 191–218.

Zuesse, E. (1979) *Ritual cosmos: the sanctification of life in African religions.* Athens: Ohio University Press.

Index